MACROMOLECULAR SELF-ASSEMBLY

Edited by

LAURENT BILLON
OLEG BORISOV

WILEY

Published by John Wiley & Sons, Inc., Hoboken, New Jersey
Published simultaneously in Canada

Library of Congress Cataloging-in-Publication Data:

Names: Billon, Laurent, 1968- editor. | Borisov, Oleg, editor.
Title: Macromolecular self-assembly / edited by Laurent Billon, Oleg Borisov.
Description: Hoboken, New Jersey : John Wiley & Sons, Inc., [2016] | Includes bibliographical references and index.
Identifiers: LCCN 2016015704 (print) | LCCN 2016021455 (ebook) | ISBN 9781118887127 (cloth) | ISBN 9781118887844 (pdf) | ISBN 9781118887974 (epub)
Subjects: LCSH: Biopolymers. | Macromolecules. | Self-assembly (Chemistry)
Classification: LCC TP248.65.P62 M325 2016 (print) | LCC TP248.65.P62 (ebook) | DDC 572–dc23
LC record available at https://lccn.loc.gov/2016015704

Set in 10/12pt, TimesLTStd by SPi Global, Chennai, India.

Printed in the United States of America

10 9 8 7 6 5 4 3 2 1

CONTENTS

LIST OF CONTRIBUTORS

Jancy Nixon Abraham, University of Geneva, Sciences II, Department of inorganic and analytical chemistry, quai Ernest Ansermet 30, 1211, Geneva 4, Switzerland

Christopher Barner-Kowollik, Preparative Macromolecular Chemistry, Institut für Technische Chemie und Polymerchemie, Karlsruhe Institute of Technology (KIT), Engesserstr. 18, 76128 Karlsruhe, Germany and Institut für Biologische Grenzflächen, Karlsruhe Institut of Technology (KIT), Hermann-von-Helmholtz-Platz 1, 76344 Eggenstein-Leopoldshafen, Germany

Laurent Billon, Institut des Sciences Analytiqueset de Physico-Chimie pour l'Environnement et les MatériauxIPREM, CNRS - UMR 5254, Université de Pau & Pays de l'Adour, 64053 Pau, France

Olga Borisova, Department of Polymer Science, Moscow State University, Leninskie Gory, Moscow 119191, Russia

Oleg Borisov, Institut des Sciences Analytiqueset de Physico-Chimie pour l'Environnement et les MatériauxIPREM, CNRS - UMR 5254, Université de Pau & Pays de l'Adour, 64053 Pau, France

Munish Chanana, ETH Zürich, Institute of Building Materials, Stefano-Franscini-Platz 3, 8093 Zürich, Switzerland, University of Bayreuth, Physical Chemistry II, Universitätsstrasse 30, 95440 Bayreuth, Germany

Franck D'Agosto, Université de Lyon, Univ. Lyon 1, CPE Lyon, CNRS UMR 5265, Laboratoire de Chimie, Catalyse, Polymères et Procédés (C2P2), LCPP group, 69616 Villeurbanne, France

Elise Deniau-Lejeune, Institut des Sciences Analytiqueset de Physico-Chimie pour l'Environnement et les MatériauxIPREM, CNRS - UMR 5254, Université de Pau & Pays de l'Adour, 64053 Pau, France

Andreas Fery, Leibniz-Institut für Polymerforschung Dresden e.V., Institute of Physical Chemistry and Polymer Physics, Technische Universität Dresden, Physical Chemistry of Polymeric Materials and Cluster of Excellence Centre for Advancing Electronics Dresden (cfaed), Hohe Strasse 6, 01069 Dresden, Germany, University of Bayreuth, Physical Chemistry II, Universitätsstrasse 30, 95440 Bayreuth, Germany

Sergey K. Filippov, Institute of Macromolecular Chemistry AS CR, Prague, Czech Republic

Martin Hrubý, Institute of Macromolecular Chemistry AS CR, Prague, Czech Republic

Matthias Karg, Heinrich Heine University Düsseldorf, Physical Chemistry I, Universitätsstrasse 1, 40225 Düsseldorf, Germany, University of Bayreuth, Physical Chemistry I, Universitätsstrasse 30, 95440 Bayreuth, Germany

Dawid Kedracki, University of Geneva, Sciences II, Department of inorganic and analytical chemistry, quai Ernest Ansermet 30, 1211, Geneva 4, Switzerland

Christian Kuttner, Leibniz-Institut für Polymerforschung Dresden e.V., Institute of Physical Chemistry and Polymer Physics, Technische Universität Dresden, Cluster of Excellence Centre for Advancing Electronics Dresden (cfaed), Hohe Strasse 6, 01069 Dresden, Germany, University of Bayreuth, Physical Chemistry II, Universitätsstrasse 30, 95440 Bayreuth, Germany

Muriel Lansalot, Université de Lyon, Univ. Lyon 1, CPE Lyon, CNRS UMR 5265, Laboratoire de Chimie, Catalyse, Polymères et Procédés (C2P2), LCPP group, 69616 Villeurbanne, France

Corinne Nardin, University of Geneva, Sciences II, Department of inorganic and analytical chemistry, quai Ernest Ansermet 30, 1211, Geneva 4, Switzerland

Dmitry V. Pergushov, Department of Chemistry, M.V. Lomonosov Moscow State University Leninskie Gory 1/3, 119991 Moscow, Russia

Olga Philippova, Physics Department, Moscow State University, 119991 Moscow, Russia

Felix A. Plamper, Institute of Physical Chemistry II, RWTH Aachen University Landoltweg 2, 52056 Aachen, Germany

Enora Prado, University of Geneva, Sciences II, Department of inorganic and analytical chemistry, quai Ernest Ansermet 30, 1211, Geneva 4, Switzerland

Jutta Rieger, UPMC Univ. Paris 6, Sorbonne Universités and CNRS, Laboratoire de Chimie des Polymères, UMR 7610, 3 rue Galilée, 94200 Ivry, France

Bernhard V. K. J. Schmidt, Materials Research Laboratory, University of California, Santa Barbara, CA, 93106, USA

Petr Štěpánek, Institute of Macromolecular Chemistry AS CR, Prague, Czech Republic

PREFACE

Macromolecular self-assembly is a generic term utilized to describe spontaneous associations of individual macromolecular species, as either identical or complementary building blocks, giving rise to a zoo of supramolecular structures. In contrast to macroscopic phase separation in polymer solution, the self-assembly process is always a result of subtle balance between attractive (i.e., driving assembly) and repulsive intermolecular forces. The latter serve as a limiting or stopping mechanism and ensure formation of supramolecular structures with well-defined shapes and sizes. In particular, self-assembly of the amphiphilic block-copolymers in aqueous environment is driven by hydrophobic attraction and counterbalanced by electrostatic repulsion or steric hindrance from ionic or non-ionic hydrophilic blocks, respectively.

The morphology of the self-assembled structures is controlled by intramolecular solvophilic/solvophobic balance, which is determined primarily by the lengths of soluble and insoluble blocks, but can be affected also by environmental conditions. As an example, spherical micelles formed in selective solvent by diblock copolymers with one soluble and another insoluble block comprise typically of the order of 10^2–10^3 individual copolymer chains and have dimensions on the order of 10^1–10^2 nm. Copolymers with longer insoluble block associate into cylindrical wormlike micelles that may reach micrometer length or bi-layer vesicles ("polymersomes") with a size of 10^2–10^3 nm.

The electrostatic attraction between oppositely charged ionic macromolecules (polyelectrolytes) provides an alternative (with respect to hydrophobic attraction) mechanism for building up supramolecular assemblies in aqueous media. The association of oppositely charged polyelectrolytes in solutions or at charged interfaces leads, respectively, to interpolyelectrolyte complexes or polyelectrolyte multi-layers. There is an evidence of a formation of soluble interpolyelectrolyte complexes involving oppositely charged macro-ions of different topologies (e.g., linear and branched polyelectrolytes). Co-micellization of a pair of oppositely charged *bis*-hydrophhilic block polyelectrolytes leads to formation of micelles with complex coacervate cores and uniformly mixed or phase-separated coronae. The latter exemplify asymmetric patchy nanoparticles capable to undergo a secondary assembly process.

At the same time, the spectrum of possible assembled structures can be enriched dramatically because of enormous diversity of macromolecular architectures—from simple diblock to multiblock copolymers, comprising of multiple blocks of different chemical nature, from linear to branched macromolecules where different topology (miktoarm stars, graft copolymers, etc.). Furthermore, (bio)nanocolloids, such as globular proteins, can be involved as elementary building blocks in the co-assembly process. This diversity of architectures of building blocks enables going beyond conventional morphologies of self-assembled aggregates and thus fabricating multicompartment, patchy, asymmetric nanoparticles, nanoworms, or nanodisks, that can serve as building blocks for more complex hierarchically assembled structures.

The hierarchical or multi-scale assembly concept assumes that in the first step individual macromolecules assemble into nanoparticles, which can undergo another assembly process into structures that in turn may serve as building blocks for larger supramolecular objects with highly complex internal organization. The success in multi-scale assembly depends crucially on proper encoding of specific properties into primary chemical sequence of the elementary building blocks and precise control and directing of the assembly on each stage. Ultimately, structures as complex as those manufactured by nature can be built up from rationally designed and properly combined macromolecular building blocks by multi-step hierarchical assembly.

In many cases the block copolymer aggregates behave as "frozen" supramolecular structures. Indeed, micellization and formation of mesophases by block copolymers in selective solvents resembles corresponding phenomena in solutions of low-molecular-weight amphiphilies (surfactants). However, the polymeric nature of the assembling species slows down dramatically the dynamics of the assembly process and the exchange rate between associated into superstructures species and individual free macromolecules (unimers) in solution. Thus special efforts are required to make the supermolecular structures capable of "dynamic" response to varied environmental conditions.

In the block copolymer self-assembly, the combination of monomer units with different properties within single functional blocks enables fine-tuning of the strength of intermolecular interactions and achieving perfect control over the thermodynamics and kinetics of the assembly process. An important effort was made and a significant progress was achieved in creating "smart" self-assembled polymer nanostructures, so called because they respond by variation in size, shape, and aggregation state to specific variation in environmental conditions (temperature, pH and ionic strength of the solution, light, etc.). This can be achieved by involving monomers with stimuli-responsive properties (e.g., pH-, themoresponsive) as constituents of the copolymer building blocks.

Moreover, in interpolyelectrolyte complexes or polyelectrolyte multi-layers, the strength of attractive electrostatic interactions can be efficiently tuned by the pH or ionic strength of the solution. Hence electrostatically assembled structures inherently exhibit pronounced stimuli-responsive features. Furthermore, the enormous diversity of possible combinations of co-assembling components, including oppositely charged ionic polymers, nucleic acids and proteins, metal/ligand complexes and inorganic nanoparticles, makes electrostatically driven assembly a very promising approach for design of novel smart functional materials.

Macromolecular self-assembly is a domain of fundamental research and, at the same time, a versatile tool in soft nanotechnology, based on the bottom-up approach, that is exploited to rationally build up structures of almost arbitrary complexity and functionality by directing the assembly routes. This strategy enables one to reach precision and complexity unattainable by top-down methods. Smart nanocontainers, colloidal nanoreactors, molecular templates for nanoelectronic devices are just a few examples of prospective applications.

A completely new and easily scaled up, at the industrial level, approach toward creating polymer nanostructures of various and well-defined morphologies assumes the assembly of amphiphilic block copolymers that occurs simultaneously with their controlled radical polymerization in aqueous medium (so-called polymerization-induced assembly).

Macromolecular assembly at interfaces is considered to be a versatile method of fabrication of ultra-thin coatings with improved (adhesive, tribological, optical, biointeractive, etc.) properties and a controllable nanopatterned structure.

In nanomedicine, self-assembled polymeric nanostructures, specifically diblock copolymer micelles are extensively explored too. A combination of proper micellar size that ensures efficient accumulation and retention in tumor tissues with high stability and the potential for controlled release of cargo through stimuli-triggered dissociation make block copolymer micelles very promising candidates for being exploited as delivery systems for drugs or radionuclides in anticancer therapy and diagnostics.

Biomedical applications give also a strong impulse to study assemblies of biohybrid macromolecules that consists of synthetic (typically hydrophobic) polymer blocks linked to blocks of biological origin (peptides, sugars, oligo- or polynucleotides, polysaccharides). Similar to the synthetic block copolymer, the biohybrid macromolecules demonstrate ability to form micellar-like aggregates, vesicles, or more complex supramolecular architectures in aqueous media. Amphiphilic block polypeptides have demonstrated the ability of stimuli-responsive assembly due to a tuneable hydrophilic/hydrophobic nature of the peptide blocks. The ability of biopolymer blocks to form intra- and intermolecular secondary structure and to take part in (bio)specific interactions opens up a fascinating perspective for the design of novel diagnostic systems or smart vectors that can deliver drugs or biologically active molecules on the basis of supramolecular assemblies of biohybrid macromolecules.

Hence macromolecular assembly is an important field where macromolecular chemistry merges with nanoscience and nanotechnology. Though many excellent books and reviews in this field have been recently published, we consider our present book as a relevant update with its focus on emergent developments in this domain.

Laurent Billon and Oleg Borisov

April 2016

1

A SUPRAMOLECULAR APPROACH TO MACROMOLECULAR SELF-ASSEMBLY: CYCLODEXTRIN HOST/GUEST COMPLEXES

BERNHARD V. K. J. SCHMIDT AND CHRISTOPHER BARNER-KOWOLLIK

Materials Research Laboratory, University of California, Santa Barbara, USA; Preparative Macromolecular Chemistry, Institut für Technische Chemie und Polymerchemie, Karlsruhe Institute of Technology (KIT), Karlsruhe, Germany; Institut für Biologische Grenzflächen, Karlsruhe Institut of Technology (KIT), Eggenstein-Leopoldshafen, Germany

1.1 INTRODUCTION

Macromolecular self-assembly is one of the key research areas in contemporary polymer science. Because complex macromolecular architectures have a significant effect on self-assembly behavior, tremendous effort has been made in the synthesis of well-defined complex macromolecular architectures [1]. The versatility of polymeric materials, such as indicated by polymer functionality, polymer composition, and polymer topology, enables the formation of materials for a broad range of applications, including hybrid materials [2], biomedical materials [3], drug/gene delivery [4], supersoft elastomers [5], and microelectronic materials [6]. In order to obtain well-defined structures, synthetic techniques are required that can provide precise control over the material properties of these structures. Among the polymerization techniques that have proved to be powerful tools for the synthesis of well-defined polymers are reversible-deactivation radical polymerization approaches, such as nitroxide-mediated radical polymerization (NMP) [7], atom transfer radical polymerization (ATRP) [8], and reversible addition-fragmentation

Macromolecular Self-Assembly, First Edition. Edited by Laurent Billon and Oleg Borisov.

chain transfer (RAFT) polymerization [9]. Especially their convenient handling and tolerance toward functional groups have led to a plethora of novel materials with precision-designed properties. Furthermore, the introduction of modular ligation chemistry has provided the opportunity to synthesize complex building blocks and architectures in a precise and efficient manner and again with high functional group tolerance [10]. Several modular ligation reactions are widely utilized in that regard, such as copper(I)-catalyzed azide-alkyne cycloaddition (CuAAc) [11], Diels–Alder reactions [12], and thiol-ene reactions [13]. Thus perfectly suited tools for the formation of materials for macromolecular self-assembly are currently available [14].

The introduction of the concept of supramolecular chemistry has influenced the entire field of chemistry significantly. Especially polymer science and the formation of complex macromolecular architectures have benefited from supramolecular chemistry [15]. New types of macromolecular architectures based on supramolecular bonds are now continually being investigated and higher level complex self-assemblies of macromolecules governed by supramolecular interactions have been formed. Several types of supramolecular interactions are used in polymer science such as hydrogen bonding [16], metal complexes [17], and inclusion complexes [18]. One of the frequently employed supramolecular motifs is cyclodextrin (CD), which forms inclusion complexes with hydrophobic guest molecules in aqueous solution. This property has been exploited readily in polymer chemistry and materials science for various applications, such as drug delivery [19], nanostructures [18b,20], supramolecular polymers [21], self-healing materials [22], amphiphiles [23], hydrogels [24], bioactive materials [25], or in polymerization reactions [26].

The incorporation of CD-based supramolecular chemistry has proved to be an elegant way for the formation of complex macromolecular architectures [14c,24a]. Reversible-deactivation radical polymerization and modular ligation techniques have emerged as effective tools for the synthesis of CD and guest functionalized building blocks. Taking the overall goal of macromolecular self-assembly into account, these building blocks can be considered as the primary structure specifying which blocks are guest and which are host functionalized. The formation of the direct supramolecular host/guest complexes can be considered the secondary structure leading to complex macromolecular architectures. The next level is the assembly of the supramolecularly formed macromolecules into higher aggregates/self-assemblies—the tertiary structure. Thus several levels of molecular complexity are available via the combination of CD host/guest chemistry and polymeric building blocks (Figure 1.1) [18a].

An interesting feature of polymer architectures governed by supramolecular interactions is modularity. The formation of a variety of architectures can be achieved by a small number of initial building blocks much like modularity in modular ligation chemistry. Thus structure–property relationships are accessible via a small amount of reactions compared to traditional material formation. Furthermore, the dynamic nature of the supramolecular bonds affords the opportunity to study systems in the bound as well as the unbound state or to dynamically change the properties of the

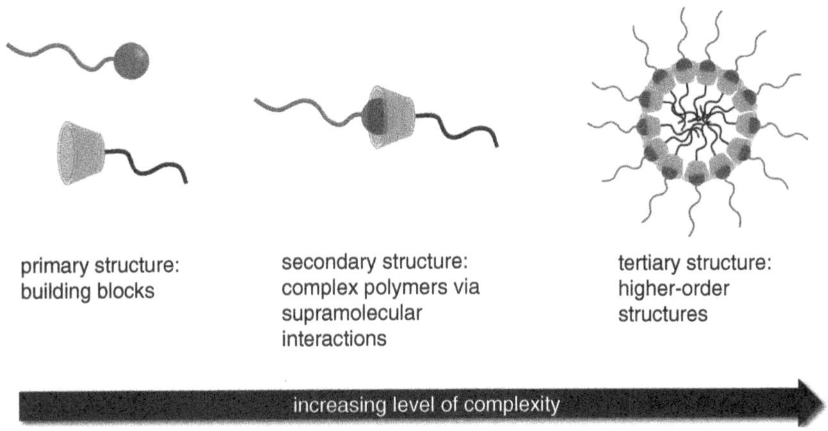

Figure 1.1 Overview over the different levels of complexity enabled via the combination of CD host/guest chemistry and macromolecular structures.

materials via external stimuli or addition of materials with competing supramolecular interactions. Especially in the case of CD host/guest chemistry, a broad range of stimuli-responsive host/guest pairs is available. Combined with stimuli-responsive polymers an extraordinary amount of combinations, and thus materials with unique properties, is accessible.

1.2 SYNTHETIC APPROACHES TO HOST/GUEST FUNCTIONALIZED BUILDING BLOCKS

1.2.1 CD Functionalization

CDs are oligosaccharides and thus contain a significant number of hydroxyl groups that can be utilized for functionalization. Hence selectivity of CD functionalization reactions is a major issue. The primary hydroxyls at C-6 are more reactive due to less steric hindrance, while the secondary hydroxyls at C-2 or C-3 are less reactive. The difference in reactivity gives the opportunity to obtain selectivity with regard to the addressed face of the CD and can be tuned with reaction conditions [27]. The selectivity toward the number of functionalized hydroxyl function remains much more challenging, yet the optimization of reaction conditions has led to several effective protocols to yield—mostly—mono functionalized CDs.

Mono tosyl CDs are the most utilized building blocks because they are readily converted into a variety of useful reactants (Figure 1.2). Several methods have been described for the synthesis of mono tosylated CDs at C-6. The most convenient route for α-CD and β-CD utilizes tosylchloride in aqueous NaOH [28], while another convenient method toward mono tosyl β-CD makes use of 1-(*p*-toluenesulfonyl)imidazole instead of tosylchloride [29]. For γ-CD, a synthesis with triisopropylphenylsulfonyl chloride has been reported in order to form a γ-CD

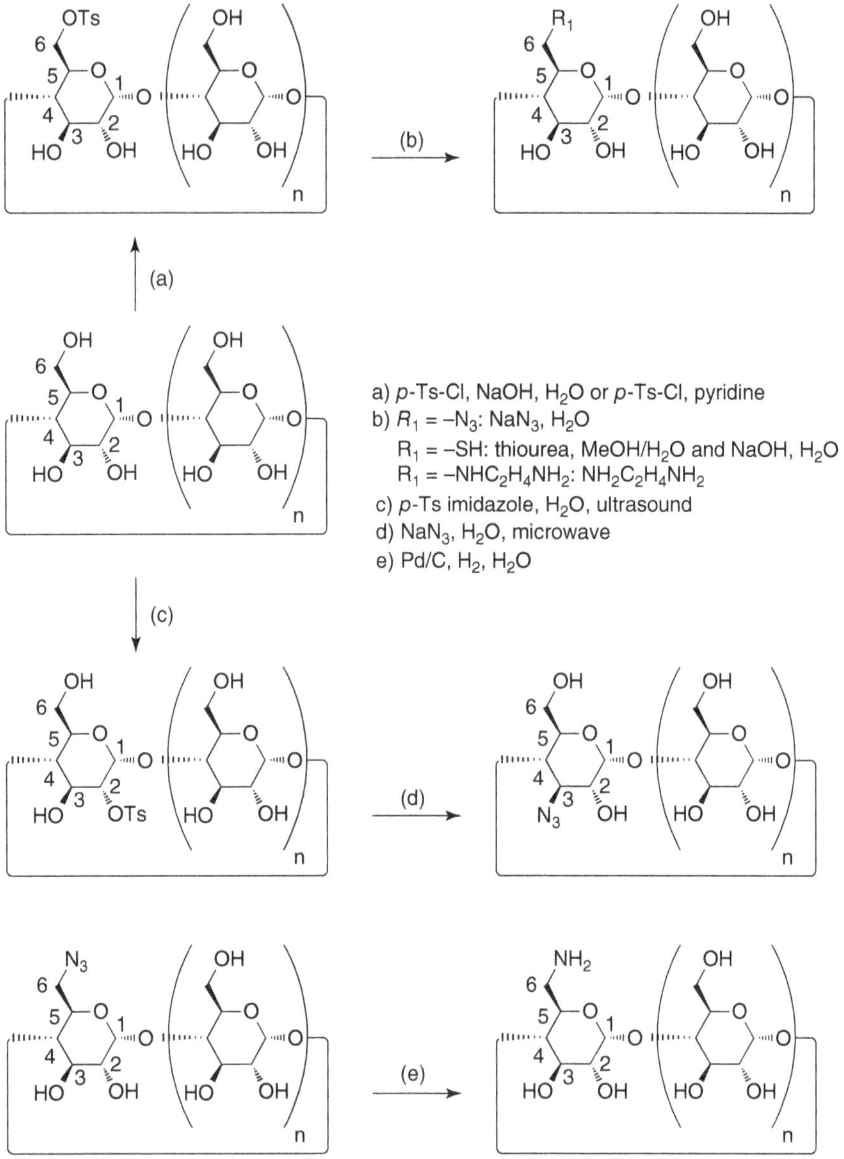

Figure 1.2 Synthesis of various mono functionalized CD derivatives [14c]. Reprinted from [14c]. Copyright 2014, with permission from Elsevier.

derivative with single leaving group [30]. Furthermore, all CD mono tosylates are available via tosylation in pyridine as well [31]. Starting from mono tosylated CD or CDs with similar leaving groups, several useful building blocks are accessible. A nucleophilic substitution with sodium azide leads to the corresponding azides that are suitable for click reactions [31], namely CuAAc. After methyl ether protection, the mono tosylates can be converted into mono alkynes via sodium propargylate, which is the complementary building block for CuAAc in addition to the well-known CD azides [32]. The azides can be further converted to amines via reduction, for example, via hydrogenolysis [31b,33] or Staudinger reduction [31a]. Another possibility to obtain mono amine functionalized CD is the substitution of the mono tosylate with an excess of a suitable diamine [34]. A thiol functionalization is amenable via substitution with thiourea and subsequent hydrolysis [35], which opens up access to thiol-ene click chemistry [36]. Less frequently utilized are C-2 or C-3 substituted CD derivatives, which is most likely due to the inconvenient and tedious synthesis of pure mono functionalized derivatives. Nevertheless, several reports on the synthesis exist [10]. Having several hydroxyl groups, CDs are, in principle, targets for esterification or etherification reactions as well, yet the selectivity in ester/ether functionalization reactions is usually low. Either full conversions of the hydroxyl groups are desired or—in the case of lower targeted substitution grades—complicated purification methods are required in order to obtain pure products. Nevertheless, the broad range of different mono functionalizations of CDs allows for the incorporation into polymers either pre- or post-polymerization. Several examples for CD functionalized polymerization mediators—the pre-polymerization incorporation—are described in the literature, for example, for NMP [37], ATRP [38], and RAFT [39]. Furthermore, post-polymerization conjugations are described as well, for example, after ATRP [38a] or RAFT polymerization [40].

1.2.2 Suitable Guest Groups

Besides functionalization with CDs, guest moieties have to be incorporated in order to form supramolecular host/guest complexes. The common guest groups do not possess a similar multifunctionality as CDs, which makes the pre- or post-polymerization functionalization straightforward. Common routes include esterification, amide formation, or several types of modular ligation reaction.

One of the most interesting features of CD complexes is their response to external stimuli, that is, the complex dissociates and/or associates reversibly due to external stimuli. The stimuli response that all guests share is temperature, namely at higher temperatures the complexes dissociate due to the usually negative association enthalpy (Figure 1.3a) [41]. A further frequently utilized stimulus is redox response based on the ferrocene/CD pair. Oxidation of ferrocene to ferrocenium leads to an increase in size that ultimately leads to complex dissociation, since the ferrocenium cation does not fit into the β-CD cavity (Figure 1.3b) [42]. Furthermore, after reduction, complexation is observed again, which can be followed via cyclovoltammetry [43]. Complexation of phenolphthalein derivatives with β-CD leads to a color change at basic pH from pink to colorless (Figure 1.3c). The lactone ring of phenophthalein forms again at higher pH due to association with β-CD, which forces the molecule into the sterically more compact structure [44]. Very recently, Harada *et al.* showed metal–ion responsive complexation based on bipyridine ligands and iron (II) or copper (II) ions (Figure 1.3d). While bipyridines are complexed with

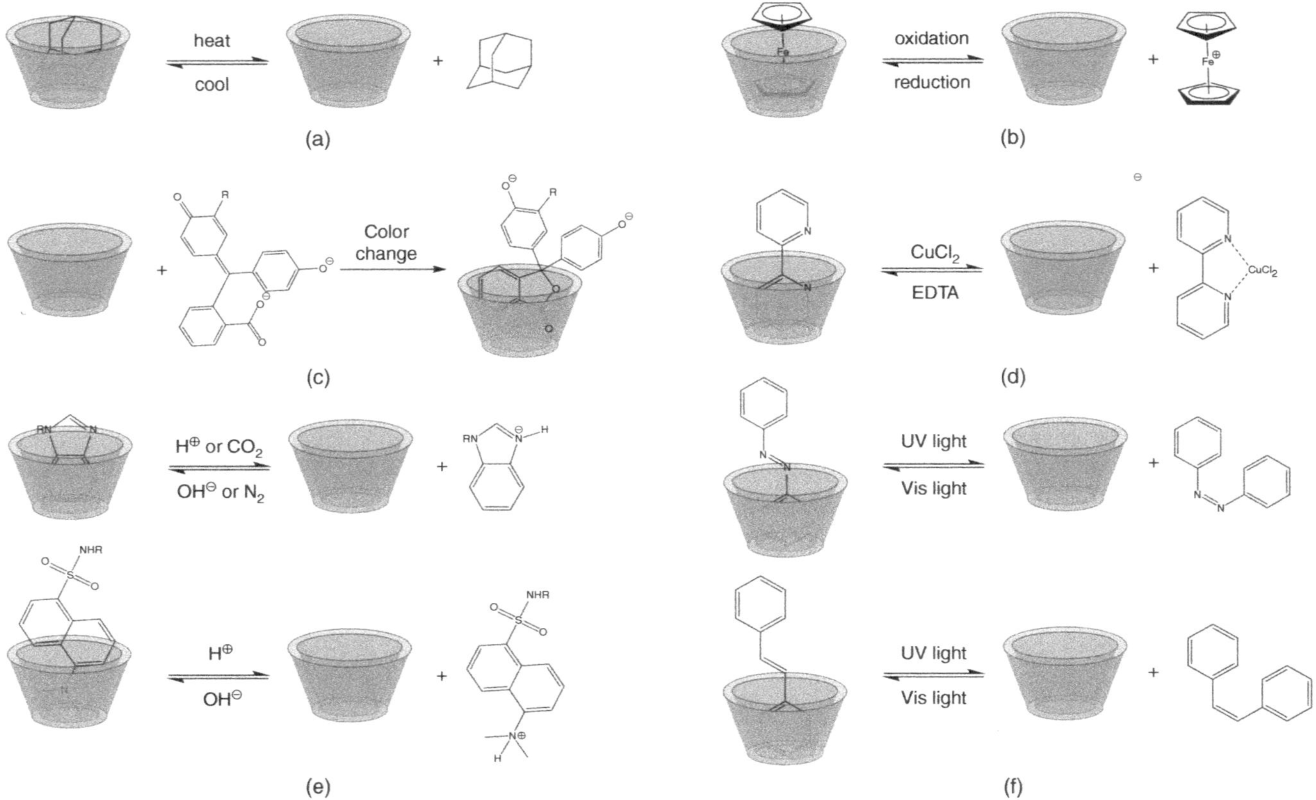

Figure 1.3 Stimuli-responsive host/guest complexation based on β-CD: (a) Thermoresponsive adamantyl complex, (b) redox-responsive ferrocene complex, (c) color changing phenolphthalein complex, (d) metal–ion-responsive bipyridine complex, (e) pH-responsive benzimidazole or dansyl complexes, and (f) light-responsive azobenzene or stilbene complexes.

β-CD in metal-ion free solutions, bipyridine/metal ion complex formation leads to an increase in size of the guest moieties and thus to decomplexation of the CD/bipyridine complex [45]. Recently, pH responsive complexes were introduced. For example, the benzimidazole/β-CD pair shows complexation/decomplexation depending on the apparent pH (Figure 1.3e) [46]. A further development of benzimidazole pH response is protonation in CO_2 enriched aqueous solution. The increased size of the protonated benzimidazole molecule leads to decomplexation, yielding a CO_2 responsive host/guest complex [47]. Dansyl groups show pH responsive complexation with β-CD as well; namely at pH below 4 the complexation is not favored [48]. A very beneficial stimulus is light as it can be controlled spatially and temporarily in a precise way. Common light responsive guest groups that lead to decomplexation upon light irradiation are azobenzenes or stilbenes. UV irradiation induces an isomerization from the thermodynamically more stable *trans* conformation to the *cis* conformation that exhibits lower complexation constants due to steric hindrance (Figure 1.3f). The situation can be reversed via irradiation with visible light, where a re-isomerization takes place and the complexes can form again. A rather biochemical stimulus is enzymatic degradation of CDs that leads to disassembly of the complexes as well, yet in an irreversible fashion [49].

1.3 SUPRAMOLECULAR CD SELF-ASSEMBLIES

After successful formation of building blocks, as described above, supramolecular interactions can be utilized to connect different building blocks in order to obtain complex architectures. Taking the manifold types of guest molecules with their various types of stimuli-responsive complexation into account, a broad range of material properties is accessible. Furthermore, the utilization of different polymer types leads to arguably unlimited possible combinations and more stimuli responses, when stimuli-responsive polymers are incorporated. In the following, several types of CD self-assemblies are presented, such as block copolymers, star polymers, and polymer brushes, leading to single macromolecules connected in a supramolecular way (Figure 1.4). CD complexes have been employed to obtain materials with special polymer functionality, polymer composition, and polymer topology. Polymer functionalities can be obtained via reversible-deactivation radical polymerization of CD and guest functionalized mediators or via modular ligation techniques. Various polymer compositions are available via CD and guest units between blocks in order to obtain supramolecular block copolymers. Complex topologies can be formed via more complex building blocks, such as multi-guest and/or CD functional building species. Complex macromolecular architectures governed by CD complexes can be constructed step by step: the polymer functionality gives rise to more complex compositions or topologies—from the primary structure to the secondary structure.

1.3.1 Linear Polymers

Linear block copolymers are a frequently studied class of CD-based macromolecular architectures. The formation of AB block copolymers is straightforward as only

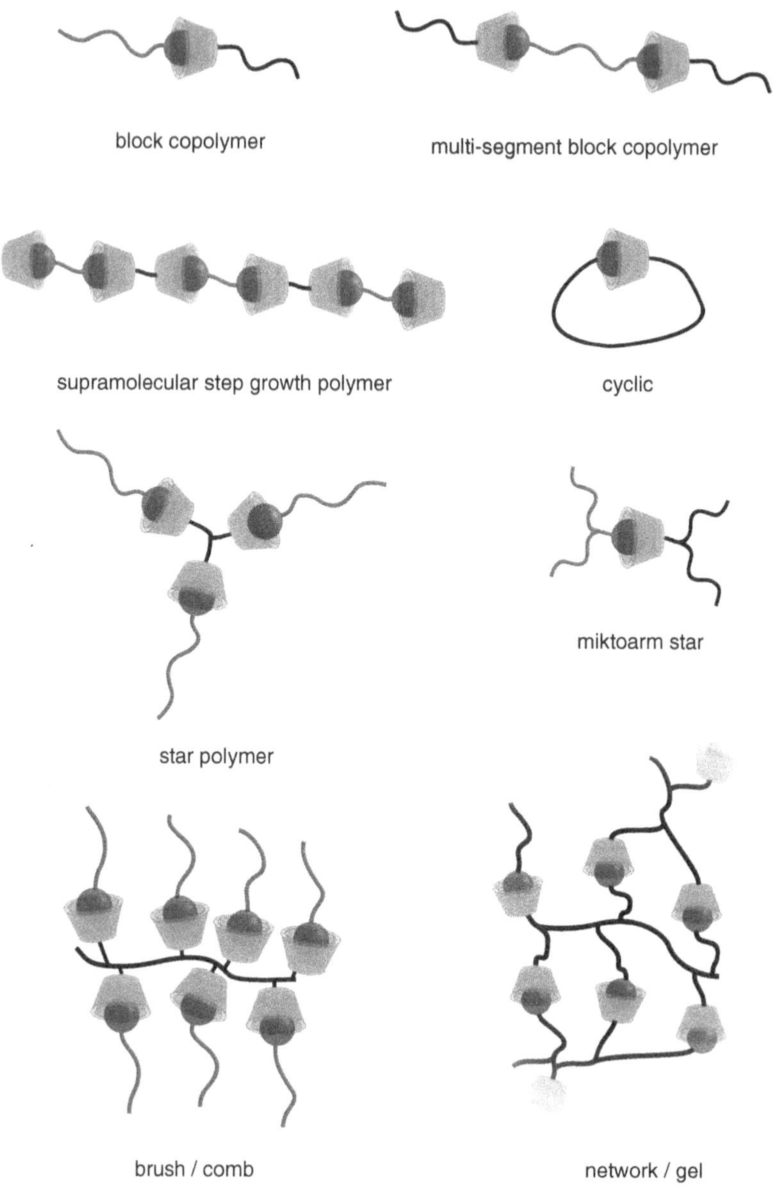

Figure 1.4 Overview of complex macromolecular architectures formed via CD host/guest complexes.

two homo polymers with guest and CD end-group, respectively, are needed. Higher block copolymers are accessible via the introduction of double functionalized middle blocks. The borderline case for higher block copolymers would be supramolecular polymers that are formed from multi-host/guest functionalized building blocks in a supramolecular step growth polymerization mechanism. The degree of polymerization is directly correlated with the number of host–guest complexes formed. This type of linear polymer is based on a step growth reaction approach. Guest and host moieties are combined in an AB- or AA/BB-type fashion to obtain supramolecular polymers.

As described in Section 1.2.3, stimuli-responsive complexation is well known with CDs, and in the following several examples of block copolymers with stimuli-responsive linkage are described. Furthermore, the respective blocks allow the incorporation of additional stimuli response, and in combination, a broad range of multi-stimuli-responsive materials is accessible, giving the opportunity to tailor the polymeric material with regard to application.

1.3.1.1 Diblock Copolymers The first example of CD-based block copolymers was described in 2008 by Zhang *et al.* (refer to Figure 1.5a) [39b]. A CD functionalized poly(4-vinylpyridine) (P4VP) and an adamantyl functionalized

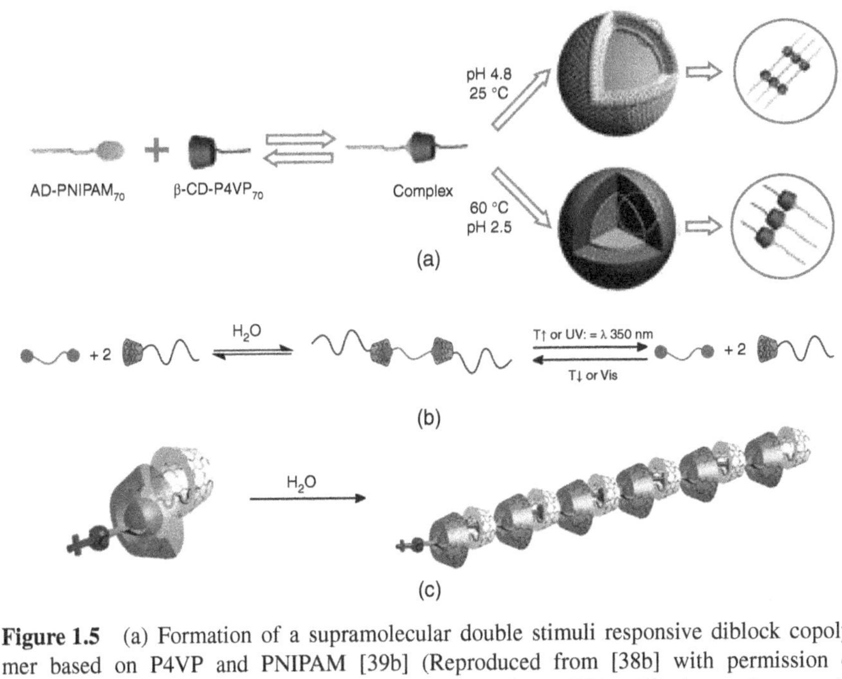

Figure 1.5 (a) Formation of a supramolecular double stimuli responsive diblock copolymer based on P4VP and PNIPAM [39b] (Reproduced from [38b] with permission of The Royal Society of Chemistry), (b) formation of an ABA triblock copolymer with temperature- and light-responsive block junctions [40b] (Adapted with permission from [39b]. Copyright 2013 American Chemical Society), and (c) formation of an AB monomer (α-CD-adamantyl/β-CD-cinnamoyl) based supramolecular alternating α-CD/β-CD copolymer [50] (Adapted with permission from [49]. Copyright 2013 American Chemical Society).

poly(*N*-isopropylacrylamide) (PNIPAM) block were synthesized via RAFT polymerization. The block copolymer was formed via supramolecular host/guest complexation and proved to be pH- and thermoresponsive, which was utilized for stimulus-induced micellization that was investigated via dynamic light scattering (DLS), static light scattering (SLS), fluorescence measurements, and transmission electron microscopy (TEM). The most frequently utilized guest moiety is adamantyl, yet it only provides a temperature responsive connection [40b]. Other examples of diblock copolymers based on CD/adamantyl complexation include poly(2-methyl oxazoline)-*b*-PNIPAM [38b], poly(*N*,*N*-dimethylaminoethyl methacrylate)-*b*-PNIPAM (PDMAEMA-*b*-PNIPAM) [38a], and poly(methyl methacrylate)-*b*-poly(hydroxyethyl acrylate) (PMMA-*b*-PHEA) [36]. A voltage/redox responsive block copolymer was presented by Yin *et al.*, where a ferrocene functionalized poly(ethylene glycol) (PEG) was connected to a β-CD functionalized poly(styrene) (PS) [51]. Vesicles were formed that were prone to disruption by an application of external current and small molecule release was probed. More recently, Yuan *et al.* presented a block copolymer of poly(lactic acid) (PLA) and PEG [52]. A pH sensitive block copolymer amphiphile was formed by He *et al.* [46]. Benzimidazole functionalized poly(ε-caprolactone) (PCL) was connected to β-CD functionalized Dextran and utilized as biodegradable drug delivery vehicle upon micelle formation in neutral aqueous solution. Drug release of Doxorubicin was studied and was supported by the difference of intra and extra cellular pH. A CO_2 responsive AB block copolymer was described by Zhao *et al.* (refer to Figure 1.8a) [47]. A β-CD functionalized Dextran was coupled to a benzimidazole functionalized poly(L-valine) in dimethylsulfoxide. Addition of water led to the formation of nanostructures depending on the degree of polymerization (DP) of the poly(L-valine) block. Vesicles were obtained for similar DP of dextran and poly(L-valine), while fiber-like structures were obtained for higher DPs of poly(L-valine). A photoresponsive block copolymer was described by Yuan *et al.* (refer to Figure 1.8b) [53]. The supramolecular block copolymer was based on PCL-*b*-poly(acrylic acid) (PCL-*b*-PAA) with azobenzene and α-CD end-groups, respectively. In aqueous solution nanotubes were formed that were disassembled upon UV irradiation. Furthermore, Rhodamine B was released from the nano tubes via light irradiation.

1.3.1.2 Higher Order Block Copolymers While diblock copolymers are described frequently, multi-block copolymers are underrepresented so far. Our team prepared an ABA triblock copolymer based on β-CD featuring thermoresponsive and light responsive connections, namely adamantyl or azobenzene guests (refer to Figure 1.5b) [40b]. Poly(*N*,*N*-dimethylacrylamide) (PDMA) and poly(*N*,*N*-diethylacrylamide) (PDEA) middle blocks were connected with biocompatible poly(*N*-2-hydroxypropyl methacrylamide) (PHPMA) outer blocks. The block formation and dissociation upon external stimuli was investigated via DLS and nuclear Overhauser enhancement spectroscopy (NOESY). Furthermore, the temperature-induced aggregation due to thermoresponsive PDEA blocks was studied via temperature sequenced DLS and turbidimetry showing a two-stage aggregation.

Another example of supramolecular ABA block copolymers was described by Zhang *et al.* [54]. PS with an adamantyl group on one end and an azobenzene group on the other end were complexed with β-CD functionalized PEG. Vesicles were formed in aqueous solution, characterized via TEM and DLS. The response of the vesicles to photo irradiation was probed as well, indicating a change in morphology toward micelles. A pH/CO_2 responsive ABC block copolymer was described by Yuan *et al.* [55]. RAFT-derived PNIPAM was subjected to aminolysis in order to obtain thiol functionalized PNIPAM. Furthermore, methacrylate and adamantyl functional hetero telechelic PCL and β-CD end-functionalized PDMAEMA were connected in one pot via a combination of a thiol-ene reaction and supramolecular complexation. Vesicles with variable size, depending on CO_2 or N_2 stimulation, were obtained via the pH responsive PDMAEMA. Furthermore, these vesicles could be transformed into micelles via heating due to the collapse of the PNIPAM block.

1.3.1.3 Supramolecular Step Growth Polymers An alternative to linear polymers based on supramolecular CD interactions is based on a step polymerization analogue [21a,56]. Host and/or guest functionalized small molecules are joined, leading to a polymer formed by multiple host/guest complexes. Either an AB or an AA/BB approach is utilized to form such polymers. Harada *et al.* utilized an α-CD functionalized with a cinnamoyl group as an AB monomer to form supramolecular polymers in the fashion of a daisy chain [57]. An AA/BB approach was undertaken by the same group [58]. A double adamantyl functionalized molecule was complexed with a β-CD dimer. Depending on the rigidity of the spacer between the adamantyl moieties, cycles, or linear polymers were obtained. The combination of different CDs was probed by Harada *et al.* as well, utilizing the strong complexations between the pairs of α-CD/cinnamoyl and β-CD/adamantyl, such as the combination of an α-CD/adamantyl and a β-CD/cinnamoyl based linker (refer to Figure 1.5c) [50]. Ritter *et al.* presented a linear supramolecular polymer based on a PDMS backbone and β-CD/adamantyl or ferrocene complexation that showed redox response [59]. A ternary supramolecular polymer was described by Liu *et al.* A naphthol functionalized β-CD was combined with an adamantyl-viologen dilinker and cucurbit[8]uril [60]. The β-CD complexes with the adamantyl moiety, while the cucurbit[8]uril complexes with the viologen and naphthol units. Thus a supramolecular polymer is formed via the utilization of two host/guest complex systems. A similar approach was performed by Zhang *et al.* [61]. Metal-complexes were utilized in the formation of supramolecular polymers as well: Tian *et al.* utilized a pyridine functionalized β-CD and a double azobenzene end-functionalized linker molecule [62]. The azobenzene functionalized linker was complexed by two pyridine containing β-CD units. Addition of a Pd (II) ethylenediamine salt led to polymer formation as the β-CD units were linked via metal-complex formation of two pyridines and a Pd (II) complex. The formation of the poly(pseudorotaxane) was evidenced by atomic-force microscopy (AFM) and NOESY. Zhang *et al.* presented a tripeptide (Phe-Gly-Gly) functionalized with an azobenzene that was complexed with cucurbit[8]uril in the ratio 2:1, which led to dimerization via complexation of two phenylalanine units [61]. The exposed azobenzenes were complexed with a β-CD dilinker, ultimately forming a supramolecular polymer. A difunctional β-CD molecule and a cationic difunctional ferrocene molecule were utilized in an AA/BB fashion to obtain redox responsive supramolecular polymers that show interesting gene vector abilities [63]. Tian *et al.* reported a photoresponsive supramolecular polymer based on γ-CD [64]. Two coumarin units were connected by a viologen unit. After addition of γ-CD in water, a ternary complex of two coumarin units originating from two different linker molecules and γ-CD was formed. Thus a step growth polymer was achieved, linked by plain γ-CD molecules. Furthermore, the coumarin units could be photo dimerized inside of the γ-CD cavity to obtain covalently bound polymers with threaded γ-CDs. A similar approach was performed by Ma *et al.* [65]. A viologen functionalized coumarin was complexed by a bis-sulfonatocalix[4]arene, which prefers the complexation of the viologen unit. Addition of γ-CD leads to the complexation of two respective coumarin units, forming a supramolecular polymer. As evidenced by TEM and DLS, the supramolecular polymer formed several hundred nanometer long fibers in solution. An interesting structure was presented by Sollogoub *et al.* who reported supramolecular polymers of α-CD azides in the solid state [66]. While α-CD C-6 mono azides lead to single-strand supramolecular polymers due to complexation of the azide by another α-CD, double azide functionalized α-CD showed higher interactions. In addition to the primary interaction, namely the complex formation of the azide and an α-CD, an azido–azido dipolar interaction is evident. Furthermore, a tertiary interaction—namely an azido hydrogen bonding —takes place. In sum, the contributions from the different interactions led to hierarchical supramolecular polymers that show a helical morphology.

1.3.2 Branched Polymers

Branched architectures have attracted significant attention in recent years. Especially the area of hydrogels has been investigated extensively, for example, with regard to their mechanical properties, stimuli response, and self-healing. As CD complexes are an ideal system for aqueous environments, prospective applications in biomedical sciences have driven progress in this field. Less pronounced branched structures are star polymers with a dominant quantity of examples of CD centered star polymers. Compared to the manifold examples of CD centered star polymers, significantly fewer reports on star architectures driven by CD host/guest complexes have been described in the literature. Nevertheless, several examples were described so far, especially with regard to stimuli responsive structures. Furthermore, supramolecular brush polymers have been added more recently to the portfolio of CD-based branched architectures.

1.3.2.1 Hydrogels By far the most studied type of CD-based branched architectures are hydrogels [24a,67], yet mostly free radical polymerization is utilized to synthesize the utilized building blocks or the branching points are distributed randomly, as in the case of α-CD/PEG crystallization driven networks [68]. A few examples utilize controlled radical polymerization techniques to obtain well-defined multiple host/guest containing polymers. Hetzer *et al.* utilized a trifunctional β-CD linker and double adamantyl end-functionalized PDMA to obtain networks

[69]. Depending on CD/guest ratios, chain length, and concentration, increasing viscosities were determined via rheological investigations. Addition of free CD or competing guest molecules led to a significant decrease in viscosity. Redox responsive networks were described by Zhang *et al.* [70] as well as Yuan *et al.* [71]. The former utilized two- or three-fold β-CD linker molecules and ferrocene functionalized PEI. The latter formed a hydrogel from β-CD and ferrocene functionalized PDMA. Photoresponsive hydrogels were described by Wang *et al.* [72]. A difunctional β-CD linker with disulfide connection was utilized to form a network with a copolymer of NIPAM and an azobenzene monomer. Thus a redox and light responsive hydrogel was formed.

1.3.2.2 Star Polymers An important type of branched architecture is the star polymer. Very frequently described are CD centered star polymers that utilize the high functionality density of CD in order to form a star core [14c,73]. In contrast, CD host/guest complexes have been employed less frequently for the formation of star polymers. Wenz *et al.* described a connection of both approaches, namely α-CD centered PMMA stars were threaded onto PEG in a rotaxane fashion with large stopper molecules [74]. Similarly Yui *et al.* investigated α-CD/PEG rotaxanes with PLA grafted α-CD [75]. A sliding rotaxane brush example was provided by Gao *et al.* [76]. An α-CD threaded PEG backbone was stoppered with β-CD. The CDs were functionalized with azides via esterification with an azide functionalized acid chloride. Subsequently, alkyne functionalized PEGs and alkyne functionalized palmitic acid (C16) were attached via CuAAc forming sliding brushes with tunable amphiphilicity. A β-CD centered PNIPAM star was synthesized by Li *et al.* [77]. Addition of a multiadamantyl functionalized PEG star led to the formation of a supramolecular multi block star polymer. Furthermore, heating above the cloud point of the PNIPAM blocks resulted in hydrogel formation.

In the area of mere star architectures driven by CD host/guest complexes, our team reported the formation of three-arm star polymers via supramolecular interactions of adamantyl end-functionalized PDMA or PDEA and a β-CD trilinker (refer to Figure 1.6a) [78]. The formation of supramolecular star polymers was probed via DLS and rotating frame nuclear Overhauser effect spectroscopy (ROESY) as well as turbidimetry measurements in the case of PDEA arms. Another three-arm star polymer was described by Wu *et al.* that connected guest functionalized oligo ethylene glycol dendrimers to a β-CD trilinker [81]. A significant effect of dendrimer generation, dendrimer end-group (ethyl or methyl), and ratio of different dendrimer types (with ethyl or methyl end group) on the thermoresponse of the supramolecular star polymers was measured. The effect of thermally induced decomposition of the supramolecular complexes as well as the effect of salt concentration on the thermoresponsive behavior was investigated, too. More complex AB_2 and A_2B_2 miktoarm star polymers were described by us (refer to Figure 1.6a) [79, 80]. Adamantyl mid functionalized and β-CD end-functionalized or adamantyl mid and β-CD mid functionalized polymers were utilized, respectively. Again, turbidimetry, DLS, and NOESY/ROESY were utilized to characterize the complexes. Similarly important is the investigation of temperature-induced micellization caused by

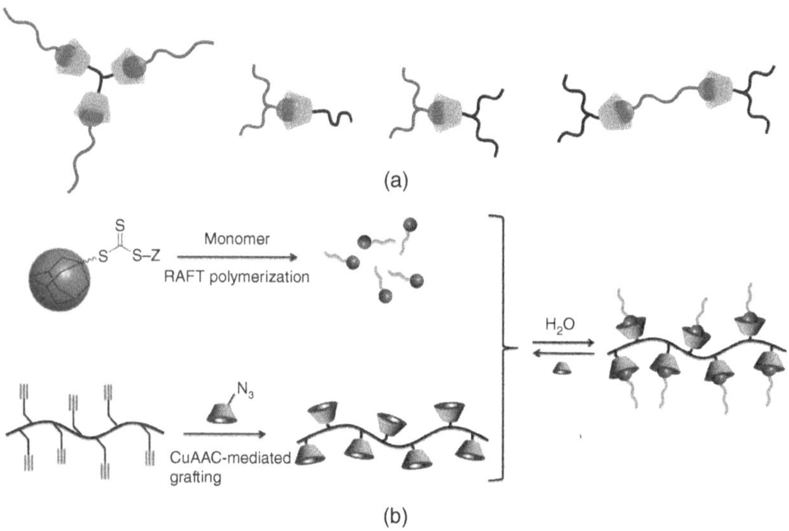

Figure 1.6 (a) Various supramolecular star architectures from left to right: Three arm stars [78], A_2B miktoarm stars [79], A_2B_2 miktoarm stars [80], and H-shape stars [80]. (b) Supramolecular brush formation via adamantyl end-functionalized PAA and CD conjugated poly(acrylate) [40a]. Reproduced from [39a] with permission of The Royal Society of Chemistry. (*See color insert for color representation of this figure*).

thermoresponsive PDEA blocks. A two-stage aggregation was detected, which was attributed to initial micelle formation close to the cloud point of the thermoresponsive block, followed by aggregation upon further heating of the complexes, which is a behavior well known in literature [38a,82]. An AB_4 miktoarm star polymer was described by Allcock *et al.* [83]. Four poly(oligo ethyleneglycol methacrylate) (POEGMA) arms were grafted from β-CD and an adamantyl end-functionalized poly(bis-(trifluoroethoxy)phosphazene) was added. Micelle formation was evident and characterized via DLS, TEM, and AFM. Even more complexity was obtained by Zhu *et al.*, who described an ABC miktoarm star polymer based on several functionalization reactions of β-CD [84]. Mono azido functionalized β-CD was mono tosylated. An alkyne functionalized PEG was attached via CuAAc and the remaining tosylate substituted with an azide. Subsequently, an ATRP initiator was incorporated via another CuAAc and DMAEMA was grafted from the β-CD core via ATRP. Then, adamantyl end-functionalized PMMA was connected to the two arm β-CD core via supramolecular interactions to afford an ABC miktoarm star polymer that formed micelles in aqueous solution due to the hydrophobic PMMA block.

The utilization of double end-functionalized polymers gives the opportunity to form more complex structures. Schmidt *et al.* described the formation of H-shape star block copolymers via double guest functionalized middle blocks and mid β-CD functionalized outer blocks [80]. Again, DLS, NOESY, and temperature-induced micellization were investigated, with results in line with previous observations, namely an aggregation that proceeds in two stages with small micelles at lower temperatures and agglomeration at higher temperatures. Doubly guest functionalized PEG and β-CD centered PNIPAM four-arm star polymers were utilized by Li *et al.* [85]. The formation of the dumbbell-shaped architecture was proved via NOESY and turbidimetry. Subsequently, the authors exchanged the middle PEG block with double adamantyl functionalized poly(propylenegylcol) (PPG), leading to a system with two different thermoresponsive polymers [82]. The temperature-dependent aggregation was studied in detail via DLS, NMR, fluorescence measurements, AFM, and TEM. In cold water, dumbbell-shaped copolymers were obtained. At temperatures over 8°C, micelle formation was observed, whereas micelle destabilization and aggregation was the case above 22°C.

Another type of branched star-like structures utilizes less well defined cores. A spherical core moiety is grafted with CD molecules and subsequently guest functionalized polymers are attached in a supramolecular fashion. In that regard mostly inorganic core material has been used so far, such as SiO_2 nanoparticles [86], CdS quantum dots [87], and gold nano particles [88]. Several guest/CD pairs were exploited to equip the formed hybrid stars with stimuli-responsive properties, such as light response [39a,87a,88b] or redox response [87b]. Furthermore, thermoresponsive blocks were added in several cases for temperature-induced hydrogelation in order to form hybrid hydrogels [87]. Polyhedral oligomeric silsesquioxane was decorated with β-CD by Li *et al.* [89]. The amphiphilic molecule self-assembled into nanospheres in aqueous solution and azobenzene end-functionalized PEG-*b*-PDMAEMA was added to form supramolecular complexes. The light-responsive azobenzene/CD complexation allowed for the formation or dissociation of the polymer shell, while the size of the micelles could be tuned via pH. Chen *et al.* presented a supramolecular star polymer generated via a combination of CD-based host/guest chemistry and a protein [90]. Concavalin A (ConA) was utilized as a core that interacts with mannopyranoside. Furthermore, a mannopyranoside/β-CD functionalized dilinker was synthesized and the complex association between ConA and the mannopyranoside was investigated via isothermal titration calorimetry. Further recognition was found when an adamantyl end-functionalized PEG was added. The CD/guest interaction led to the formation of a supramolecular four-arm star polymer. Again, isothermal titration calorimetry was performed to study the complexation. Addition of α-CD led to the formation of hydrogels via α-CD/PEG complex formation. Ritter *et al.* combined CD end-functionalized hyperbranched polyglycerols with azobenzene end-functionalized PNIPAM [91].

1.3.2.3 Brush Polymers Bottle brush polymers are another type of branched architectures that have found CD-based supramolecular analogues. Two major factors have to be considered in the formation of brush polymers. First, the formation has to be via a grafting-to approach, which results from the supramolecular nature of the grafting. Thus one has to decide whether the CD units should be bound to the backbone or as end groups of the grafts. Second, the grafting density depends strongly on the steric hindrance induced by large CD molecules. Furthermore, the grafting density depends on the association constants of the utilized CD/guest pair. Bernard *et al.* formed supramolecular brushes of adamantyl functionalized PAA and a poly(acrylate) backbone with adjacent CD units that were attached via CuAAc (refer to Figure 1.6b) [40a]. DLS and NOESY were utilized to prove the complex formation. The group of Ritter in cooperation with our team described the formation of a supramolecular brush polymer that shows changes in optical properties upon complex and thus brush formation [44a]. A CD end-functionalized PDEA was added to phenolphthalein functionalized PDMA. Phenolphthalein—a well-known pH indicator —changes its color from pink to colorless at pH > 8.4 via the opening of its lactone-group. In the presence of β-CD, the lactone ring reforms—even at higher pH—due to the formation of a host/guest complex that forces the molecule into the sterically more compact lactonoid structure. The complex formation could be followed via DLS, NOESY, and UV-VIS. Furthermore, the complex formation was visible to the naked eye. Redox-responsive brush polymers were described by Ritter *et al.* [43] as well as by Yuan *et al.* [92]. The former utilized a ferrocene functionalized acrylamide and DMA derived copolymers in conjunction with CD end-functionalized PDEA. The redox response of the formed supramolecular brushes was probed via cyclovoltammetry. Yuan *et al.* utilized PEG-*b*-poly(glycidyl methacrylate) (PEG-*b*-PGMA) that was reacted with mono thiol functionalized β-CD to obtain a β-CD functionalized block [92]. Moreover, ferrocene end-functionalized PCL was utilized to form supramolecular brush copolymers, which was probed via NOESY and UV-Vis. Self-assembly into micelles was investigated via TEM and DLS also with regard to an electrochemical redox response. Further, Yuan *et al.* prepared a photoresponsive brush polymer [93]. Ethyl cellulose was grafted with PCL and CD was conjugated to the PCL end-groups. The supramolecular brush was formed via addition of azobenzene functionalized PEG. In aqueous solution micelles formed with hydrophobic ethyl cellulose core and PEG corona. The structures were studied via TEM and DLS also with regard to their photoresponse. A rather rare polymer type was utilized by Allcock who utilized poly(phosphazene) with multiple β-CD grafts and formed brush polymers with adamantyl end-functionalized poly(phosphazene) [83].

A supramolecular dendronized polymer brush was described by Zhang *et al.* [81]. Adamantyl functionalized oligo ethylene glycol dendrons were attached onto a CD functionalized backbone and the thermoresponsive behavior was studied. A broad range of LCSTs from 34°C to 56°C was accessible via changes in dendrimer generation or hydrophobicity of dendrimer end groups. The dehydration and collapse of the oligo ethylene glycol chains leads to disassembly of the host/guest complexes as indicated by NMR and isothermal titration calorimetry data. Frey *et al.* studied the brush formation of CD functionalized PHPMA with adamantyl functionalized PEG-*b*-hyperbranched poly(glycerol) [94]. The efficiency of supramolecular brush polymer formation was studied in detail via ITC and DOSY. It was evidenced that the association constants depend strongly on the degree of branching of the poly(glycerol) block as well as the presence and the length of introduced PEG spacers. Nevertheless, the association constants have a final value independent of the degree of branching already at moderate molecular weights of the poly(glycerol) block.

1.3.3 Cyclic Polymer Architectures

Rather uncommon structures are cyclic polymers with a supramolecular CD-based ring closure. Harada *et al.* showed a CD-based macrocycle based on the PEG platform with azobenzene guests that showed the ring opening depending on temperature and light irradiation [95]. Recently Willenbacher *et al.* presented a β–CD/adamantane based macrocycle utilizing RAFT derived polymers [96]. Although only PDMA was probed in the present study, a broad range of polymer macrocycles is, in principle, available via utilization of different water soluble monomers and RAFT polymerization. The ring formation was proved via NOESY, DLS, and diffusion-ordered NMR spectroscopy (DOSY). Furthermore, the temperature response of the ring closure was probed.

1.4 HIGHER ORDER ASSEMBLIES OF CD-BASED POLYMER ARCHITECTURES TOWARD NANOSTRUCTURES

A variety of supramolecular-based macromolecular architectures has been described so far, as shown in the previous section. Starting from various building blocks—the primary structure—different assemblies were obtained—the secondary structure. Yet the obtained assemblies can be considered as single macromolecules if strong association is operational. The next step to more complex materials would be an assembly of CD host/guest bound macromolecules forming a tertiary structure. Ultimately, these ensembles of CD complex governed macromolecules yield a higher molecular level of complexity and a new level of properties as well. Again, a broad range of stimuli responsive CD/guest pairs and polymers is available rising the opportunity to form a significant amount of novel materials. Technically hydrogels belong to this category as well and the reader is referred to Section 1.3.2.1 for a brief description.

1.4.1 Micelles/Core-Shell Particles

The simplest higher order structures are micelles or core-shell particles. In the same way as covalently bound polymers, amphiphilic supramolecular block copolymers can be utilized to form micelles. As micelles are of particular interest in drug delivery, in that delivery of hydrophobic drugs/cargoes, supramolecular block copolymers, are the object of research to perform drug delivery tasks. Especially when triggered release of drug cargo is considered, stimuli-responsive block connections or stimuli-responsive blocks seem to be a tool of choice.

Zhang *et al.* described a supramolecular connected micelle in 2008 [39b]. The adamantyl/β-CD interaction was utilized to form a diblock copolymer of P4VP and PNIPAM that was able to form micelles upon pH or temperature stimuli. Similar strategies have been employed for a variety of supramolecular diblock copolymers [36,38a,46]. Micelles were also formed from supramolecular—mainly miktoarm—star polymers, such as a hydrophilic β-CD centered four-arm POEGMA complexed with a hydrophobic adamantyl end-functionalized P(bis-(trifluoroethoxy)phosphazene) [83]. Another example is a β-CD conjugated with PEG and PDMAEMA that was complexed with an adamantyl end-functionalized PMMA yielding an amphiphilic structure [84]. A micelle-forming brush copolymer was described by Yuan *et al.*, who utilized ferrocene end-functionalized PCL as hydrophobic and biodegradable block and a PEG-*b*-PGMA block conjugated with β-CD, yielding redox-responsive micelles [92]. Micelles were also formed from PCL grafted ethyl cellulose with attached β-CD that was complexed with azobenzene end-functionalized PEG [93]. A β-CD centered seven-arm poly(L-glutamic acid) with the drug *cis*-dichlorodiamine platinum (II) attached to the arms was described by Liu *et al.* [99]. Furthermore, supramolecular complexes with adamantyl functionalized PEG were formed. The formed supramolecular miktoarm star polymers assembled into micelles that were capable of drug release. A brush polymer forming micelles has been described by Jiang as well (refer to Figure 1.7a) [97]. A poly(*N*-vinyl-2-pyrrolidone) copolymerized with β-CD containing monomers was grafted with adamantyl end-functionalized PCL. Characterization via TEM, DLS, and AFM revealed multicore micelles in the case of preparation under nonequilibrium conditions, whereas usual core-shell micelles were obtained under equilibrium conditions. Li *et al.* utilized *O*-isopropylidenation of several hydroxyl groups in α-CD or β-CD in order to protect the hydroxyl groups as acetals and alter the properties of the molecules [100]. Furthermore, α-CD or β-CD polymers were formed by means of epichlorohydrin-induced polycondensation under basic conditions in water and the obtained polymeric hydroxyls could be converted into acetals as well. In the case of acetalated poly(β-CD), addition of an adamantyl functionalized PEG led to the formation of micelles or cylindrical assemblies in water, which could be controlled via the PEG/poly(β-CD) ratio. Due to the acid labile acetal protection, the assemblies showed a pH-dependent disintegration. The

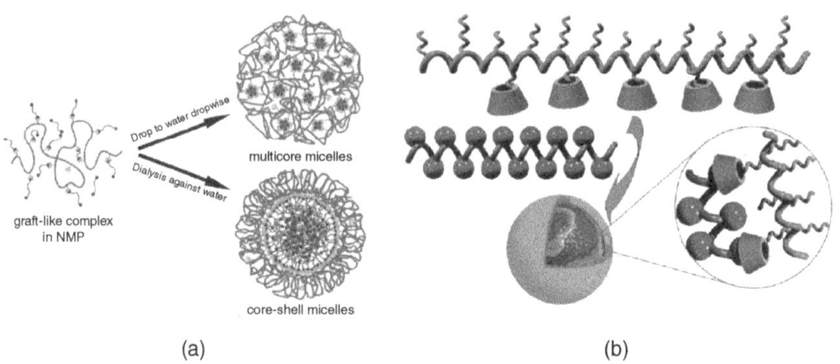

Figure 1.7 (a) Formation of micelles based on poly(*N*-vinyl-2-pyrrolidone) with β-CD side chains and adamantyl end-functionalized PCL leading to different micelle architectures depending on the preparation technique [97] (Reproduced from [96] with permission of John Wiley and Sons) and (b) formation of micelles based on β-CD conjugated poly(ethylene imine) and poly(β-benzyl-L-aspartate) [98] (Reproduced with permission from [97]. Copyright 2010 American Chemical Society). (*See color insert for color representation of this figure*).

rate of hydrolysis was controlled via the degree of acetylation. A micellar aggregate was described by Jiang *et al.*, too [101]. A poly(methacrylate) with β-CD in the side chains was complexed with a copolymer of poly(*tert*-butyl acrylate) (PTBA) and an adamantyl containing monomer. The hydrophobic adamantyl containing PTBA block formed the core that was bound to the shell-forming hydrophilic CD containing block via supramolecular interactions. Furthermore, the shells were cross-linked and the core removed in order to obtain β-CD containing nanocages and the surface of the micelles/nanocages could be modified via guest addition, for example, cationic or anionic adamantyl derivatives. A similar approach was described by Ma *et al.* (refer to Figure 1.7b) [98]. A β-CD conjugated poly(ethylene imine) (PEI) was combined with a poly(β-benzyl-L-aspartate) and core-shell particles were formed via the dialysis method, leading to a complex formation of the benzyl groups and β-CD. Furthermore, drug-loading/release was studied as well as plasmid DNA (pDNA) loading and delivery *in vitro*.

1.4.2 Vesicles

More complex macromolecular assemblies are vesicles. While micelles have two interfaces, namely the interface between the blocks and the interface shell/solvent, vesicles have three interfaces, namely the interface between the blocks and the interface between inner shell/solvent and the interface between outer shell/solvent. Thus supramolecular-based vesicles are less frequently described in the literature. Nevertheless, vesicles are very well suited for cargo delivery/release, especially of hydrophilic molecules, and thus are an important target structure for polymer chemists.

Vesicles were described for several systems, such as the supramolecular diblock copolymer. A β-CD functionalized PS and a ferrocene functionalized PEG were utilized to form vesicles that could be disrupted via a redox stimulus [51]. A supramolecular diblock copolymer of β-CD functionalized dextran and benzimidazole functionalized poly(L-valine) formed vesicles for similar DP of dextran

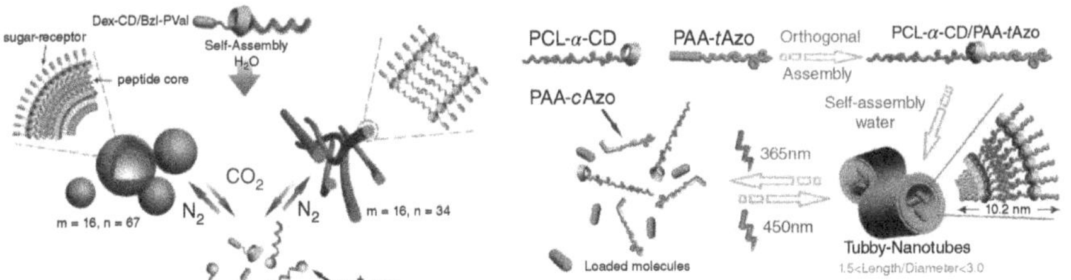

Figure 1.8 (a) Formation of pH/CO_2 responsive vesicles or fibers based on β-CD/benzimidazole interactions [47] (Reprinted with permission from [46]. Copyright 2014 American Chemical Society) and (b) formation of supramolecular light-responsive nanotubes based on azobenzene/α-CD interactions [53] (Reproduced from [52] with permission of The Royal Society of Chemistry). (*See color insert for color representation of this figure*).

and poly(L-valine), while fiber-like structures were obtained for higher DPs of poly(L-valine) (refer to Figure 1.8a) [47]. Jiang *et al.* prepared vesicles with doubly β-CD end-functionalized poly(ether imide) [102]. Furthermore, the inner and outer surface could be modified via host/guest complexes with adamantyl functionalized PEG. Li *et al.* presented recently a β-CD fully functionalized at C-6 with hydrophobic ortho esters that formed vesicles or spherical nanoparticles after complexation with adamantyl end-functionalized PEG [103]. The size and morphology of the aggregates could be adjusted via the chain lengths of pendant aliphatic chains attached to the ortho esters and feed ratio of the components. Furthermore, the surface of the aggregates could be modified via addition of ionic guest compounds as shown via zeta potential measurements. Due to the incorporation of acid labile ortho esters, the aggregates were dePEGylated at pH of 7.4 and finally disassembled completely at pH 6.4. Yan *et al.* presented a Janus-type hyperbranched polymer [104]. A β-CD centered hyperbranched polyglycerol and a hyperbranched poly(3-ethyl-3-oxetanemethanol) with an azobenzene at the apex were complexed via azobenzene/β-CD host/guest interaction. The complex formed vesicles in aqueous solution that could be disassembled into unimers after irradiation with UV light. The example of an ABA block copolymer with a hetero bifunctional middle block by Zhang *et al.* makes use of a PS with an adamantyl group on one end and an azobenzene group on the other end [54]. Complexation with β-CD functionalized PEG leads to vesicles, yet the morphology changes upon light irradiation due to the change in the block composition from ABA to AB after dissociation of the azobenzene/β-CD complexes. Another system that transforms from vesicles to micelles after application of an external stimulus—namely temperature—consists of a PNIPAM-*b*-PCL that is connected to PDMAEMA via adamantyl/β-CD interaction [55]. While the vesicle size could be tuned via CO_2 or N_2 pressure, the change in morphology toward micelles could be achieved via heating and collapse of the PNIAPM block. Ravoo *et al.* formed vesicles of hydrophobically decorated β-CD and added adamantyl functionalized maltose or lactose to modify the surface of the vesicles [105]. Furthermore, addition of ConA or peanut agglutinin led to agglutination of the vesicles due to interactions with either the lactose or maltose units on the surface of the vesicles.

1.4.3 Nanotubes and Fibers

Compared to the isotropic morphologies of micelles and vesicles, anisotropic structures like nanotubes and fibers are much more difficult to achieve. Therefore examples in the literature are rare. One example was described in Sections 1.3.11 and 1.4.2, fibers of a dextran and poly(L-valine) supramolecular block copolymer were obtained for high DPs of poly(L-valine) (refer to Figure 1.8a) [47]. Another example was described earlier as well, where nanoparticles of acetalated β-CD derived from an emulsion method lead to cyclindrical assemblies in water after addition of adamantyl end-functionalized PEG [100]. A photoresponsive example was reported by Yuan *et al.* (refer to Figure 1.8b). Supramolecular nanotubes with a length of 220 nm and a diameter of 90 nm were obtained from a block copolymer of azobenzene end-functionalized PCL and α-CD end-functionalized PAA [53]. The light-responsive linkage could be exploited for the triggered release of Rhodamine B.

1.4.4 Nanoparticles and Hybrid Materials

The formation of nano particles via supramolecular interactions requires the careful adjustment of the utilized building blocks, yet significant potential toward applications like drug-delivery exists for supramolecular CD-based nanoparticles. Furthermore, the combination of polymer-based CD host/guest chemistry with biological motifs or inorganic materials, such as DNA and metal oxide nanoparticles, has found increasing interest in the last years. Thus a broad range of materials with enhanced or novel properties is available, especially when the stimuli-responsive nature of several CD complex types is taken into account.

Recently, Huskens *et al.* described a supramolecular nanoparticle that does not rely on the amphiphilicity of the employed polymers [106]. Poly(isobutyl-*alt*-maleic acid) was grafted with amino functionalized β-CD or *tert*-butyl aniline via amidation. To prevent the system from forming a hydrogel, adamantyl end-functionalized PEG was added as capping agent, yet it was found that the addition of the stopper effected the nanoparticle formation in water only slightly. This fact was attributed to the poly anionic nature of the poly(isobutyl-*alt*-maleic acid) that prevents the system from aggregation due to electrostatic repulsion. Nevertheless, under acidic conditions or at high ionic strengths aggregation was observed. A polycationic system was utilized by Tseng *et al.* [107]. Adamantyl-grafted poly(amidoamine) dendrimers, β-CD-grafted branched PEI, adamantyl functionalized PEG, and adamantyl-grafted $Zn_{0.4}Fe_{2.6}O_4$ superparamagnetic nanoparticles were assembled into nanoparticles in aqueous solution. Furthermore, Doxorubicin molecules were added to perform drug delivery. The release of the drug molecules was triggered via an external magnetic field that interacts with the embedded superparamagnetic nanoparticles and the release could be measured via fluorescence of the Doxorubicin molecules. Finally, the effects of the drug release were studied *in vitro* as well *in vivo*. In a similar manner a small interfering ribonucleic acid (siRNA) delivery system was presented by Davis *et al.* [108], which is described in detail in the next section. A DNA-based hybrid material based on CD/guest chemistry has been described by Xu *et al.* [109]. A β-CD centered star polymer, namely PGMA reacted with ethanolamine and a linear PGMA reacted with adamantyl amine, were synthesized. Both polymers were used to form a supramolecular brush and pDNA was condensed into the brushes. Finally, low cell viability and enhanced gene transfection *in vitro* was found. A similar approach made use of a β-CD centered four-arm PDMAEMA with a disulfide moiety between polymer and core [110]. Furthermore, adamantyl end-functionalized poly(poly(ethylene glycol)ethyl ether methacrylate) was synthesized and a supramolecular miktoarm star polymer formed. Addition of pDNA led to the formation of DNA/polymer hydrid particles that were studied with regard to redox-triggered release due to disulfide breakage, toxicity, and gene transfection *in vitro* and *in vivo*. A β-CD centered star polymer was employed as vector for siRNA by ElSayed *et al.* [111]. Copolymerization of hexyl methacrylate, DMAEMA, and DMAEMA methyl ammonium salt via

ATRP led to the desired star polymer. Notably, the polymers were coupled to the core via acid-labile hydrazone linkage to enable hydrolytic degradation of the star polymer after siRNA delivery. The star polymers were condensed with siRNA and the uptake into HeLa cancer cells was investigated. Feiters and coworkers presented a β-CD-based polymer/enzyme conjugate [112]. Vesicles from β-CD functionalized PS were formed in aqueous solution and adamantyl-PEG-functionalized horseradish peroxidase was conjugated in a supramolecular fashion, while the catalytic activity of the enzyme was retained.

In the case of hybrid materials, incorporating CDs and inorganic materials several examples have been described in the literature, such as inorganic nanoparticle centered stars from SiO_2 nanoparticles [86], CdS quantum dots [87], and gold nanoparticles [88], Kaifer *et al.* presented β-CD functionalized gold nanoparticles that could be aggregated via ferrocene functionalized dilinkers. Furthermore, the aggregation process could be reversed via redox stimulus [113]. In a similar way, Ravoo *et al.* showed a light-responsive aggregation of β-CD decorated SiO_2 nanoparticles via the addition of small molecule diazobenzene linkers [114]. A polyhedral oligomeric silsesquioxane grafted with β-CD was utilized by Li *et al.* and azobenzene end-functionalized PEG-*b*-PDMAEMA was attached in a photoresponsive fashion [89]. Yang *et al.* grafted azobenzenes onto mesoporous silica nanoparticles and Rhodamine B was loaded [115]. To encapsulate the model compound entirely, the pores were capped with a β-CD functionalized linear PGMA. Furthermore, the release of the model compound was studied with regard to light and temperature stimuli as well as the addition of competing binding agents. Thus the β-CD functionalized polymers acted as "nanovalves." Another mesoporous silica hybrid was described by Zhao *et al.* [116]. β-CD was attached to the silica surface via disulfide bonds and Doxorubicin was added as cargo. Subsequently, adamantyl end-functionalized PEG and folic acid/adamantyl hetero bifunctional PEG were added, forming a supramolecular complex. The particles were utilized for drug delivery, where the folic acid acted as targeting unit toward cancer cells and the PEG units as antifouling barrier and stabilizer to extend the circulation period. After cell internalization, reductive glutathione inside the cell led to disulfide bond cleavage, enhancing the drug release due to removal of the CD capping on the mesoporous silica carrier.

1.4.5 Planar Surface Modification

Surface functionalization plays a significant role in contemporary polymer science and supramolecular interactions provide powerful abilities to advance surface chemistry. Mono layers of CDs on surfaces have been entitled "molecular printboards" by Reinhoudt and Huskens [117]. A variety of grafted structures was presented with that concept in mind, such as on gold [117], Si [118] or SiO_2 [119]. In addition, patterned surfaces have been utilized in order to obtain spatially controlled host grafting and subsequently spatially controlled supramolecular complexation, for example, via micro contact printing [119, 120]. A variety of guest functional moieties have been grafted on surfaces in a supramolecular fashion, such as proteins [121], fluorescent dyes [120b], and Eu^{3+} luminescent complexes [120a]. Remarkably, azobenzene functionalized cell recognition peptides were grafted onto an α-CD functionalized gold surface that showed reversible and photocontrollable cell attachment [122].

1.5 APPLICATIONS

Taking the previous chapters into consideration, a broad variety of architectures and more complex assemblies are available. Especially when the stimuli responsive nature of CD/guest complexes is taken into account, there certainly are applications for CD-based materials. As described in the last section a very important application for CD-based materials is drug delivery, in particular because CD/host guest chemistry is usually occurring in aqueous solutions and plain CDs are utilized to great extent in drug delivery already [19, 123].

An example for drug delivery via CD-based macromolecular architectures was presented by Zhuo *et al.* [124]. A dilinker containing α-CD and β-CD was employed to connect a PNIPAM and a PCL block. Cell-targeting ligands were attached as well as PEG units in order to protect the core-shell assemblies of the supramolecular diblock copolymer in the body. After formation of noncovalently connected micelles [125], drug delivery experiments were performed that showed tumor-triggered release of loaded Doxorubicin molecules. Davis *et al.* presented a CD-based nano particle for siRNA delivery (refer to Figure 1.9) [108, 126]. A polymer with CD units in the backbone was synthesized via a condensation copolymerization approach. Thus a difunctionalized β-CD monomer and the difunctional dimethylsuberimidate monomer were utilized to obtain a polycationic polymer (CDP). Furthermore, imidazole endgroups were introduced. In order to stabilize the delivery vehicle, adamantyl end-functionalized PEG was added (AD-PEG), yet an adamantyl and targeting ligand (human transferrin, Tf) functionalized PEG was added as well in order to engage cell receptors and induce endocytosis (AD-PEG-Tf). Adding a mixture of the three components to a siRNA solution led to the formation of PEG stabilized polycationic nanoparticles with targeting ligands on the outside and

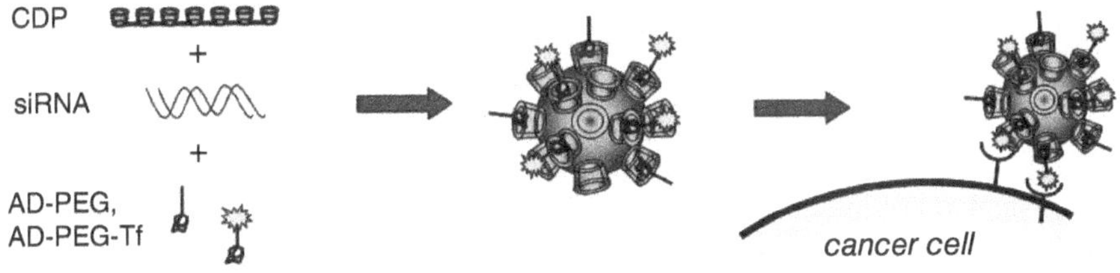

Figure 1.9 Formation of a CD-based supramolecular nano particle for siRNA delivery via mixing of a CD containing copolymer (CDP), siRNA, adamantyl end-functionalized PEG (AD-PEG) and adamantyl end-functionalized PEG with targeting ligand (AD-PEG-Tf) [108]. Reprinted with permission from [107]. Copyright 2009 American Chemical Society.

siRNA in the inside. Nanoparticles with diameters ranging from 60 to 80 nm were obtained. *In vivo* delivery was probed for DNAzymes, pDNA, and siRNA. The delivery system proved to be very efficient and eventually went to clinic trials. The examples above show that CD complexes are a very efficient tool for the formation of macromolecular assemblies with regard to drug delivery applications.

Apart from drug delivery other applications have emerged as well, such as in sensing. Reed *et al.* formed a self-assembled monolayer of β-CD on a silicon nanowire field-effect transistor [127]. The sensor was capable of detecting thyroxin molecules in solution and even to discriminate between the respective enantiomers. Furthermore, the streptavidin–biotin interactions could be probed via addition of orthogonal linker molecules, such as a double adamantyl functionalized biotin derivative. In addition, the sensing was reversible and the sensor could be regenerated for further use of the device.

Harada *et al.* introduced a new concept and type of materials in 2011—the so-called macroscopic self-assembly (refer to Figure 1.10a) [128]. Small centimeter-size blocks containing either CD functionalized gels or guest functionalized gels were brought together on a petri dish in water. Shaking led to recognition of the respective CD and guest functionalized blocks. For visualization different dyes were added to the respective gel blocks in order to show that the attachment of the blocks proceeds selectively. This concept was exploited for a variety of examples, for example, different guests and different CDs [128, 131] and stimuli-responsive complexation [48, 132]. Recently, Yan *et al.* utilized poly(ionic liquid) membrane strips functionalized with either β-CD or ferrocene moieties in the formation of a supramolecular velcro (refer to Figure 1.10b) [129]. The guest and CD functionalized strips were pressed together—zipped, so to speak—and showed strong adhesion under air und in water with lap shear adhesion strengths of around 90 kPa. Furthermore, oxidation of the ferrocenes via application of an external current or addition of oxidizer led to disassembly of the membrane connection—the unzipping. The zipping–unzipping could be performed several times with only small losses in adhesion performance. In another macroscopic example Xie *et al.* utilized three types of macroscopic gels to generate complex stimuli responsive assemblies: A β-CD and carboxylic acid containing gel (RH), a β-CD containing gel (NRH), and a ferrocene containing gel (NRG) (refer to Figure 1.10c) [130]. Due to the incorporation of carboxylic acids, the RH gel showed swelling/deswelling depending on ionic strength or pH. A combination of the RH and the NRG gel led to adhesion, and a change in pH led to bending of the assembly due to swelling of only one-half of the assembly, namely the RH gel. Addition of the NRH gel led to further possibilities, such as 3D assemblies that showed extraordinary bending structures. Due to their reversible nature, supramolecular interactions are well suited for the construction of self-healing materials. Harada *et al.* recently showed an example of a β-CD/ferrocene based self-healing hydrogel (Figure 1.10d) [22a]. A hydrogel was formed from a β-CD containing acrylamido polymer and a ferrocene containing acrylamido polymer. Cutting of the gel and rejoining led to reformation of the initial structure after 24 hours. Stress–strain measurements showed a recovery of 84% of the initial strength after self-healing. Furthermore, treating the cut surface with an

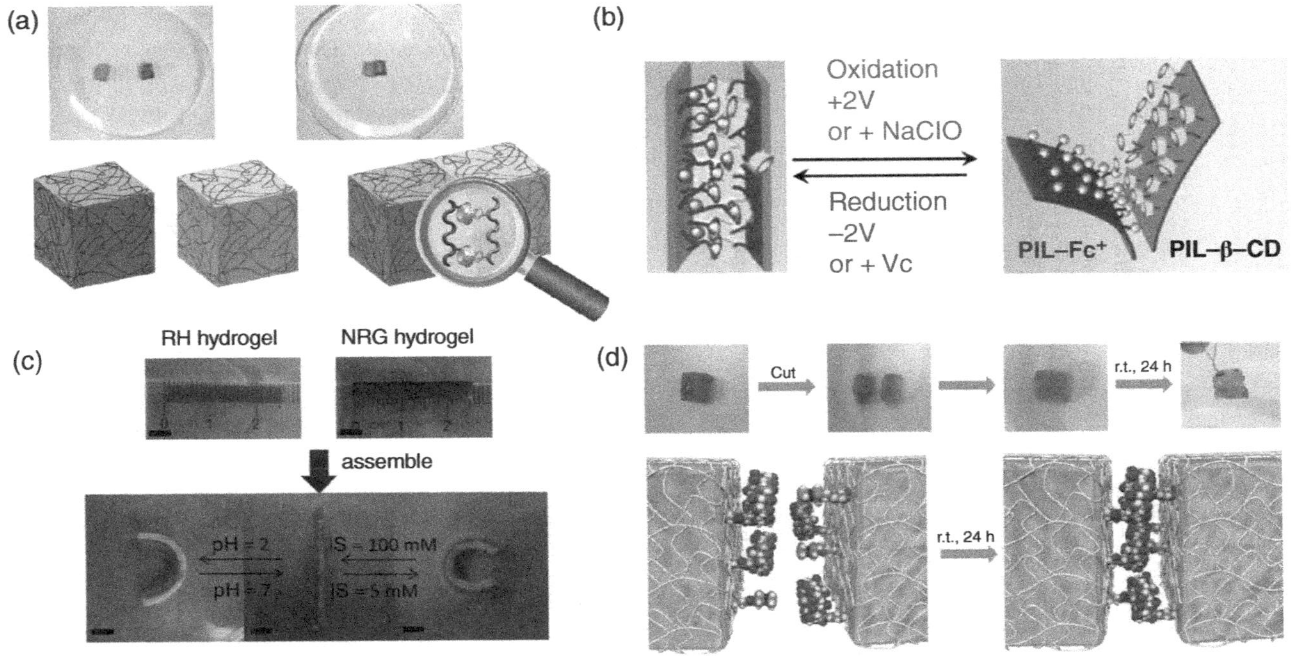

Figure 1.10 (a) Macroscopic self-assembly of an adamantyl containing hydrogel cube (green) and a β-CD containing hydrogel cube (red) [128] (Reprinted by permission from Macmillan Publishers Ltd: *Nat. Chem.* [126], copyright 2011), (b) β-CD/ferrocene based supramolecular velcro with reversible zipping/unzipping depending on the ferrocene oxidation state [129] (Reproduced from [127] with permission of The Royal Society of Chemistry), (c) combination of a β-CD and carboxylic acid containing gel (RH hydrogel) and a ferrocene containing gel (NRG hydrogel) with pH or ionic strength depending bending [130] (Adapted from [128] with permission of John Wiley and Sons), and (d) β-CD/ferrocene based self-healing hydrogel [22a] (Reprinted by permission from Macmillan Publishers Ltd: *Nat. Commun.* [21a], copyright 2011). (*See color insert for color representation of this figure*).

oxidant showed no self-healing, as oxidized ferrocenes do not complex with β-CDs, whereas subsequent treatment with a reductant led to self-healing of the cut.

1.6 CONCLUSION AND OUTLOOK

Supramolecular governed macromolecular architectures and CD-based complex macromolecular architectures, in particular, have a significant impact on polymer science. Relying on effective protocols for the synthesis of building blocks containing CD or guest moieties, CD complexation has found its way into macromolecular self-assemblies. The control over the functionalities of the utilized polymers, i.e. the primary structure, can be transferred into macromolecular assemblies. These assemblies can be considered as single polymers as well, assuming strong supramolecular association, resembling novel polymer compositions or topologies, namely the secondary structure. Thus complex macromolecular architectures such as block copolymers, star polymers, and brush polymers can be formed in a supramolecular fashion. A higher level of complexity can be achieved when these new supramolecular assembled macromolecules undergo further self-assembly into multi-polymer assemblies, namely the tertiary structure. Supramolecular governed macromolecular architectural polymers were utilized to form a variety of multi-polymer assemblies, such as micelles, vesicles, nanotubes, and hybrid materials. Combined with the large number of stimuli-responsive CD/guest complexes and stimuli-responsive polymers, a large variety of functional materials is accessible. Especially with regard to prospective applications, the utilized polymer systems can be tuned in a precise and convenient way via efficient chemistries and because of the modular character of supramolecular building blocks. Several potential applications have been investigated so far showing promising results, for example, drug-delivery, sensing, adhesives, and self-healing materials. In summary, we conclude that CD host/guest chemistry had significant impact on polymer and materials science and will play a key role in future developments. Certainly more complex structures and prospective applications will be investigated and presented in the near future.

REFERENCES

1. (a) A. H. Nikos Hadjichristidis, Yasuyuki Tezuka, Filip Du Prez, John Wiley & Sons (Asia) Pte Ltd, Singapore, **2011**, p. 856; (b) C. J. Hawker, K. L. Wooley, *Science* **2005**, *309*, 1200–1205; (c) A. Gregory, M. H. Stenzel, *Prog. Polym. Sci.* **2012**, *37*, 38–105.
2. G. J. A. A. Soler-Illia, O. Azzaroni, *Chem. Soc. Rev.* **2011**, *40*, 1107–1150.
3. H. Tian, Z. Tang, X. Zhuang, X. Chen, X. Jing, *Prog. Polym. Sci.* **2012**, *37*, 237–280.
4. S. M. Grayson, W. T. Godbey, *J. Drug Target.* **2008**, *16*, 329–356.
5. D. Neugebauer, Y. Zhang, T. Pakula, S. S. Sheiko, K. Matyjaszewski, *Macromolecules* **2003**, *36*, 6746–6755.
6. J. L. Hedrick, T. Magbitang, E. F. Connor, T. Glauser, W. Volksen, C. J. Hawker, V. Y. Lee, R. D. Miller, *Chem., Eur. J.* **2002**, *8*, 3308–3319.

7. (a) C. J. Hawker, A. W. Bosman, E. Harth, *Chem. Rev.* **2001**, *101*, 3661-3688; (b) J. Nicolas, Y. Guillaneuf, C. Lefay, D. Bertin, D. Gigmes, B. Charleux, *Prog. Polym. Sci.* **2013**, *38*, 63–235.
8. (a) M. Ouchi, T. Terashima, M. Sawamoto, *Chem. Rev.* **2009**, *109*, 4963-5050; (b) K. Matyjaszewski, *Macromolecules* **2012**, *45*, 4015–4039.
9. (a) C. Barner-Kowollik, *Wiley-VCH*, Weinheim, Germany, **2008**, p. 556; (b) G. Moad, E. Rizzardo, S. H. Thang, *Aust. J. Chem.* **2012**, *65*, 985–1076; (c) C. Barner-Kowollik, S. Perrier, *J. Polym. Sci., Part A: Polym. Chem.* **2008**, *46*, 5715–5723.
10. (a) H. C. Kolb, M. G. Finn, K. B. Sharpless, *Angew. Chem., Int. Ed.* **2001**, *40*, 2004–2021; (b) C. Barner-Kowollik, F. E. Du Prez, P. Espeel, C. J. Hawker, T. Junkers, H. Schlaad, W. Van Camp, *Angew. Chem., Int. Ed.* **2011**, *50*, 60–62; (c) K. Kempe, A. Krieg, C. R. Becer, U. S. Schubert, *Chem. Soc. Rev.* **2012**, *41*, 176–191; (d) W. H. Binder, R. Sachsenhofer, *Macromol. Rapid Commun.* **2007**, *28*, 15–54.
11. J. F. Lutz, *Angew. Chem., Int. Ed.* **2007**, 46, 1018–1025.
12. M. A. Tasdelen, *Polym. Chem.* **2011**, *2*, 2133–2145.
13. (a) C. E. Hoyle, C. N. Bowman, *Angew. Chem., Int. Ed.* **2010**, *49*, 1540–1573; (b) A. B. Lowe, *Polym. Chem.* **2010**, *1*, 17–36.
14. (a) H. Nandivada, X. Jiang, J. Lahann, *Adv. Mater.* **2007**, *19*, 2197–2208; (b) A. S. Goldmann, M. Glassner, A. J. Inglis, C. Barner-Kowollik, *Macromol. Rapid Commun.* **2013**, *34*, 810–849; (c) B. V. K. J. Schmidt, M. Hetzer, H. Ritter, C. Barner-Kowollik, *Prog. Polym. Sci.* **2014**, *39*, 235–249.
15. T. Aida, E. W. Meijer, S. I. Stupp, *Science* **2012**, *335*, 813–817.
16. (a) A. J. Wilson, *Soft Matter* **2007**, *3*, 409-425; (b) A. Bertrand, F. Lortie, J. Bernard, *Macromol. Rapid Commun.* **2012**, *33*, 2062–2091.
17. D. G. Kurth, M. Higuchi, *Soft Matter* **2006**, *2*, 915–927.
18. (a) J. M. Zayed, N. Nouvel, U. Rauwald, O. A. Scherman, *Chem. Soc. Rev.* **2010**, *39*, 2806–2816; (b) G. Chen, M. Jiang, *Chem. Soc. Rev.* **2011**, *40*, 2254–2266; (c) B. Zheng, F. Wang, S. Dong, F. Huang, *Chem. Soc. Rev.* **2012**, *41*, 1621–1636; (d) J. Hu, S. Liu, *Acc. Chem. Res.* **2014**, *47*, 2084–2095.
19. J. Zhou, H. Ritter, *Polym. Chem.* **2010**, *1*, 1552–1559.
20. A. F. Hirschbiel, B. V. K. J. Schmidt, P. Krolla-Sidenstein, J. P. Blinco, C. Barner-Kowollik, *Macromolecules* **2015**, *48*, 4410–4420.
21. (a) A. Harada, Y. Takashima, H. Yamaguchi, *Chem. Soc. Rev.* **2009**, *38*, 875–882; (b) X. Ma, H. Tian, *Acc. Chem. Res.* **2014**, *47*, 1971–1981.
22. (a) M. Nakahata, Y. Takashima, H. Yamaguchi, A. Harada, *Nat. Commun.* **2011**, *2*, No. 511; (b) A. Harada, Y. Takashima, M. Nakahata, *Acc. Chem. Res.* **2014**, *47*, 2128–2140.
23. X. Zhang, C. Wang, *Chem. Soc. Rev.* **2011**, *40*, 94–101.
24. (a) S. Tan, K. Ladewig, Q. Fu, A. Blencowe, G. G. Qiao, Macromol. *Rapid Commun.* **2014**, *35*, 1166–1184; (b) T. Murakami, B. V. K. J. Schmidt, H. R. Brown, C. J. Hawker, *Macromolecules* **2015**, *48*, 7774–7781.
25. Y. Chen, Y. Liu, *Chem. Soc. Rev.* **2010**, *39*, 495–505.
26. (a) B. V. K. J. Schmidt, M. Hetzer, H. Ritter, C. Barner-Kowollik, *Macromolecules* **2011**, *44*, 7220–7232; (b) M. Hetzer, B. V. K. J. Schmidt, C. Barner-Kowollik, H. Ritter, *J. Polym. Sci., Part A: Polym. Chem.* **2013**, *51*, 2504–2517; (c) H. S. Köllisch, C. Barner-Kowollik, H. Ritter, *Chem. Commun.* **2009**, 1097–1099; (d) L. Ding, Y. Li, J. Deng, W. Yang, *Polym. Chem.* **2011**, *2*, 694–701.

27. A. R. Khan, P. Forgo, K. J. Stine, V. T. D'Souza, *Chem. Rev.* **1998**, *98*, 1977–1996.
28. (a) B. V. K. J. Schmidt, *Novel Macromolecular Architectures via a Combination of Cyclodextrin Host/Guest Complexation and RAFT Polymerization*, Springer International Publishing, **2014**; (b) S. Amajjahe, S. Choi, M. Munteanu, H. Ritter, *Angew. Chem., Int. Ed.* **2008**, 47, 3435–3437; (c) G. Tripodo, C. Wischke, A. T. Neffe, A. Lendlein, *Carbohydr. Res.* **2013**, *381*, 59–63.
29. H.-S. Z. Byun, Ning; Robert Bittman, *Org. Syn*, **2000**, *77*, 225.
30. R. Palin, S. J. A. Grove, A. B. Prosser, M.-Q. Zhang, *Tetrahedron Lett.* **2001**, *42*, 8897–8899.
31. (a) W. Tang, S.-C. Ng, *Nat. Protoc.* **2008**, *3*, 691–697; (b) L. D. Melton, K. N. Slessor, *Carbohydr. Res.* **1971**, *18*, 29–37.
32. J. A. Faiz, N. Spencer, Z. Pikramenou, *Org. Biomol. Chem.* **2005**, *3*, 4239–4245.
33. K. Hamasaki, H. Ikeda, A. Nakamura, A. Ueno, F. Toda, I. Suzuki, T. Osa, *J. Am. Chem. Soc.* **1993**, *115*, 5035–5040.
34. R. Bonomo, V. Cucinotta, F. Allessandro, G. Impellizzeri, G. Maccarrone, E. Rizzarelli, G. Vecchio, *J. Inclusion Phenom. Macrocyclic Chem.* **1993**, *15*, 167–180.
35. (a) A. Kuzuya, T. Ohnishi, T. Wasano, S. Nagaoka, J. Sumaoka, T. Ihara, A. Jyo, M. Komiyama, *Bioconjugate Chem.* **2009**, *20*, 1643–1649; (b) J. Wang, L. Kong, Z. Guo, J. Xu, J. Liu, *J. Mater. Chem.* **2010**, *20*, 5271–5279; (c) K. Fujita, T. Ueda, T. Imoto, I. Tabushi, N. Toh, T. Koga, *Bioorg. Chem.* **1982**, *11*, 72–84.
36. F. Yhaya, S. Binauld, M. Callari, M. H. Stenzel, *Aust. J. Chem.* **2012**, *65*, 1095–1103.
37. C. Giacomelli, V. Schmidt, J.-L. Putaux, A. Narumi, T. Kakuchi, R. Borsali, *Biomacromolecules* **2009**, *10*, 449–453.
38. (a) H. Liu, Y. Zhang, J. Hu, C. Li, S. Liu, *Macromol. Chem. Phys.* **2009**, *210*, 2125–2137; (b) J. Stadermann, H. Komber, M. Erber, F. Däbritz, H. Ritter, B. Voit, *Macromolecules* **2011**, *44*, 3250–3259.
39. (a) T. Cai, W. J. Yang, Z. Zhang, X. Zhu, K.-G. Neoh, E.-T. Kang, *Soft Matter* **2012**, *8*, 5612–5620; (b) J. Zeng, K. Shi, Y. Zhang, X. Sun, B. Zhang, *Chem. Commun.* **2008**, 3753–3755.
40. (a) A. Bertrand, M. Stenzel, E. Fleury, J. Bernard, *Polym. Chem.* **2012**, *3*, 377–383; (b) B. V. K. J. Schmidt, M. Hetzer, H. Ritter, C. Barner-Kowollik, *Macromolecules* **2013**, *46*, 1054–1065.
41. K. A. Connors, *Chem. Rev.* **1997**, *97*, 1325–1358.
42. L. Peng, A. Feng, M. Huo, J. Yuan, *Chem. Commun.* **2014**.
43. F. Szillat, B. V. K. J. Schmidt, A. Hubert, C. Barner-Kowollik, H. Ritter, *Macromol. Rapid Commun.* **2014**, *35*, 1293–1300.
44. (a) M. Hetzer, C. Fleischmann, B. V. K. J. Schmidt, C. Barner-Kowollik, H. Ritter, *Polymer* **2013**, *54*, 5141–5147; (b) C. Fleischmann, H. Wöhlk, H. Ritter, *Beilstein J. Org. Chem.* **2014**, *10*, 2263–2269.
45. T. Nakamura, Y. Takashima, A. Hashidzume, H. Yamaguchi, A. Harada, *Nat. Commun.* **2014**, *5*, no. 4622.
46. Z. Zhang, J. Ding, X. Chen, C. Xiao, C. He, X. Zhuang, L. Chen, X. Chen, *Polym. Chem.* **2013**, *4*, 3265–3271.
47. Q. Yan, H. Zhang, Y. Zhao, *ACS Macro Lett.* **2014**, *3*, 472–476.
48. Y. Zheng, A. Hashidzume, A. Harada, *Macromol. Rapid Commun.* **2013**, *34*, 1062–1066.

49. (a) B. V. K. J. Schmidt, M. Hetzer, H. Ritter, C. Barner-Kowollik, *Macromol. Rapid Commun.* **2013**, *34*, 1306–1311; (b) A. Fetzner, S. Böhm, S. Schreder, R. Schubert, *Eur. J. Pharm. Biopharm.* **2004**, *58*, 91–97; (c) C. Park, H. Kim, S. Kim, C. Kim, *J. Am. Chem. Soc.* **2009**, *131*, 16614–16615.
50. M. Miyauchi, A. Harada, *J. Am. Chem. Soc.* **2004**, *126*, 11418–11419.
51. Q. Yan, J. Yuan, Z. Cai, Y. Xin, Y. Kang, Y. Yin, *J. Am. Chem. Soc.* **2010**, *132*, 9268–9270.
52. L. Peng, A. Feng, H. Zhang, H. Wang, C. Jian, B. Liu, W. Gao, J. Yuan, *Polym. Chem.* **2014**, *5*, 1751–1759.
53. Q. Yan, Y. Xin, R. Zhou, Y. Yin, J. Yuan, *Chem. Commun.* **2011**, *47*, 9594–9596.
54. L. Liu, L. Rui, Y. Gao, W. Zhang, *Polym. Chem.* **2014**, *5*, 5453–5460.
55. B.-W. Liu, H. Zhou, S.-T. Zhou, H.-J. Zhang, A.-C. Feng, C.-M. Jian, J. Hu, W.-P. Gao, J.-Y. Yuan, *Macromolecules* **2014**, *47*, 2938–2946.
56. F. Huang, O. A. Scherman, *Chem. Soc. Rev.* **2012**, *41*, 5879–5880.
57. M. Miyauchi, Y. Kawaguchi, A. Harada, *J. Inclusion Phenom. Macrocyclic Chem.* **2004**, *50*, 57–62.
58. K. Ohga, Y. Takashima, H. Takahashi, Y. Kawaguchi, H. Yamaguchi, A. Harada, *Macromolecules* **2005**, *38*, 5897–5904.
59. B. Knudsen, B. E. Kergl, H. Paulsen, V. Durnev, H. Ritter, *J. Polym. Sci., Part A: Polym. Chem.* **2013**, *51*, 2472–2482.
60. Q. Wang, Y. Chen, Y. Liu, *Polym. Chem.* **2013**, *4*, 4192–4198.
61. Q. Song, F. Li, X. Tan, L. Yang, Z. Wang, X. Zhang, *Polym. Chem.* **2014**.
62. L. Zhu, M. Lu, Q. Zhang, D. Qu, H. Tian, *Macromolecules* **2011**, *44*, 4092–4097.
63. R. Dong, Y. Su, S. Yu, Y. Zhou, Y. Lu, X. Zhu, *Chem. Commun.* **2013**, *49*, 9845–9847.
64. Q. Zhang, D.-H. Qu, J. Wu, X. Ma, Q. Wang, H. Tian, *Langmuir* **2013**, *29*, 5345–5350.
65. Q. Zhang, X. Yao, D.-H. Qu, X. Ma, *Chem. Commun.* **2014**, *50*, 1567–1569.
66. M. Ménand, S. Adam de Beaumais, L.-M. Chamoreau, E. Derat, S. Blanchard, Y. Zhang, L. Bouteiller, M. Sollogoub, *Angew. Chem.* **2014**, *126*, 7366–7370.
67. (a) J. Li, in *Inclusion Polymers*, Vol. *222* (Ed.: G. Wenz), Springer Berlin Heidelberg, **2009**, pp. 175–203; (b) E. A. Appel, J. del Barrio, X. J. Loh, O. A. Scherman, *Chem. Soc. Rev.* **2012**, *41*, 6195–6214.
68. R. Bleta, S. Menuel, B. Leger, A. Da Costa, E. Monflier, A. Ponchel, *RSC Adv.* **2014**, *4*, 8200–8208.
69. M. Hetzer, B. V. K. J. Schmidt, C. Barner-Kowollik, H. Ritter, *Polym. Chem.* **2014**, *5*, 2142–2152.
70. Y.-F. Wang, D.-L. Zhang, T. Zhou, H.-S. Zhang, W.-Z. Zhang, L. Luo, A.-M. Zhang, B.-J. Li, S. Zhang, *Polym. Chem.* **2014**, *5*, 2922–2927.
71. H. Zhang, L. Peng, Y. Xin, Q. Yan, J. Yuan, *Macromol. Symp.* **2013**, *329*, 66–69.
72. Y. Guan, H.-B. Zhao, L.-X. Yu, S.-C. Chen, Y.-Z. Wang, *RSC Adv.* **2014**, *4*, 4955–4959.
73. (a) M. Stenzel-Rosenbaum, T. P. Davis, V. Chen, A. G. Fane, *J. Polym. Sci., Part A: Polym. Chem.* **2001**, *39*, 2777–2783; (b) K. Ohno, B. Wong, D. M. Haddleton, *J. Polym. Sci., Part A: Polym. Chem.* **2001**, *39*, 2206–2214.
74. C. Teuchert, C. Michel, F. Hausen, D.-Y. Park, H. W. Beckham, G. Wenz, *Macromolecules* **2012**, *46*, 2–7.
75. K. Nagahama, R. Aoki, T. Saito, T. Ouchi, Y. Ohya, N. Yui, *Polym. Chem.* **2013**, *4*, 1769–1773.

76. J. Wu, C. Gao, *Macromolecules* **2010**, *43*, 7139–7146.
77. Z.-X. Zhang, K. L. Liu, J. Li, *Angew. Chem., Int. Ed.* **2013**, *52*, 6180–6184.
78. B. V. K. J. Schmidt, T. Rudolph, M. Hetzer, H. Ritter, F. H. Schacher, C. Barner-Kowollik, *Polym. Chem.* **2012**, *3*, 3139–3145.
79. B. V. K. J. Schmidt, M. Hetzer, H. Ritter, C. Barner-Kowollik, *Polym. Chem.* **2012**, *3*, 3064–3067.
80. B. V. K. J. Schmidt, C. Barner-Kowollik, *Polym. Chem.* **2014**, *5*, 2461–2472.
81. J. Yan, X. Zhang, X. Zhang, K. Liu, W. Li, P. Wu, A. Zhang, *Macromol. Chem. Phys.* **2012**, *213*, 2003–2010.
82. Z.-X. Zhang, K. L. Liu, J. Li, *Macromolecules* **2011**, *44*, 1182–1193.
83. Z. Tian, C. Chen, H. R. Allcock, *Macromolecules* **2014**, *47*, 1065–1072.
84. X. Huan, D. Wang, R. Dong, C. Tu, B. Zhu, D. Yan, X. Zhu, *Macromolecules* **2012**, *45*, 5941–5947.
85. Z.-X. Zhang, X. Liu, F. J. Xu, X. J. Loh, E.-T. Kang, K.-G. Neoh, J. Li, *Macromolecules* **2008**, *41*, 5967–5970.
86. M. Guo, M. Jiang, S. Pispas, W. Yu, C. Zhou, *Macromolecules* **2008**, *41*, 9744–9749.
87. (a) J. Liu, G. Chen, M. Guo, M. Jiang, *Macromolecules* **2010**, *43*, 8086–8093; (b) P. Du, J. Liu, G. Chen, M. Jiang, *Langmuir* **2011**, *27*, 9602–9608.
88. (a) K. Wei, J. Li, J. Liu, G. Chen, M. Jiang, *Soft Matter* **2012**, *8*, 3300–3303; (b) C. Luo, F. Zuo, Z. Zheng, X. Cheng, X. Ding, Y. Peng, *Macromol. Rapid Commun.* **2008**, *29*, 149–154.
89. J. Li, Z. Zhou, L. Ma, G. Chen, Q. Li, *Macromolecules* **2014**, *47*, 5739–5748.
90. K. Wei, J. Li, G. Chen, M. Jiang, *ACS Macro Lett.* **2013**, *2*, 278–283.
91. G. Maatz, A. Maciollek, H. Ritter, *Beilstein J. Org. Chem.* **2012**, *8*, 1929–1935.
92. A. Feng, Q. Yan, H. Zhang, L. Peng, J. Yuan, *Chem. Commun.* **2014**, *50*, 4740–4742.
93. C.-M. Jian, B.-W. Liu, X. Chen, S.-T. Zhou, T. Fang, J.-Y. Yuan, *Chin. J. Polym. Sci.* **2014**, *32*, 690–702.
94. C. Moers, L. Nuhn, M. Wissel, R. Stangenberg, M. Mondeshki, E. Berger-Nicoletti, A. Thomas, D. Schaeffel, K. Koynov, M. Klapper, R. Zentel, H. Frey, *Macromolecules* **2013**, *46*, 9544–9553.
95. Y. Inoue, P. Kuad, Y. Okumura, Y. Takashima, H. Yamaguchi, A. Harada, *J. Am. Chem. Soc.* **2007**, *129*, 6396–6397.
96. J. Willenbacher, B. V. K. J. Schmidt, D. Schulze-Suenninghausen, O. Altintas, B. Luy, G. Delaittre, C. Barner-Kowollik, *Chem. Commun.* **2014**, *50*, 7056–7059.
97. S. Ren, D. Chen, M. Jiang, *J. Polym. Sci., Part A: Polym. Chem.* **2009**, *47*, 4267–4278.
98. J. Zhang, H. Sun, P. X. Ma, *ACS Nano* **2010**, *4*, 1049–1059.
99. D. Yong, Y. Luo, F. Du, J. Huang, W. Lu, Z. Dai, J. Yu, S. Liu, *Colloids Surf., B.* **2013**, *105*, 31–36.
100. J. Zhang, Y. Jia, X. Li, Y. Hu, X. Li, *Adv. Mater.* **2011**, *23*, 3035–3040.
101. J. Wang, M. Jiang, *J. Am. Chem. Soc.* **2006**, *128*, 3703–3708.
102. M. Guo, M. Jiang, G. Zhang, *Langmuir* **2008**, *24*, 10583–10586.
103. R. Ji, J. Cheng, T. Yang, C. C. Song, L. Li, F.-S. Du, Z.-C. Li, *Biomacromolecules* **2014**.
104. Y. Liu, C. Yu, H. Jin, B. Jiang, X. Zhu, Y. Zhou, Z. Lu, D. Yan, *J. Am. Chem. Soc.* **2013**, *135*, 4765–4770.

105. J. Voskuhl, M. C. A. Stuart, B. J. Ravoo, *Chem. – Eur. J.* **2010**, *16*, 2790–2796.

106. L. Grana Suarez, W. Verboom, J. Huskens, *Chem. Commun.* **2014**, *50*, 7280–7282.

107. J.-H. Lee, K.-J. Chen, S.-H. Noh, M. A. Garcia, H. Wang, W.-Y. Lin, H. Jeong, B. J. Kong, D. B. Stout, J. Cheon, H.-R. Tseng, *Angew. Chem., Int. Ed.* **2013**, *52*, 4384–4388.

108. M. E. Davis, *Mol. Pharmaceutics* **2009**, *6*, 659–668.

109. Y. Hu, M. Y. Chai, W. T. Yang, F. J. Xu, *Bioconjugate Chem.* **2013**, *24*, 1049–1056.

110. Y. Hu, W. Yuan, N.-N. Zhao, J. Ma, W.-T. Yang, F.-J. Xu, *Biomaterials* **2013**, *34*, 5411–5422.

111. Y. Y. Durmaz, Y.-L. Lin, M. E. H. ElSayed, *Adv. Funct. Mater.* **2013**, *23*, 3885–3895.

112. M. Felici, M. Marzá-Pérez, N. S. Hatzakis, R. J. M. Nolte, M. C. Feiters, *Chem., Eur. J.* **2008**, *14*, 9914–9920.

113. J. Liu, S. Mendoza, E. Román, M. J. Lynn, R. Xu, A. E. Kaifer, *J. Am. Chem. Soc.* **1999**, *121*, 4304–4305.

114. J. A. Krings, B. Vonhoren, P. Tegeder, V. Siozios, M. Peterlechner, B. J. Ravoo, *J. Mat. Chem. A* **2014**, *2*, 9587–9593.

115. Q.-L. Li, L. Wang, X.-L. Qiu, Y.-L. Sun, P.-X. Wang, Y. Liu, F. Li, A.-D. Qi, H. Gao, Y.-W. Yang, *Polym. Chem.* **2014**, *5*, 3389–3395.

116. Q. Zhang, F. Liu, K. T. Nguyen, X. Ma, X. Wang, B. Xing, Y. Zhao, *Adv. Funct. Mater.* **2012**, *22*, 5144–5156.

117. J. Huskens, M. A. Deij, D. N. Reinhoudt, *Angew. Chem., Int. Ed.* **2002**, *41*, 4467–4471.

118. D. Abt, B. V. K. J. Schmidt, O. Pop-Georgievski, A. S. Quick, D. Danilov, N. Y. Kostina, M. Bruns, W. Wenzel, M. Wegener, C. Rodriguez-Emmenegger, C. Barner-Kowollik, *Chem., Eur. J.* **2015**, *21*, 13186–13190.

119. T. Auletta, B. Dordi, A. Mulder, A. Sartori, S. Onclin, C. M. Bruinink, M. Péter, C. A. Nijhuis, H. Beijleveld, H. Schönherr, G. J. Vancso, A. Casnati, R. Ungaro, B. J. Ravoo, J. Huskens, D. N. Reinhoudt, *Angew. Chem., Int. Ed.* **2004**, *43*, 369–373.

120. (a) S.-H. Hsu, M. D. Yilmaz, C. Blum, V. Subramaniam, D. N. Reinhoudt, A. H. Velders, J. Huskens, *J. Am. Chem. Soc.* **2009**, *131*, 12567–12569; (b) A. González-Campo, S.-H. Hsu, L. Puig, J. Huskens, D. N. Reinhoudt, A. H. Velders, *J. Am. Chem. Soc.* **2010**, *132*, 11434–11436.

121. (a) M. J. W. Ludden, A. Mulder, R. Tampé, D. N. Reinhoudt, J. Huskens, *Angew. Chem., Int. Ed.* **2007**, *46*, 4104–4107; (b) D. A. Uhlenheuer, D. Wasserberg, C. Haase, H. D. Nguyen, J. H. Schenkel, J. Huskens, B. J. Ravoo, P. Jonkheijm, L. Brunsveld, *Chem., Eur. J.* **2012**, *18*, 6788–6794.

122. Y.-H. Gong, C. Li, J. Yang, H.-Y. Wang, R.-X. Zhuo, X.-Z. Zhang, *Macromolecules* **2011**, *44*, 7499–7502.

123. T. Loftsson, P. Jarho, M. Másson, T. Järvinen, *Expert Opin. Drug Deliv.* **2005**, *2*, 335–351.

124. C.-Y. Quan, J.-X. Chen, H.-Y. Wang, C. Li, C. Chang, X.-Z. Zhang, R.-X. Zhuo, *ACS Nano* **2010**, *4*, 4211–4219.

125. M. Guo, M. Jiang, *Soft Matter* **2009**, *5*, 495–500.

126. M. E. Davis, J. E. Zuckerman, C. H. J. Choi, D. Seligson, A. Tolcher, C. A. Alabi, Y. Yen, J. D. Heidel, A. Ribas, *Nature* **2010**, *464*, 1067–1070.

127. X. Duan, N. K. Rajan, D. A. Routenberg, J. Huskens, M. A. Reed, *ACS Nano* **2013**, *7*, 4014–4021.

128. A. Harada, R. Kobayashi, Y. Takashima, A. Hashidzume, H. Yamaguchi, *Nat. Chem.* **2011**, *3*, 34–37.

129. J. Guo, C. Yuan, M. Guo, L. Wang, F. Yan, *Chem. Sci.* **2014**, *5*, 3261–3266.

130. C. Ma, T. Li, Q. Zhao, X. Yang, J. Wu, Y. Luo, T. Xie, *Adv. Mater.* **2014**, *26*, 5665–5669.

131. H. Yamaguchi, R. Kobayashi, Y. Takashima, A. Hashidzume, A. Harada, *Macromolecules* **2011**, *44*, 2395–2399.

132. H. Yamaguchi, Y. Kobayashi, R. Kobayashi, Y. Takashima, A. Hashidzume, A. Harada, *Nat. Commun.* **2012**, *3*, 603.

2

POLYMERIZATION-INDUCED SELF-ASSEMBLY: THE CONTRIBUTION OF CONTROLLED RADICAL POLYMERIZATION TO THE FORMATION OF SELF-STABILIZED POLYMER PARTICLES OF VARIOUS MORPHOLOGIES

MURIEL LANSALOT, JUTTA RIEGER, AND FRANCK D'AGOSTO
Université de Lyon and UPMC Université-Paris 6, Sorbonne Universités, CNRS, France

2.1 INTRODUCTION

The assembly of block copolymers in solution is an important area in the field of macromolecular self-assembly. Assembly occurs because of the specific affinity of the blocks with different media and with each other. Among these macromolecules, those incorporating blocks with antagonistic properties are particularly interesting in that they allow phase separation either in bulk or in a selective solvent for one of the blocks [1]. The synthesis of block copolymers [2] usually requires the use of living/controlled polymerization techniques. They allow the synthesis of well-defined polymer chains with reactive end-functionalities in a first polymerization step, chains that may then be extended with a second monomer. An alternative scheme that often also requires the use of living polymerization techniques is based on efficient coupling reactions between two pre-formed well-defined polymer chains carrying appropriately chosen reactive end-groups [3, 4]. Among the

Macromolecular Self-Assembly, First Edition. Edited by Laurent Billon and Oleg Borisov.

variety of possible block copolymers, one of the most attractive classes consists in amphiphilic structures composed of a hydrophilic and a hydrophobic block. These macromolecules can self-assemble in the most benign and easy-to-handle solvent: water. The self-assembly of such macromolecules was first developed by Eisenberg [5–7] who prepared solutions of the desired amphiphilic block copolymers in a common solvent for both blocks and then induced their self-assembly by slowly adding a selective nonsolvent for one of the blocks (i.e., water) in accord with a solvent displacement method. Various morphologies, such as spherical aggregates, rods, worms, fibers and vesicles, could be achieved by adjusting the molar masses of both the hydrophilic and hydrophobic blocks (Figure 2.1), but extremely dilute conditions and long equilibration times were typically required. The resulting objects are expected to contribute to a large development in high-performance materials [6, 8]. This growing demand is encouraging the development of a large variety of amphiphilic block copolymer structures, along with methods to assemble and isolate the corresponding objects more rapidly and in higher concentration.

Polymerization in aqueous dispersed systems is a well-known means to obtain micro- to nano-objects that are colloidally stable in a dispersant (water or water-based mixtures) and in which the polymer chains can be confined to spherical particles. Among the associated processes, emulsion polymerization is an industrial tool used to produce hydrophobic polymer particles in water with the aid of a molecular surfactant [9]. An emulsion system contains a water-soluble initiator, the hydrophobic monomer (mostly located in large droplets and small surfactant micelles), and the surfactant. Hydrosoluble radicals are generated in water and because of a certain, albeit very low, water solubility of the hydrophobic monomer, they start to add monomer units in the aqueous phase and generate oligoradicals. Beyond a critical degree of polymerization, the growing oligomeric species become insoluble and migrate into the monomer-swollen surfactant micelles, where the polymerization continues. Swollen by the monomer and fed by monomer diffusion from large monomer droplets through

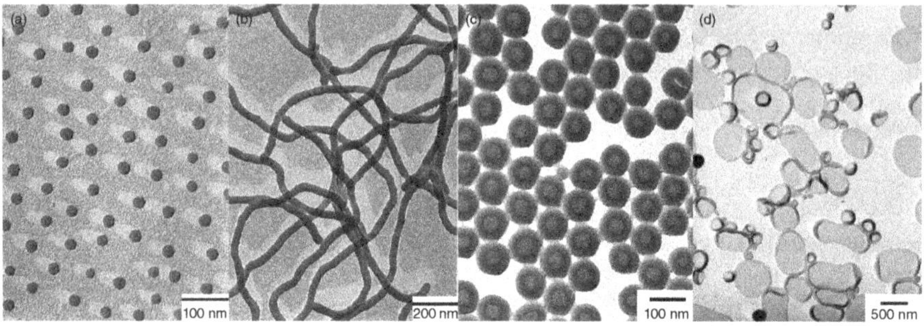

Figure 2.1 An example of "pure" morphologies obtained by self-assembling preformed poly(acrylic acid)-*b*-polystyrene (PAA-*b*-PSt) amphiphilic block copolymers in water. (a) PAA_{21}-*b*-PSt_{200}—spheres; (b) PAA_{20}-*b*-PSt_{190}—worms; (c) PAA_{13}-*b*-PSt_{410}—vesicles; (d) PAA_{20}-*b*-PSt_{132}—lamellae. The degree of polymerization (DP_n) for each block is indicated in the subscript. Reprinted with permission from [7]. Copyright 1999 NRC Canada.

the water phase, the growing particles can continue to consume monomers and incorporate the polymer chains created in the water phase, leading to the final spherical nanoparticles. The stability of these objects is thus achieved by a key component in the process: the surfactant. However, the surfactant is known to have detrimental effects on the latex stability when frozen or subjected to high shear, and diminishes the properties of the resulting films when exposed to water or high humidity.

Since the 1990s, with the development of controlled radical polymerization techniques, the advances in polymer chemistry have accelerated in combining the desirable features of living polymerization techniques already known mainly from ionic polymerizations with the simplicity of a free radical process [10]. Myriad new architectures have been achieved, and the most prominent are the amphiphilic block copolymers. This concept was also envisaged for polymerizations in dispersed media with the goal of reproducing the same degree of topological control over the macromolecular structures as in solution polymerizations [11, 12].

However, implementing this approach proved unsuccessful in the first years of research due to an inability to restrict the location of the controlling agent to the site of polymerization in the dispersed systems (first in water then in the particles) [13]. This issue was eventually overcome by exploiting the ability of CRP to produce amphiphilic block copolymers, which can then assume the function of the surfactant that is typically employed in a conventional emulsion-polymerization system [14]. Based on the same concept as the original work on living anionic dispersion polymerization in organic solvents [15], the hydrophilic living polymer chain was grown in a first step and further chain-extended in water as a hydrophobic monomer was slowly fed into the reactor to prevent the formation of monomer droplets (and their potential nucleation). The resulting amphiphilic block copolymer proved then to be able to self-assemble into nanosized self-stabilized particles [16]. The successful realization of this aim constituted the first example of surfactant-free emulsion polymerization based on self-assembled amphiphilic block copolymers formed *in situ*, and this was later coined polymerization-induced micellization [17] or polymerization-induced self-assembly (PISA) (Figure 2.2) [18, 19].

Although PISA has not replaced conventional surfactant-containing emulsion polymerization in producing industrial-scale polymer latexes, it has rapidly evolved into a versatile tool capable of producing amphiphilic block copolymers directly in water at high solids contents [20]. In addition to emulsion polymerization, PISA has been applied to dispersion-polymerization systems, in which the monomers

Figure 2.2 Polymerization-induced self-assembly (PISA) using CRP in water. (*See color insert for color representation of this figure*).

are initially soluble but still form a solid polymer phase. Both in emulsion and in dispersion, it was quickly recognized that under optimized conditions, PISA could directly yield the same self-assembled morphologies (spheres, rods, fibers, vesicles) previously obtained by the solvent-displacement method using preformed amphiphilic block copolymers, but at much higher solids contents (up to 40–50%) and with significantly less experimental effort (a few hours of polymerization) [21].

PISA thus provides several advances: first, it allows direct amphiphilic block copolymer synthesis in water; second, it permits the formation of various micellar architectures in the absence of added cosolvents; and third, it corresponds to a breakthrough in surfactant-free controlled radical emulsion polymerization with the formation of well-defined and self-stabilized particles. It has been the topic of excellent recent reviews by us and other leaders in the field.

The aim of this chapter is to give an accurate and up-to-date (as of end of 2014) overview of the recent work involving the PISA process performed in water and water-based systems using different controlled polymerization techniques.

2.2 PRELIMINARY COMMENTS UNDERLYING CONTROLLED RADICAL POLYMERIZATION

When discussing the origins of the PISA process, it is important to recognize that it is not restricted to controlled radical polymerization techniques, and that any living polymerization technique that allows the reactivation of a soluble polymer with the formation of a second, insoluble polymer block could be referred to as a PISA process. Work published almost 40 years ago by Barrett [15], and more recently by Okay *et al.* [22] and Kim *et al.* [23], exploited the livingness of poly(*t*-butyl styryl)-lithium chains to initiate the dispersion polymerization of styrene (St) [15, 23] or divinylbenzene [22] in *n*-hexane, and indeed pioneered the concept of what is today called a PISA process. PISA can, in essence, be defined as the self-assembly of (amphiphilic) block copolymers during polymerization. However, as it will be detailed below, some of the soluble living polymers used for the polymerization in dispersed media (emulsion or dispersion) are in some cases amphiphilic and able to self-assemble before the polymerization starts [18, 24]. In both cases, however, particle formation relies on an efficient micellar nucleation. Both types of living precursors will thus be described in this chapter.

Since controlled radical polymerization techniques are central to the PISA process, they will be briefly described in the following sections.

2.2.1 Introduction

Free-radical polymerization is the most important method for the production of synthetic polymers in large-scale industrial production and in the manifold applications. Despite the wide range of functional monomers that can be polymerized by a radical mechanism and the great variety of statistical copolymers with many structures and properties [25] that can be produced by free-radical polymerization,

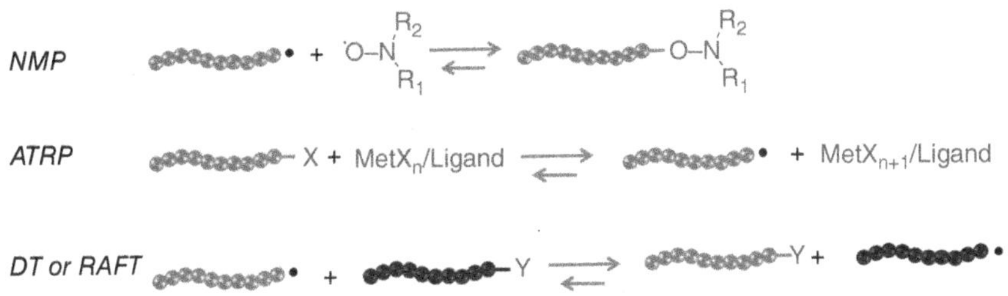

Figure 2.3 Commonly used techniques to control the free radical polymerization. Nitroxide-mediated polymerization (NMP), atom transfer radical polymerization (ATRP, Met = metal atom, X = halogen atom), degenerative-transfer polymerization (DT, −Y= −I, $-TeCH_3$) or reversible addition–fragmentation chain transfer polymerization (RAFT, −Y = −SC(=S) − Z ; Z = activating group).

the unavoidable occurrence of fast radical–radical termination reactions leads to a lack of control over the resulting polymers' molar masses, molar mass distributions, chain-end functionalities and the macromolecular architecture. The pathway found to get around this is the emergence of the so-called living or controlled radical polymerization techniques (IUPAC: reversible-deactivation radical polymerization, RDRP) that has opened a new era for the free-radical polymerization. The CRP techniques can be categorically subdivided into two groups. They are either based on a *reversible termination reaction* or on a *reversible chain transfer reaction* (Figure 2.3). In both cases macromolecular radicals, called "active species," undergo reversible deactivation—that is, successive activation–deactivation cycles—by being capped by a specific group under a form called dormant. Only a very small fraction of chains are simultaneously active and can propagate, but as a consequence of the rapid exchange of the radical between active and dormant macromolecular chains, they grow simultaneously during the whole polymerization period. The main feature of CRP is that the number-average molar masses increase linearly with monomer conversion and the corresponding molar mass distributions are narrow. This requires that a fast exchange between active and dormant chains takes place. After the monomer has been consumed, the formed chains still bear the mediating function at the chain end and can thus be further extended with either the same or another monomer in a subsequent polymerization step. This paves the way for the synthesis of block copolymers and other more complex architectures.

2.2.2 Major Methods Based on a Reversible Termination Mechanism

Nitroxide-mediated polymerization [26] and atom transfer radical polymerization [27] represent the two most commonly used methods of controlled radical polymerization based on a reversible termination reaction. In these polymerization systems, the dormant species are either an alkoxyamine for NMP or an alkyl halide for ATRP (Figure 2.3). In NMP, the homolytic cleavage of the alkoxyamine is thermally activated, which generally requires elevated temperatures, whereas in ATRP, the activation of the alkyl halide chain ends is achieved through a redox reaction with a transition metal complex under a broader range of conditions. In both cases the polymerization kinetics are governed by the activation–deactivation equilibrium and by the persistent radical effect [28]. The theoretical DP_n can be calculated by the ratio of the initial monomer concentration and the initiator (i.e., alkoxyamine or alkyl halide) concentration, multiplied by the monomer conversion.

2.2.2.1 ***Nitroxide-Mediated Controlled Radical Polymerization (NMP)*** Nitroxides are stable radicals that are able to trap carbon-centered radicals at a nearly diffusion-controlled rate. While at low temperatures the formation of the alkoxyamine is kinetically irreversible, the formed C–O bond may undergo homolytic cleavage at elevated temperatures, regenerating a propagating radical and the nitroxide. This equilibrium between propagating radicals and dormant macroalkoxyamines governs the NMP. The initiation step can be performed in two different ways: either in a bicomponent system using a classical radical initiator in combination with the free nitroxide or in a monocomponent system based on a preformed alkoxyamine. The latter is nowadays the most popular one because it allows the kinetics and the molar masses of the obtained polymer chains to be tuned very precisely. The typically used nitroxides are 2,2,6,6-tetramethylpiperidine-1-oxyl (TEMPO), which was successfully applied for St and derivatives, and *N-tert*-butyl-*N*-(1-diethyl phosphono-2,2-dimethyl propyl) nitroxide (SG1) and 2,2,5-trimethyl-4-phenyl-3-azahexane-3-oxyl (TIPNO) (Figure 2.4), which have significantly expanded the range of monomers amenable for use in NMP and reduced the required polymerization times and temperatures.

2.2.2.2 ***Atom-Transfer Radical Polymerization (ATRP)*** ATRP [27] is based on the reversible transfer of a halogen atom between a dormant alkyl halide and a transition metal catalyst in a redox reaction (Figure 2.3). The alkyl halide is reduced to a growing radical while the transition metal is oxidized. In most of the studied reactions, Cu(I) species complexed by multidentate amine ligands constitute the activator and the corresponding Cu(II) species are the deactivator. The initiation step can be performed in different ways, starting either from Cu(I) or from Cu(II). With Cu(I), the process is called direct ATRP. An alkyl halide initiator is employed. With Cu(II), there are multiple initiation possibilities, including the reverse ATRP process [27],

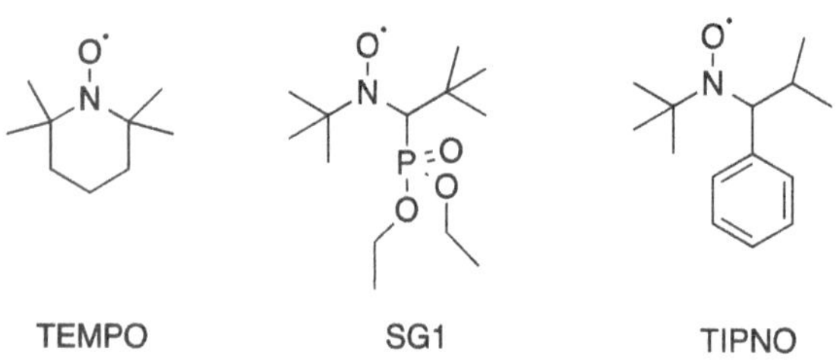

Figure 2.4 Chemical structures of commonly used nitroxides in NMP.

which uses a classical radical initiator; the simultaneous reverse and normal initiation (SR&NI) method [29], which combines an alkyl halide along with a small fraction of a classical radical initiator (the latter produces radicals able to reduce Cu(II) to Cu(I)); the activator (re)generated by electron transfer (A(R)GET) technique [30], which uses an alkyl-halide initiator in combination with a reducing agent to turn Cu(II) into Cu(I); and e-ATRP [31], which utilizes the concept of ARGET-ATRP via an electrochemical stimulus to provide enhanced polymerization control. ATRP can be successfully applied to a broad variety of monomers using a multitude of available mono- or multifunctional initiators and catalysts that allow the polymerization and polymer characteristics to be fine-tuned.

2.2.3 Major Methods Based on a Reversible Transfer Mechanism

For the methods proceeding according to a reversible transfer mechanism, a bimolecular reaction between an active and a dormant macromolecule leads to the transfer of the functional end-group (Figure 2.3). This can be a direct exchange as in the so-called degenerative transfer (DT) technique where an iodine atom is interchanged ((reverse) iodine transfer polymerization, (R)ITP) [32–34]. The organo-tellurium mediated CRP (TERP) follows a similar mechanism [35], with the exchange of a terminal $-TeCH_3$ group. This reaction proceeds via both thermal dissociation and degenerative transfer. However, when an external source of free radicals is used at low temperatures, only degenerative transfer takes place. Another technique is the well-known reversible addition–fragmentation chain transfer (RAFT) [36, 37] polymerization in which chains are end-functionalized by an unsaturated group (typically the C=S bond from a dithioester, a dithiocarbonate, a dithiocarbamate, or a trithiocarbonate). The radical is transferred through an addition–fragmentation step with a tertiary carbon-based radical as the intermediate. A reversible transfer system requires the use of a conventional radical source in addition to a reversible chain transfer agent RY (Figure 2.3). The first step can be considered as a conventional transfer reaction to RY, creating new chains with a R group at one end and a Y atom or functional group at the other end. The next step is the transfer of Y from an end-functionalized chain to a propagating macroradical, which is a thermodynamically neutral (i.e., degenerative) process. This second step does not create new chains but contributes to the extension and continuous redistribution of the active function amongst the existing chains. If the initial concentration of the radical initiator is low with respect to the initial concentration of the transfer agent, a large majority of the macromolecules have the same RAFT agent-derived end-groups. The concentration of growing macromolecular chains becomes constant and close to the initial concentration of the RAFT agent, once the latter has been completely consumed. Only at that stage, and when the reversible transfer takes place, can a linear increase of the number-average molar mass with monomer conversion be observed (the DP_n can be approximated by the ratio of the initial monomer concentration divided by the chain transfer agent concentration, multiplied by the monomer conversion). The main requirement is that the rate constant of the transfer reaction to RY should be large, which is the case for the RAFT process.

2.3 PISA VIA CRP BASED ON REVERSIBLE TERMINATION

As mentioned above, CRP techniques based on reversible termination such as NMP or ATRP do not require an additional source of radicals because the radicals are either produced by thermal activation of an alkoxyamine, in the case of NMP, or by activation of the halogen–carbon bond with a metallic complex, in the case of ATRP. In theory, this represents a considerable advantage compared with other techniques, particularly in regard to block copolymer synthesis because the formation of the second block will not be accompanied by the formation of a homopolymer side product that can deteriorate the efficiency of the self-assembly.

2.3.1 PISA Using NMP

2.3.1.1 Starting with a Poly(acrylic acid)-Based Macroalkoxyamine In 2005 Couvreur *et al.* [38] managed to control for the first time the polymerization of acrylic acid (AA) in dioxane using an alkoxyamine based on SG1 nitroxide (Figure 2.4). The resulting low molar mass PAA macroalkoxyamines could be easily dissolved in water after treatment with NaOH. Delaittre *et al.* [39, 40] first exploited the use of these water-soluble macromolecular controlling agents synthesized by NMP to polymerize styrene and *n*-butyl acrylate (BA) under batch emulsion conditions in water. The polymerization was conducted under basic conditions (pH value around 8) to preserve the integrity of SG1 that is known to degrade under acidic conditions. The polymerization was performed at 120°C under a nitrogen pressure of 3 bar and with a targeted solids content of 20 wt%. The analyses of the samples withdrawn during the polymerization of the hydrophobic monomers showed a linear increase of the molar masses with conversion and low molar mass dispersity, demonstrating the formation of well-defined block copolymers. However, higher molar masses than theoretically expected were obtained. This was consistent with a low initiation efficiency (estimated to be 60% for BA and 35% for St) of the PAA-SG1 macroalkoxyamine, meaning that a considerable amount of PAA chains remained in the aqueous phase. The addition of free SG1 during the growth of the hydrophobic block only slightly increased the initiation efficiency without altering the molar mass distribution. The colloidal characterization showed the presence of isometric and spherical particles with diameters between 40 and 110 nm. Given the advantage of PISA over the solvent displacement method, the solids content could even be increased up to 40 wt% with the same level of control over the block copolymer synthesis and with the same colloidal stability of the latex.

Building on this first example of controlled emulsion polymerization using an *ab initio* batch process, Delaittre *et al.* further extended this strategy to dispersion polymerization, in order to produce thermosensitive nanogels [42, 43]. Self-stabilized nanoparticles with a temperature-responsive poly(*N*,*N*-diethylacrylamide-*co*-*N*,*N*′-methylenebisacrylamide) (P(DEAAm-*co*-MBAAm)) microgel core were obtained using PAA-SG1 macroalkoxyamine in basic conditions. First non-crosslinked particles were obtained in absence of MBAAm at 112°C under a nitrogen pressure of 3 bar at pH = 10. The solids content had been fixed to 20 wt%. The temperature-induced phase transition of PDEAAm at its lower critical solution temperature (LCST, around 32°C) led to a redissolution of the obtained particles composed of PAA-*b*-PDEAAm block copolymers when the temperature was decreased at the end of the polymerization. Upon re-increasing the temperature, a latex formed again, but the formed particles had a diameter significantly larger than the original ones, showing the difference of assembling block copolymers *in situ* during their formation or post-assembling them afterward. At low temperatures and pH values, "inverse particles" could be formed, composed of a PAA core and a hydrophilic PDEAAm shell. Again, the molar mass analysis of the obtained block copolymers showed higher molar mass values than expected, indicating a low initiation efficiency of PAA at high pH values. The incorporation of a small amount of MBAAm (<3 mol% with respect to the monomer concentration) during the formation of the PDEAAm block allowed locking of the particle morphology by crosslinking the core, leading to microgel particles. Higher amounts of MBAAm gave rise to macrogelation. Nevertheless, higher extents of crosslinking could be obtained by adding the MBAAm after the nucleation of the particles.

Taking advantage of the controlled growth of the PDEAAm block, physical crosslinking was also investigated [43] by adding St units during the DEAAm polymerization step (after 50% conversion) and targeting a solids content of 30 wt%. The conversion plateaued after 70% and 22 hours. Upon cooling, the obtained latex remained stable and the particles exhibited a diameter of 95 nm at 50°C and 120 nm at 15°C. Although SEC analysis of the resulting block copolymers turned out to be complicated, ^{13}C-NMR spectroscopy showed that the block incorporating the St units contained a high fraction of $-(St)_2$-DEAAm triads. However, DSC measurements were not conclusive concerning the potential physical crosslinking of the particles. Ultimately, the study by Delaittre *et al.* demonstrated the versatility of the technique in producing well-defined and custom-designed microgel particles.

As mentioned in the introduction, the simultaneous *in situ* formation of amphiphilic block copolymers and their self-assembly in water should give rise to the formation of morphologies that have already been observed when self-assembling preformed block copolymers in a selective solvent. Indeed, Delaittre *et al.* [21] employed PAA-SG1 macroalkoxyamine (poly(sodium acrylate), M_n = 2000 g mol^{-1}) at pH = 11 to polymerize 4-vinylpyridine (4VP) in water (20 wt%) at 120°C under a nitrogen pressure of 3 bar. The polymerization was fast (3–4 h) and reached a conversion of 90%. The control over the polymerization of the P4VP block was shown by the gradual shift of the SEC chromatograms toward higher molar masses with conversion. A similar initiation efficiency to the one obtained when St was used as the hydrophobic monomer was obtained here (around 30%). Dynamic light scattering (DLS) and transmission electron microscopy (TEM) analyses were consistent with the presence of particle morphologies different from the spherical morphology obtained in the previously mentioned study employing a PAA-SG1 macroalkoxyamine in a PISA system. In addition to spherical particles with a diameter of 60 nm, a broad range of morphologies were observed including wormlike micelles of the same size, spherical vesicles, mixed wormlike/vesicular micelles, elongated vesicles and multicompartment vesicles when a DP_n of 880 was targeted for the P4VP block. TEM images taken at different times during the polymerization of 4VP showed that small spherical particles first formed up to a monomer conversion of 33%. At a conversion of 58%, the other morphologies started to arise. The conversion of the spherical micelles to larger assemblies was assumed to be driven by the swelling of the hydrophobic core by the monomer. The low initiation efficiency observed with poly(sodium acrylate) was also suspected to induce a continuous creation of amphiphilic block copolymers of very different compositions and molar masses that may lead to these various morphologies. Eventually, decreasing the pH value to 2 led to a complete dissolution of the particles due to the pH responsiveness of the P4VP block. Increase the pH value again induced the self-assembly, but the formed particles had spherical shapes (diameter 85 nm). This result was interpreted by the formation of out-of-equilibrium morphologies during the polymerization process.

2.3.1.2 Improving the Macroalkoxyamine Initiation Efficiency: Hydrosoluble Poly(methacrylate)-Type Macroalkoxyamines In all the above-mentioned studies, the poor initiation efficiency associated with the use of PAA-SG1 macroalkoxyamine at high pH values was partly due to the low dissociation constant at 120°C, the temperature at which the polymerizations were performed. This led the researchers to consider the alternative of water-soluble macroalkoxyamines with higher dissociation rates. In this context, an ideal candidate would be poly(methacrylic acid) capped with SG1 as it would mimic the well-known BlocBuilder® structure that shows a high initiating efficiency [44, 45] due to a comparably high dissociation rate constant. Although the control of methacrylic monomer had been found to be complicated because of the low stability of the alkoxyamine chain end and the side reaction that this group can undergo, Charleux *et al.* showed that the use of a small amount of a co-monomer such as St or acrylonitrile in the polymerization of methacrylates could alleviate these issues and allowed the formation of living polymethacrylates, including poly(methacrylic acid-*co*-styrene) (P(MAA-*co*-St)-SG1), which can dissociate at low temperature (70–90°C) [26]. This new type of macroalkoxyamine was then used in alkaline conditions in water for the copolymerization of methyl methacrylate (MMA) and St (up to 8.8 mol% with respect to MMA) [46] at 90°C. As anticipated, 100% initiation efficiency was achieved leading to well-defined amphiphilic block copolymers that self-assemble into spherical nanoparticles of less than 100 nm. Still, the colloidal characteristics of the final latex were less satisfactory (increase of particle dispersity and size observed during the polymerization due to aggregation), since the hydrophilicity of the P(MAA-*co*-St)-SG1 was partly impacted by the quite hydrophobic nature of the polymethacrylic backbone and the presence of St units. In the same vein, spherical nanoparticles decorated by poly(ethylene oxide) hydrophilic moieties were obtained using hydrosoluble macroalkoxyamines based on PEO methyl ether methacrylate (PEOMA) and MAA [17, 47]. Interestingly, Qiao *et al.* [47] designed P(PEOMA-*co*-St)-SG1 macroalkoxyamines with two PEOMA blocks of distinct molar masses (300 g mol^{-1} and 950 g mol^{-1}). The resulting macroalkoxyamines were used to copolymerize *n*-butyl methacrylate (BMA) with a small amount of St in an emulsion system according to a PISA process performed at 85°C. The low cloud point of the macroalkoxyamine incorporating a low molar mass PEOMA prevented the formation of stable particles. In contrast, well-defined amphiphilic block copolymers were obtained by varying the concentration of the starting macroalkoxyamine comprising the high molar mass PEOMA and their self-assembly led to spherical particles with diameters between 60 and 300 nm at a pH value of 4.2. Increasing the pH value from 4.2 to 6.7 drastically affected the obtained morphologies (change to vesicles and anisotropic objects; Figure 2.5). This was rationalized by the salting-out effect induced by a concomitant increase of the ionic strength and confirmed by the formation of fibers and vesicles when increasing

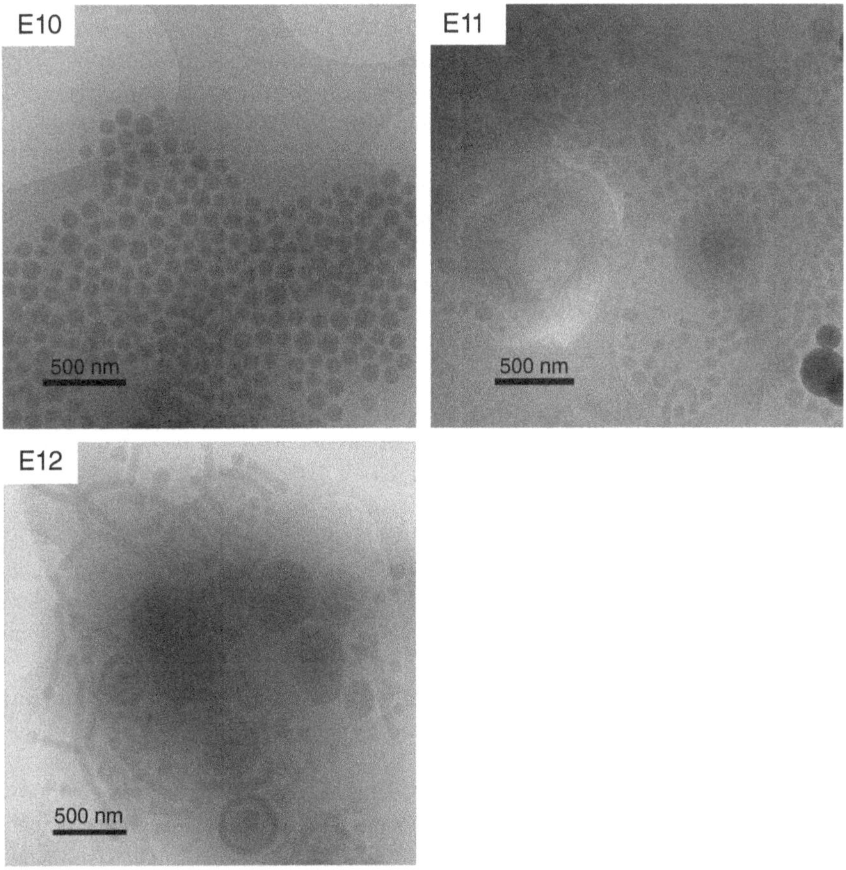

Figure 2.5 Cryo-TEM images of the final P(PEOMA-*co*-St)-*b*-P(BMA-*co*-St) latex particles obtained by surfactant-free emulsion polymerization of BMA and St in the presence of P(PEOMA-*co*-St)-SG1 with different concentrations of NaCl. E10: 10^{-3} mol L^{-1}; E11: 10^{-2} mol L^{-1}; E12: 10^{-1} mol L^{-1}. Reprinted with permission from [47]. Copyright 2013 American Chemical Society.

the amounts of sodium chloride when the PISA process was performed at a pH value of 4.2.

With the aim of preparing stimuli-responsive materials, Darabi *et al.* [48] investigated the bulk synthesis and the use of poly((2-diethylaminoethyl methacrylate)-*co*-styrene) macroalkoxyamine (P(DEAEMA-*co*-St)-SG1) in its protonated form to copolymerize MMA and St in aqueous emulsion at 90°C via a PISA process. The obtained macroalkoxyamines exhibit a quantitative initiation efficiency and a stable latex was obtained from the self-assembly of the resulting well-defined P(DEAEMA-*co*-St)-*b*-P(MMA-*co*-St) amphiphilic block copolymers. Spherical particles were found that had diameters between 70 and 90 nm. Their redispersion in carbonated water after neutralization with NaOH (causing precipitation) was, however, difficult.

Eventually, Yoshida also investigated the photopolymerization of various methacrylic monomers in the presence of TEMPO or derivatives, of an azoinitiator and of photoacid generators such as bis(alkylphenyl)iodonium hexafluorophosphate (BAI), to produce radicals [49, 50]. The mechanism remained unclear as the role of the photoacid generators was not elucidated [26]. Nevertheless, the author claimed that the polymerization of methacrylic acid [50] and methyl methacrylate [49] had been controlled at room temperature without the need of supporting co-monomer such as St or acrylonitrile as mentioned and accepted in the literature so far. Vesicles were claimed to be prepared by using a PMAA-TEMPO-OMe macroalkoxyamine (M_n = 17 100 g mol^{-1}, Đ = 1.76), produced in methanol after 76% conversion of MAA, to reinitiate the copolymerization of MMA and the remaining MAA at room temperature in a mixture of methanol and water (4/1 by volume) [51]. PMAA-*b*-P($MAA_{0.183}$-*co*-$MMA_{0.817}$) amphiphilic block copolymers (M_n = 50 500 g mol^{-1}, Đ = 1.70) formed and self-assembled into micrometer-sized giant vesicles. Then [52], an increasing content of MAA units was incorporated in the P(MMA-*co*-MAA) statistical hydrophobic block (P($MAA_{0.117}$-*co*-$MMA_{0.883}$), M_n = 50 100 g mol^{-1}, Đ = 1.85; P($MAA_{0.218}$-*co*-$MMA_{0.782}$), M_n = 69 700 g mol^{-1}, Đ = 1.66; P($MAA_{0.299}$-*co*-$MMA_{0.701}$), M_n = 98 100 g mol^{-1}, Đ = 1.71), which provoked the switch from vesicles to membranes. This was attributed to a change in the packing parameters of the block copolymers [53] when increasing the hydrophilicity of the hydrophobic block [54], but without commenting on the concomitant increase of the block's molar mass. A similar study was further performed [55] in which only the molar mass parameter was varied. In this case, fission of the already obtained giant vesicles was observed.

2.3.1.3 Toward a Whole Process in Water: The One-Pot PISA In order to improve the stability of the final particles while maintaining a quantitative initiation efficiency, Brusseau *et al.* attempted to substitute St by sodium styrene sulfonate (NaSS) as a co-monomer for MAA in the synthesis of the hydrosoluble macroalkoxyamine [56]. The use of a pair of totally hydrosoluble comonomers was thought to additionally enable the possibility to perform the synthesis of the macroalkoxyamine in water and to subsequently add the monomers St and MMA in the same reactor, thus producing amphiphilic block copolymer particles in a one-pot process. However, as mentioned

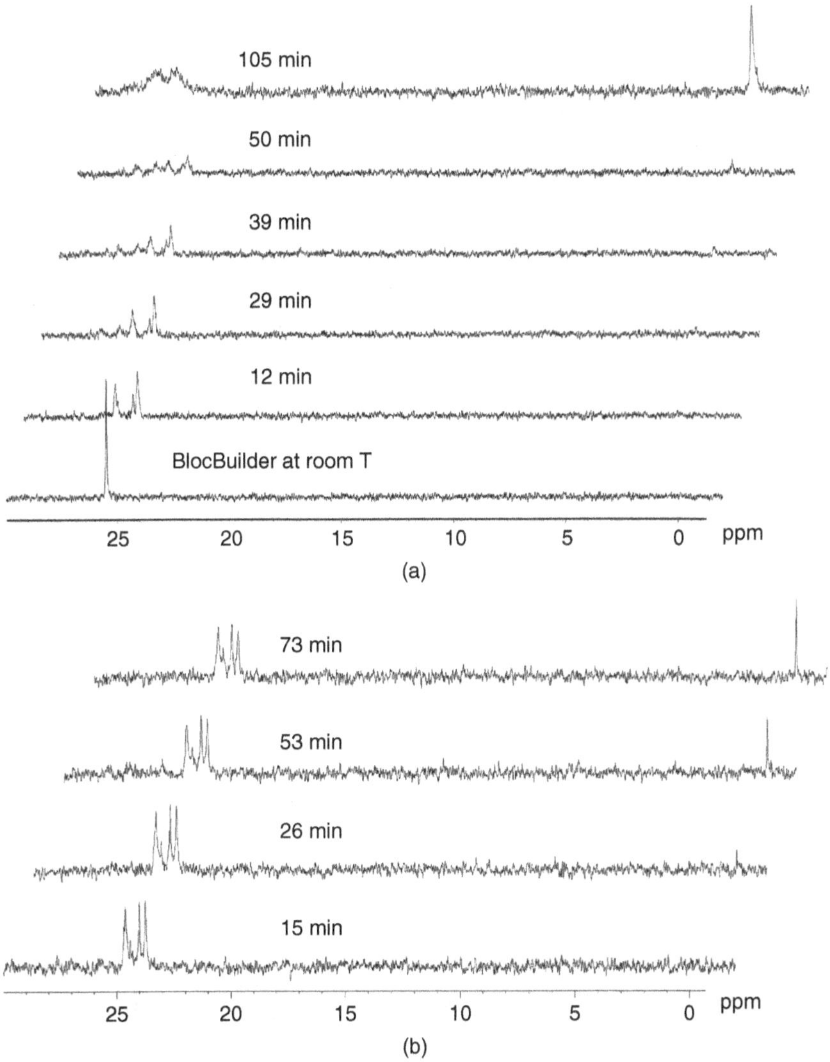

Figure 2.6 Evolution of the ^{31}P-NMR spectra with polymerization time during the SG1-mediated copolymerization of MAA with NaSS performed at 76°C (a) in D_2O and (b) in DMSO-d_6. Reprinted with permission from [56]. Copyright 2011 American Chemical Society.

earlier, the stability of the SG1 nitroxide is known to be poor at low pH values such as the one of aqueous solution of MAA and NaSS. Indeed, it was shown (Figure 2.6) that within short reaction times (less than 15 min) and at a low temperature (76°C), living chains of P(MAA-*co*-NaSS)-SG1 could be generated that can successfully be chain extended with MMA and St in the same reactor.

Although unreacted hydrophilic monomers remained at the end of this first step (about 50% conversion), they were supposed to be quickly consumed in the emulsion polymerization step (when MMA and St were added and when the polymerization temperature was raised to 90°C). Well-defined amphiphilic block copolymers were formed and simultaneously self-assembled by PISA in a one-pot process to produce spherical nanoparticles with diameters of less than 100 nm.

After optimizing both the initiation efficiency by switching from a polyacrylate- to a polymethacrylate-type hydrosoluble macroalkoxyamine and the colloidal stability by a careful choice of the controlling styrenic co-monomer, the formation of morphologies different from spheres in a one-pot process could be investigated. This was indeed achieved using the same P(MAA-*co*-NaSS)-SG1 macroalkoxyamine and by varying the molar mass of the P(MMA-*co*-St) hydrophobic block [57]. Using a starting P(MAA-*co*-NaSS)-SG1 macroalkoxyamine with a M_n of 5600 g mol^{-1}, small spheres (25 nm), a mixture of small spheres and short fibers, fibers and vesicles, were obtained when the M_n value of the final P(MAA-*co*-NaSS)-*b*-P(MMA-*co*-St) block copolymers was 24570 g mol^{-1}, 31900 g mol^{1}, 34200 g mol^{-1}, 46300 g mol^{-1}, respectively, showing that the window to obtain a pure morphology such as fibers remains narrow (Figure 2.7).

2.3.2 Using ATRP

The only examples for the use of a PISA strategy to self-assemble amphiphilic block copolymers in water in the absence of additional surfactant via ATRP have all made use of a poly(ethylene glycol) (PEG)-based macroinitiator end-functionalized with a labile bromine atom (PEG-Br). These water-soluble chains were then

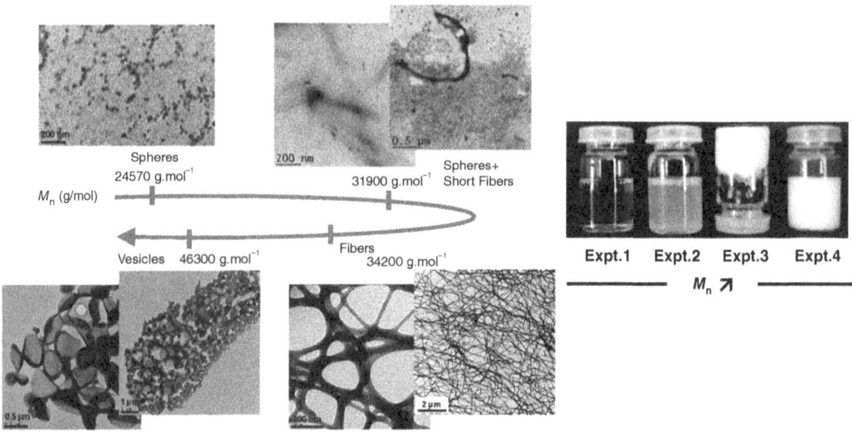

Figure 2.7 TEM images of the final nano-objects (left) and photographs of the latexes obtained with the same macroinitiator P(MAA-*co*-NaSS)-SG1 (5600 g mol^{-1}) for different molar masses of the hydrophobic block. Reprinted with permission from [57]. Copyright 2012 American Chemical Society.

extended with *N*-isopropylacrylamide (NIPAM) [58], 4VP [59], dimethylaminoethyl methacrylate (DMAEMA) [60], and 2-(methacryloyloxy)ethyl phosphorylcholine (MPC) [61] according to a dispersion–polymerization process. The only example performed in pure water was presented by Kim *et al.* [58] who performed the NIPAM polymerization at 50°C using PEG-Br in the presence of CuBr and tris-(2-dimethylaminoethyl)amine (M_6TREN) as a ligand. Block copolymer micelles were obtained as the polymerization temperature was set above the LCST of PNIPAM (32°C). To preserve the particles at lower temperatures, *N,N′*-ethylenebisacrylamide (EBAAm, 10 mol% with respect to NIPAM) was used as a crosslinker and allowed the formation of a nanogel with a diameter of 67.5 nm. This paper [58] constitutes one of the earliest reports on PISA in aqueous dispersion and the first one reporting the synthesis of a hydrogel using this strategy. Similarly, in a mixture of ethanol and water (1:1 by volume), the polymerization of 4VP was carried out using a PEG-Br macroinitiator, CuBr and *N,N,N′,N′,N″*-pentamethyldiethylenetriamine (PMDETA) as a ligand, and MBAAm as the crosslinker (5 mol% with respect to 4VP) [59]. The obtained nanogels with sizes of around 30 nm were further used to complex $HAuCl_4$ onto the strongly bonding 4VP unit. After reduction with $NaBH_4$, composite nanoparticles incorporating gold nanoparticles in their core could be formed. Eventually, Sugihara *et al.* [60, 61] employed an alcohol-rich solvent (isopropanol:water = 9:1 by volume) to design and self-assemble block copolymers based on MPC from a PEG-Br macroinitiator using CuBr with bipyridine (Bpy) as the ligand. In a first approach, PEG-Br was first chain-extended with DMAEMA at 40°C, and in a second step, with MPC. The resulting triblock copolymer underwent spontaneous self-assembly to PMPC-core micelles *in situ* as a result of the cononsolvency behavior of PMPC chains in alcohol/water mixtures. Shell-crosslinked micelles were further obtained by adding 1,2-bis(2-iodo-ethoxy)ethane (BIEE), which led to the quaternization of the DMAEMA units (Figure 2.8).

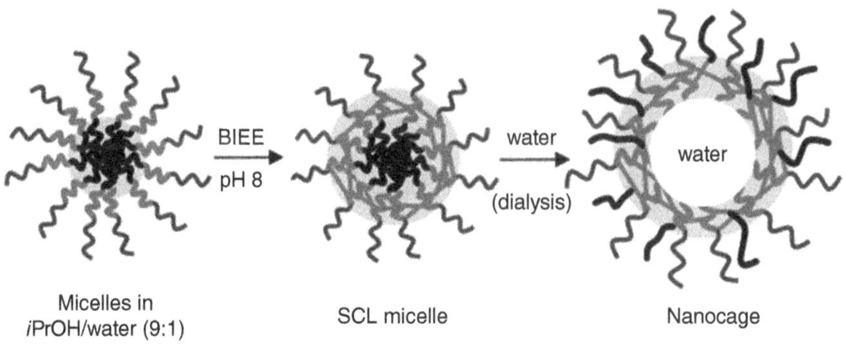

Figure 2.8 Strategy employed by Sugihara *et al.* to design shell-crosslinked micelles at a solids content of 10 wt% using PEG-*b*-PDMAEMA-*b*-PMPC (red: PEG, blue: PDMAEMA, black: PMPC). Reprinted with permission from [60]. Copyright 2010 Wiley-VCH Verlag GmbH & Co. KGaA, Weinheim.

A similar strategy was employed in a second approach to chain-extend PEG-Br directly with MPC. The resulting micelles in alcoholic solvent were further crosslinked to obtain nanogels after dialysis of water. The fine-tuning of the molar mass of the PMPC block and the degree of crosslinking led to gel particles with different sizes exhibiting a high swelling capacity in water.

A last example that is worth to be mentioned is based on the recent progress made in the field of Cu(0)-mediated controlled radical polymerization in pure water [62–64]. Zhang *et al.* [65] took advantage of the excellent control achieved under optimized conditions (water, 0°C, M_6TREN, CuBr, dihydroxypropyl-2-bromo-2methylpropanoate used as the water-soluble initiator) to first synthesize a hydrophilic block composed of unprotected mannose glycomonomer units (ManA). This PManA block was used as the macroinitiator for the polymerization of dimethyleneglycol acrylate (DEGA) in a methanol/water mixture. The resulting well-defined doubly hydrophilic block copolymer PManA-*b*-PDEGA did not self-assemble in water at the polymerization temperature (0°C) but exhibited an LCST below room temperature that gave rise to its self-assembly at room temperature, and facilitated the TEM analysis of the spherical micelles that form under these conditions. Although strictly it cannot be considered a true PISA process, the last example shows that a careful choice of the monomer (including hydrophobic ones) may pave the way for the use of Cu(0)-mediated CRP via PISA. The general technique has now been developed and will undoubtedly help expand the range of strategies to implement PISA in pure water.

2.4 PISA VIA CRP BASED ON REVERSIBLE TRANSFER

In contrast to the polymerization techniques presented before, which were based on a reversible termination mechanism, PISA using RAFT (or TERP) requires the addition of an external radical source, added to the continuous phase (typically water). These PISA techniques may therefore suffer from competition between the desired "self-assembly nucleation" and unwanted homogeneous nucleation, that is, nucleation by precipitation of solvophobic oligomers before the chain transfer reaction can take place. Homogeneous nucleation may result from a too low transfer efficiency to the RAFT agent, a too low RAFT-agent concentration (in the continuous phase), or a too high concentration of initiator. It may lead to a low control over the RAFT polymerization with residual macroRAFT agent still being present in the serum of the resulting dispersions and to the formation of purely solvophobic homopolymers. As detailed in the next sections, these possible drawbacks have been successfully overcome and RAFT has become the most used and versatile CRP technique to conduct PISA, either in emulsion or dispersion.

2.4.1 Using RAFT in Emulsion Polymerization

2.4.1.1 PISA by RAFT: The First Steps in Emulsion PISA using RAFT in emulsion was pioneered by the work of Hawkett *et al.* [14, 16]. For the first time, good control over both the colloidal stability and the molar masses of the produced polymer chains could be achieved. Oligomeric poly(acrylic acid) (PAA_5) macro-RAFT agents were used to mediate the polymerization of BA in semi-continuous feed conditions. Small particles (60 nm) composed of well-defined PAA-*b*-PBA amphiphilic block copolymers were formed. Particles of poly(methyl acrylate) (PMA) as well as core–shell PBA-*b*-PSt particles were also reported [16]. Particle formation was originally thought to occur via the formation of rigid micelles [14], but deeper investigation showed that in the early stages of the polymerization when the block copolymer chains are still short, the micelles are in fact dynamic structures. Macromolecular chains are mobile enough to migrate between micelles, and the micelles are prone to coalescence, explaining the high aggregation number and the relatively large size of the final particles (Figure 2.9a) [16, 66–68]. The proposed

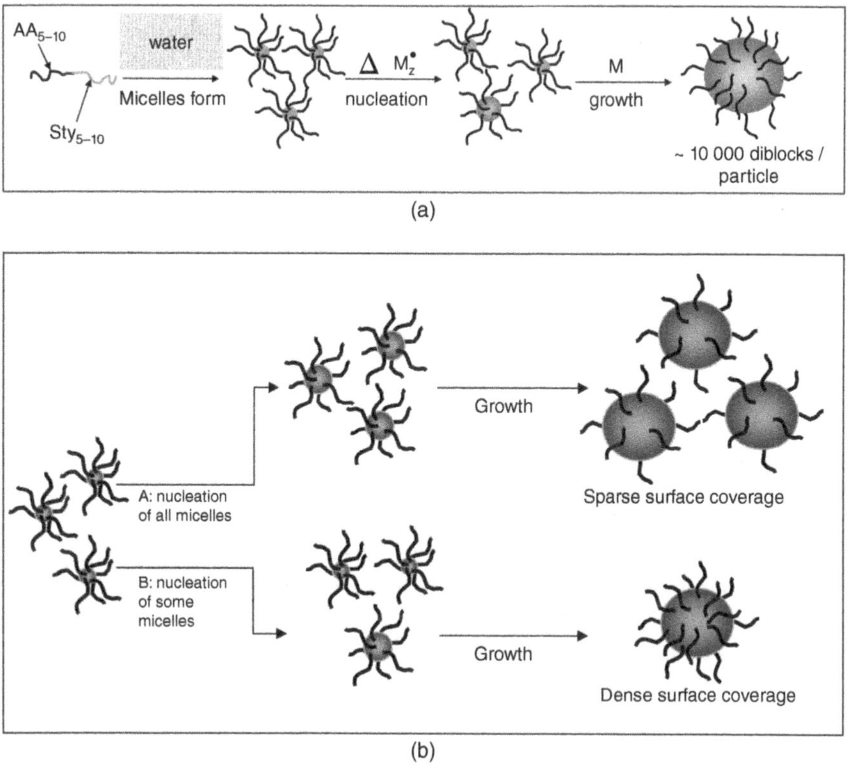

Figure 2.9 (a) Mechanism for particle formation in emulsion systems under RAFT control. (b) Elementary steps of the suggested mechanism for particle formation in emulsion polymerization stabilized by amphiphilic RAFT-capped diblock copolymers employed in concentrations above their CMC (Route A: micelles act as a seed. Route B: some micelles are nucleated and others serve as a diblock-copolymer reservoir). Reproduced with permission from reference [67]. Copyright 2007 American Chemical Society.

mechanism turned out to be especially well suited to describe systems with the particular PAA macroRAFT agents used in these studies. In contrast, the use of less mobile amphiphilic macroRAFT agents (i.e., PAA-*b*-PSt copolymers) led to the nucleation of an increased number of micelles (up to complete nucleation, depending on the concentrations of initiator and macroRAFT agent, Figure 2.9b) [67]. If the macromolecular RAFT agent is not hydrophobic enough and/or employed below its critical micelle concentration (CMC), homogeneous nucleation may actually be favored.

PAA and PAA-*b*-PSt macroRAFT agents have later been used for the emulsion polymerization of isoprene (I) and butadiene [69]. Despite a limited control over the polymerization, the resulting PI-based copolymers assumed interesting nanostructures with likely PI-rich nanodomains being incorporated in a PSt-rich matrix. Using a similar semi-continuous process, Božović-Vukić *et al.* obtained stable poly(styrene-*co*-acrylonitrile) particles using a P4VP macroRAFT agent [70]. More recently, Chen *et al.* obtained amphiphilic gradient copolymers of 2,2,2-trifluoroethyl methacrylate and AA using a short PAA macroRAFT agent in a mixture of water and acetone (5 wt%) to raise the solubility of the PAA segments [71]. All the preceding syntheses were carried out using relatively short macroRAFT agents and a controlled feed of hydrophobic monomer to avoid the formation of droplets that were thought to favor partitioning of the macroRAFT agents and thus negatively impact the course of the polymerization. Significant efforts were then devoted to the development of PISA under batch conditions, using hydrophilic macroRAFT agents with higher molar masses.

To this end, a trithiocarbonate PEO-based macroRAFT agent (M_n = 2000 g mol^{-1}) carrying a hydrophobic dodecyl chain was successfully used in *ab initio* batch emulsion polymerizations of styrene and butyl acrylate [72, 73]. This macroRAFT agent, deployed above its CMC, undoubtedly favors chain extension in the preformed micelles. Stable latexes and controlled molar masses were obtained in both cases (Figure 2.10a and b), but with a significantly higher polymerization rate in the case of BA. The particle size of the PBA latexes could be easily tuned, varying the molar mass of the PEO block (1000, 2000, or 5000 g mol^{-1}—the longer PEO block leading to the smaller particles) or by mixing two PEO macroRAFT agents of different molar masses while targeting the same molar mass for the PBA segment [73]. Moreover, P(BA-*co*-MMA) particles were obtained with good control over the polymerization as long as the MMA content was kept below 75 mol%. For both types of latexes, the size of the particles (typically higher than 100 nm) may indicate the incorporation of a part of the PEO-based block copolymers inside the particles, as previously observed with similar macroRAFT agents [74]. This may result from limited aggregation of small block copolymer particles at around 20–30% conversion and was supported by the observation of nanostructured particles with PEO-rich domains (Figure 2.10c and d). Recently, the same PEO macroRAFT agents with M_n of 2000 g mol^{-1} or 6000 g mol^{-1} could also be successfully used for the synthesis of industrially relevant poly(vinylidene chloride-*co*-methyl acrylate) (P(VDC-*co*-MA)) latexes [75].

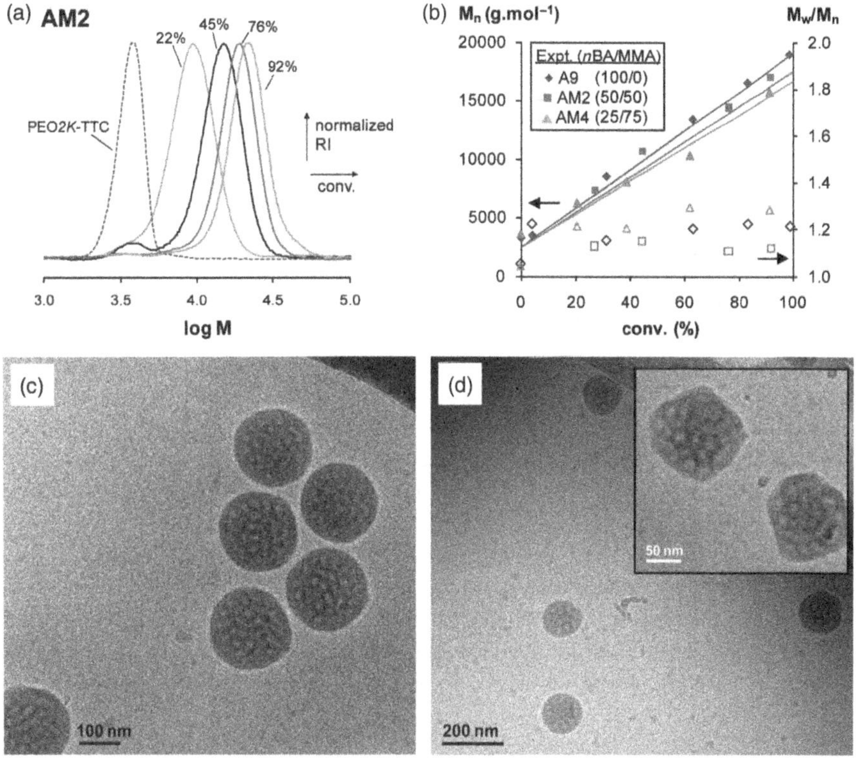

Figure 2.10 Surfactant-free, batch emulsion homopolymerization of BA (A9) and copolymerizations of BA with MMA (BA:MMA molar ratio 50:50, AM2 and 25:75, AM4) in water at 70°C: (a) evolution of the size exclusion chromatograms with conversion for experiment AM2; (b) number-average molar mass, M_n, and dispersity, $Đ = M_w/M_n$, determined by SEC as a function of the conversion. The straight lines correspond to the theoretically expected M_n versus conversion functions. Cryo-TEM micrographs of particles obtained in (c) experiment A9 and (d) experiment AM4. Reproduced with permission from reference [73]. Copyright 2009 American Chemical Society. (*See color insert for color representation of this figure*).

The PEO segment was then replaced by poly(*N*,*N*-dimethylacrylamide) (PDMAAm) of various molar masses and structures, obtained using different trithiocarbonate-type RAFT agents [76]. PBA particles composed of amphiphilic di- or triblock copolymers were obtained with a solids content of up to 40%. A modification of the structure of the RAFT agent, such as a variation of the length of the Z-group alkyl chain or the switch from an asymmetric to a symmetric RAFT agent, leading to BAB triblock copolymers, did not significantly impact the polymerization characteristics. In contrast, a variation of the polymers' molar masses did have a significant effect: longer PDMAAm blocks led to more stable latexes but also to broader molar mass distributions. Stable PDMAAm-*b*-PSt particles were also reported.

A polyacrylamide macroRAFT agent was used by Ji *et al.* [77] in the batch emulsion polymerization of styrene, preceded by a preliminary sonication step aiming at reducing the size of the monomer droplets. Small spherical particles were indeed formed (58 nm), according to the authors mainly by the nucleation of the small monomer droplets obtained after ultrasonication.

Well-defined copolymers of AA and PEO methyl ether acrylate (PEOA) were synthesized using a trithiocarbonate RAFT agent (P(AA-*co*-PEOA)) by Boissé *et al.* [78] and then used for emulsion polymerization of styrene. Different parameters such as the pH value, the copolymer composition, and the concentration of salts were varied. Whereas PAA or PPEOA homopolymers both led to spherical particles, nanofibers together with small vesicles and spherical particles could be observed with the brush-like copolymers at acidic pH values (3 or 6) or high $NaHCO_3$ concentrations (at pH = 8) (Figure 2.11). Under these conditions, the volume fraction of the hydrophilic block was likely decreased (via the formation of a hydrophobic complex between the protonated carboxylic acid groups of AA and the ethylene oxide units, or via the screening of the electrostatic repulsion between the hydrophilic segments), leading to the observed nonspherical morphologies. The monitoring of the particle morphology during the polymerization clearly showed morphological transitions with increasing conversion: spherical particles gradually changed to worm-like nano-objects (with a few vesicles), demonstrating the influence of the molar mass of the PSt block. The formation of worms was also accompanied by an increase of the viscosity of the polymerization medium. All these experiments clearly point out the

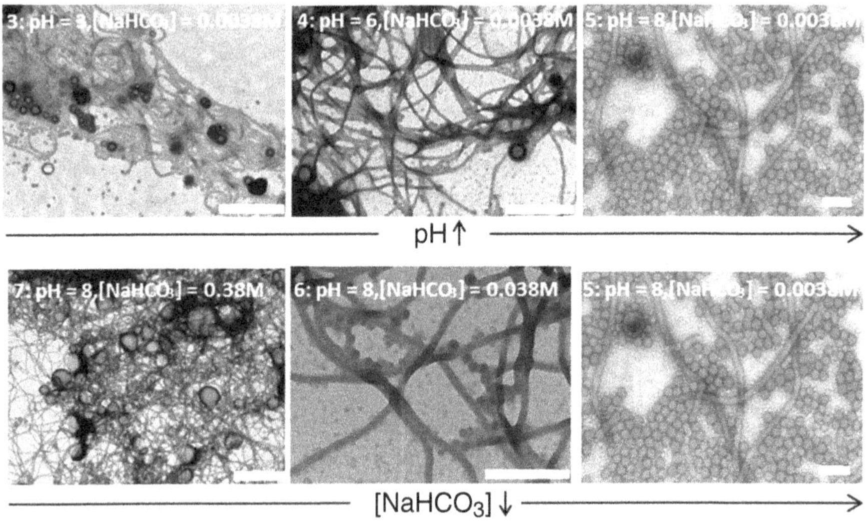

Figure 2.11 Influence of pH value and salt concentration on the morphologies obtained for the emulsion polymerization of St using a P(AA-*co*-PEO_8A) macroRAFT agent (AA:PEO_8A = 50:50). Reproduced with permission from reference [78]. Copyright 2010 Royal Society of Chemistry.

decisive influence of the volume fraction of each block on the final morphology. The same authors also investigated the influence of the stirring speed and the concentration of the divalent salt $CaCl_2$ on the resulting nano-objects' morphologies [79].

The previously outlined studies constitute the first example of nonspherical objects using the PISA approach in emulsion polymerization according to a RAFT process. However, residual macromolecular RAFT agent was usually observed [78, 79] and this was partly assigned to the acrylic nature of the macroRAFT agent, which is poorly suited to efficiently re-initiate the RAFT block copolymerization of styrene. Similar investigations were thus undertaken with a methacrylate macroRAFT homolog, P(MAA-*co*-PEO_8MA) using, however, a thiopropyl Z group instead of a thiododecyl one [80]. The polymerization was in fact faster, particularly at acidic pH values, with a very good control over the molar masses of the PSt block. No residual macroRAFT agent was detected at pH = 5. Independent of the pH value, mainly small spherical particles (ca. 20 nm) were obtained. The structure, concentration, and molar mass of the macroRAFT agent; the pH value; and the concentration of styrene were also varied. The crucial role of the pH value was evidenced again: only spheres were obtained at pH = 3.5 using a P(MAA-*co*-PEO_{19}MA) macroRAFT agent, while different morphologies in dependence on the length of the PSt block could be observed at pH = 5. Various morphologies were also obtained when the same macroRAFT agent was used in the polymerization of benzyl methacrylate (BzMA), which, depending on the composition of the initial water/organic solvent mixture, proceeds either via an emulsion or a dispersion polymerization mechanism [81]. The use of a cosolvent (either ethanol or dioxane) notably proved to be an interesting tool to broaden the formation domain of fibers previously observed in pure water at pH = 5, showing again the impact on the resulting morphology of the respective affinity of the individual blocks to their environment.

While most of the time the trigger to achieve different morphologies had been the variation of the average molar mass of the hydrophobic block, a recent study has investigated the influence on the morphology of the hydrophilic segment. Three different types of PEO-based brush macroRAFT agents were synthesized by the RAFT (co)polymerization of poly(ethylene oxide) methyl ether vinyl phenyl (PEOV) using a symmetrical trithiocarbonate, in order to study the effect of the brush copolymer sequence and chemical composition on the morphology of the particles obtained in the emulsion polymerization of styrene: PPEOV, P(PEOV-*co*-St), and PPEOV-*b*-PSt-*b*-PPEOV [82]. A pronounced effect of the brush copolymer structure on the kinetics and the molar mass control was found, the P(PEOV-*co*-St) statistical copolymer being the most efficient one. This was ascribed by the authors to a fast and efficient swelling of the initial micelles by styrene combined with a lower steric hindrance from the PEO side chains to the entry of the polymer chains generated in water. Only spherical particles were generated in the first part of the study. The authors then focused on morphological transitions, varying the absolute concentrations of the two copolymers as well as their ratio (increasing amount of styrene). The morphological transition was also monitored with increasing conversion (Figure 2.12). In this last case, the key parameter was found to be the molar mass of the PSt block rather than the structure of the hydrophilic one. The same authors

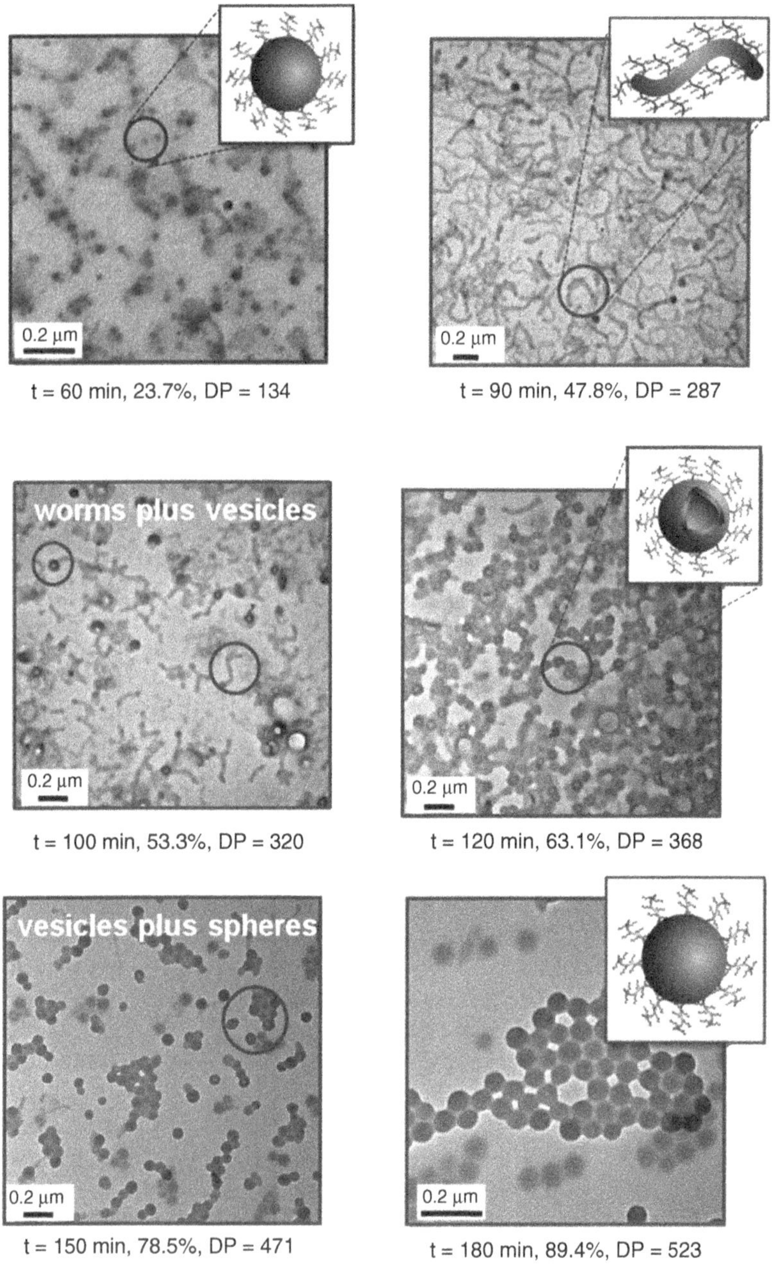

Figure 2.12 TEM images of the nano-objects formed after different times (and monomer conversion) in the emulsion RAFT polymerization of St in the presence of the P(PEOV$_{12}$-*co*-St$_{25}$)-TTC macroRAFT agent ([St]$_0$:[MacroRAFT]$_0$:[V50]$_0$ = 2400:4:1). Reproduced with permission from reference [82]. Copyright 2013 Wiley-VCH Verlag GmbH & Co.

also reported the use of a poly[*N*-(4-vinylbenzyl)-*N,N*-dibutylamine hydrochloride] macroRAFT agent for the emulsion polymerization of styrene, but only spheres were formed in the range of the investigated experimental conditions [83].

Recently, a 2-dimethylaminoethyl methacrylate (DMAEMA)-based macroRAFT agent was used for the synthesis of cationic PMMA particles. During the synthesis of this macroRAFT agent in water at pH = 7, a minor fraction (up to 7%) of the DMAEMA hydrolyzed to yield a copolymer incorporating MAA units, P(DMAEMA-*co*-MAA), that was subsequently used for the emulsion polymerization of MMA at pH = 6. The spherical cationic nanoparticles were readily adsorbed onto negatively charged cellulose model surfaces in aqueous solutions without any post-modification of the latex (Figure 2.13) [84].

The macroRAFT agents (defined here as macromolecular or oligomeric RAFT agents with a DP_n higher than 5) used in the studies depicted so far were mostly trithiocarbonates, which usually show better control over the molar masses and molar mass distributions (presumably related to their lower sensitivity to hydrolysis) [85] than RAFT agents with other common functional groups. All macromolecular RAFT agents had previously been synthesized in an organic solvent. A more straightforward strategy was recently reported by Chaduc *et al.* [20] for the synthesis of PSt particles using P(M)AA and P(MAA-*co*-PEOMA) macroRAFT agents in a two-step one-pot process. The macroRAFT agent is formed directly in water, providing the opportunity to skip the often time-consuming step of purification of the hydrophilic macroRAFT agent synthesized in an organic solvent. Subsequently the emulsion polymerization is carried out in the same reaction vessel. This strategy requires a very good control over the polymerization of the hydrophilic monomer(s) up to complete conversion,

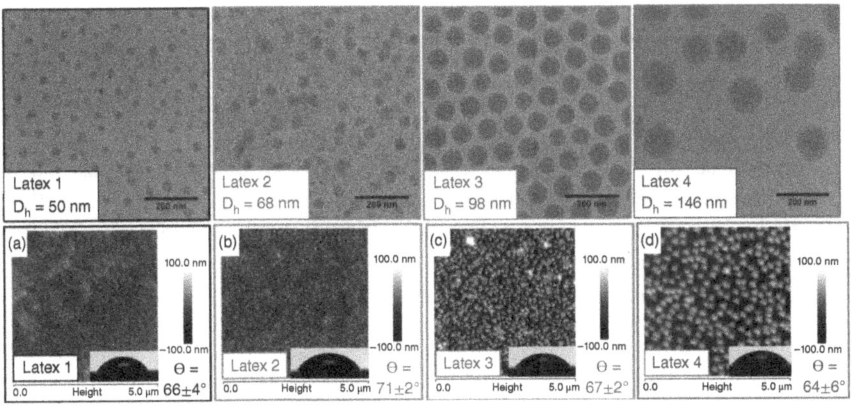

Figure 2.13 *Top*: TEM images and hydrodynamic diameter from DLS of cationic P(DMAEMA-*co*-MAA)-*b*-PMMA latexes, obtained by RAFT-mediated surfactant-free emulsion polymerization at pH = 6. *Bottom*: AFM images of the same latexes adsorbed onto cellulose model surfaces formed on QCM crystals. The inlay pictures show the contact angles of water. Adapted with permission from reference [84]. Copyright 2014 Royal Society of Chemistry.

which was indeed the case for the three considered macroRAFT agents. The use of P(MAA-*co*-PEOMA) macroRAFT agents in this one-pot approach was then investigated systematically. First, the macroRAFT composition (molar ratio of MAA and $PEOMA_{19}$) and concentration were varied in styrene emulsion polymerizations at pH = 3.5, and spheres (sometimes with the presence of holes indicative of partial burial of the macroRAFT agent) were obtained [86]. Apart from TEM and DLS, these particles were characterized in a very innovative way: the molar mass and the molar mass distribution of the particles were determined by electrospray-charge detection mass spectrometry [87, 88]. Another series of polymerizations was undertaken at pH = 5 [89]. The molar mass, composition and concentration of the macroRAFT agent were varied resulting in different morphologies (spheres, fibers, and vesicles). Together with the pH value, the molar masses of both hydrophilic and hydrophobic blocks were again shown to be the main parameters governing the final morphology. If crosslinked, the PSt fibers showed a good thermal stability, resistance to solvent, and an interesting dynamic behavior [90]. PMMA and P(MMA-*co*-St) were then chosen as core-forming polymers [91]. First, MMA polymerizations were performed at pH = 3.5, 5, and 7. Regardless of the pH, the morphologies evolved from spheres to fibers, and finally vesicles with increasing segment molar mass, but the reaction was only well controlled at pH = 3.5, with lower blocking efficiency for higher pH values. If part of the MMA was substituted by St targeting a DP_n, which would lead to the formation of fibers in the case of pure PMMA, only spheres were formed, for all compositions of the MMA/St mixture. This shows the crucial influence of the initial chain growth steps occurring in the aqueous phase on the further self-assembly of the growing polymer chains. These early chain-growth steps are governed by the water-solubility of the monomer(s), the rate constant of propagation, and the chain transfer constant to the macroRAFT agent. Suspensions of both PMMA and PSt nanofibers were characterized by asymmetric flow field-flow fractionation coupled to static light scattering, giving access to their average length and length distribution [92]. The viscoelastic properties of PMMA and PSt nanofibers have also been investigated and were shown to depend mainly on their aspect ratio rather than on their nature and composition [93].

The syntheses of both PMAA [94, 95] and PAA [96] and their use as macroRAFT agents have been thoroughly investigated using the one-pot approach. Both hydrophilic macroRAFT agents were employed for the synthesis of PSt particles and the effect of different parameters was studied (initial pH value, concentration and molar mass of the macroRAFT, amount of styrene). It was found that the pH value plays a key role in determining the type of particles obtained: at acidic pH values, small particles of about 30 nm were obtained, comprised of polymer chains with controlled and narrowly distributed molar masses (Figure 2.14). In that case, the particles formed through the self-assembly of amphiphilic P(M)AA-*b*-PSt macroRAFT agents generated during the polymerization. Increasing the pH value still led to stable particles (50 nm and not uniform in size with PMAA, 150 nm and uniform in size for PAA; Figure 2.14) and in both cases, residual unreacted macroRAFT agent was present at alkaline pH values. In the case of PMAA, the control over the molar mass was, however, lost. Hydrolysis of the trithiocarbonate end-group (shown at pH = 8.0),

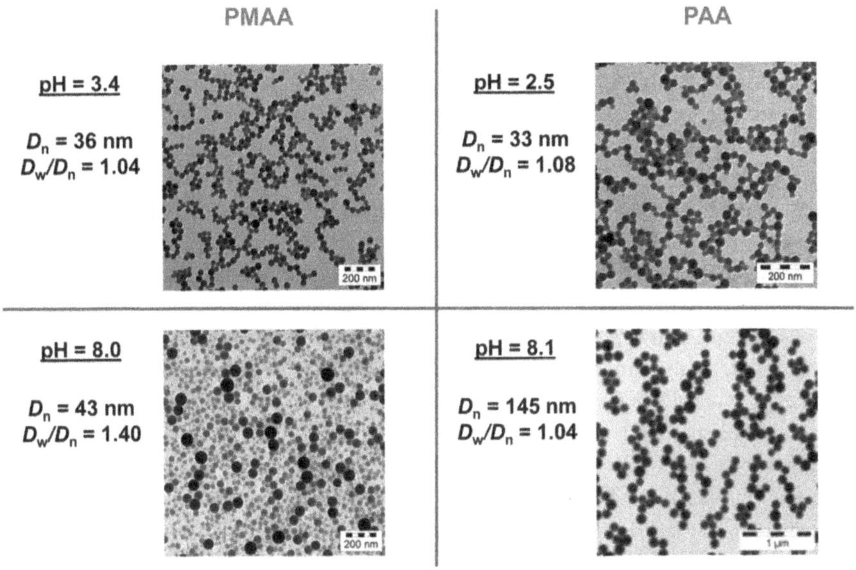

Figure 2.14 TEM images and hydrodynamic diameter of PSt particles obtained by RAFT-mediated surfactant-free emulsion polymerization using PMAA (*left*) or PAA (*right*) macroRAFT agents. Reprinted and adapted with permission from [95, 96]. Copyright 2012 and 2013 American Chemical Society.

together with the more open water-swollen structure at pH $\geq$ 6.5 exhibited by PMAA chains would offer a less favorable environment for the growth of the PSt segment, and homogeneous nucleation would instead be favored. The presence of residual PAA chains was also observed. However, no degradation of the chain end was noticed over the course of the polymerization, and the good control of the styrene polymerization was to some extent maintained. The presence of remaining but still active PAA chains was rationalized by their ionization at alkaline pH, which impeded an efficient re-initiation of PAA macroRAFT agent (lowering the chain transfer coefficient). The formation of amphiphilic block copolymer nanoparticles of PMAA-*b*-PMMA, PMAA-*b*-PBA, PMAA-*b*-P(BA-*co*-St), and also of PMAA-*b*-PSt-*b*-PBA have also been reported at acidic pH [95].

These two systems using PAA and PMAA under acidic conditions proved to be extremely robust, forming well-defined block copolymers particles that self-assemble into spherical nano-objects regardless of the molar masses of either the hydrophilic block (2400 g mol^{-1} < M_n < 13000 g mol^{-1}) or the hydrophobic block (200 < DP_n < 800). This is in sharp contrast to the switch in morphologies that has been experimentally obtained in other systems and suggested by the expected evolution of the packing parameter of the formed amphiphilic block copolymers.

As mentioned in the introduction, many studies have employed the use of amphiphilic diblock copolymers as starting macroRAFT agents for the emulsion polymerization of various hydrophobic monomers. In these cases, self-assembly/micellization of the copolymers (usually used above their CMC) likely occurred before the onset of polymerization. However, to provide a complete overview of the field, it is worth mentioning the following studies, which offer insight into the mechanism of particle formation. Indeed, as described above, Ganeva *et al.* used amphiphilic PAA-*b*-PSt to study the influence of the macroRAFT structure on the nucleation step for St or St/BA polymerization [67]. Similar copolymers were studied by Wang *et al.* who showed the strong impact of pH on the type of nucleation and quality of control [97]. Mirroring Chaduc *et al.*'s observations for pure PAA macroRAFT [96], the control was effective at acidic pH albeit with some stability issues, whereas a higher degree of ionization of the PAA segment when the pH was increased to 6.3 led to stable latexes but poor control over the growth of the PSt segment. In the latter case, the systems contained a higher number of micelles (i.e., fewer macroRAFT chains per micelle), and prolonged nucleation was observed with some macroRAFT agent remaining unreacted. The influence of the molar mass of each block (15 to 30 AA units and 0 to 7 St units) was also investigated [98]. PAA-*b*-PSt macroRAFT agents were employed for the synthesis of various kinds of particles: PSt-*b*-PBA-*b*-PSt thermoplastic elastomers [99], polybutadiene-based latexes [69, 100], film-forming latexes of P(St-*co*-BA) [101], high glass transition temperature (T_g) poly[styrene-*co*-(γ-methyl-α-methylene-γ-butyrolactone)] particles [102], and poly(styrene-*co*-acrylonitrile) [103] particles. The PPEOV, P(PEOV-*co*-St), and PPEOV-*b*-PSt-*b*-PPEOV copolymers discussed above [82] are also relevant to this section, as they were all used above their CMC for the emulsion polymerization of styrene. P(VDC-*co*-MA) latexes have also been reported, using either PEO-*b*-P(VDC-*co*-MA) [75] or PAA-*b*-PSt [104] macroRAFT agents.

2.4.1.2 PISA by RAFT: A Powerful Tool The robustness of the PISA process has opened up many possibilities for polymer chemists to play with each of the components of the system and access new particle compositions. Playing with the miscibility of the core-forming monomer(s) with the continuous phase, an interesting study by Ratcliffe *et al.* reports the (co)polymerization of water-miscible 2-hydroxyethyl methacrylate (HEMA) and water-immiscible 4-hydroxybutyl methacrylate (HBMA) mediated in water using a poly(glycerol monomethacrylate) (PGMA) macroRAFT agent [105]. This macroRAFT agent has been extensively studied by the same team for RAFT aqueous dispersion polymerization of 2-hydroxypropyl methacrylate (HPMA) (see Section 2.4.2.1). As expected, preliminary studies on the homopolymerization of HEMA generated water soluble amphiphilic copolymers, whereas homopolymerization of HBMA operated via a emulsion polymerization process, providing spherical particles whose sizes increased with the length of the PHBMA block. Considering the water miscibility of both monomers, the polymerization of a 1/1 (molar ratio) mixture of HEMA/HBMA was expected, and then experimentally verified, to proceed via an emulsion polymerization mechanism. For this 1/1 monomer ratio the particle core is effectively isomeric with PHPMA, and could therefore offer an emulsion-based alternative to the use of HPMA in dispersion. Different degrees of polymerization were targeted

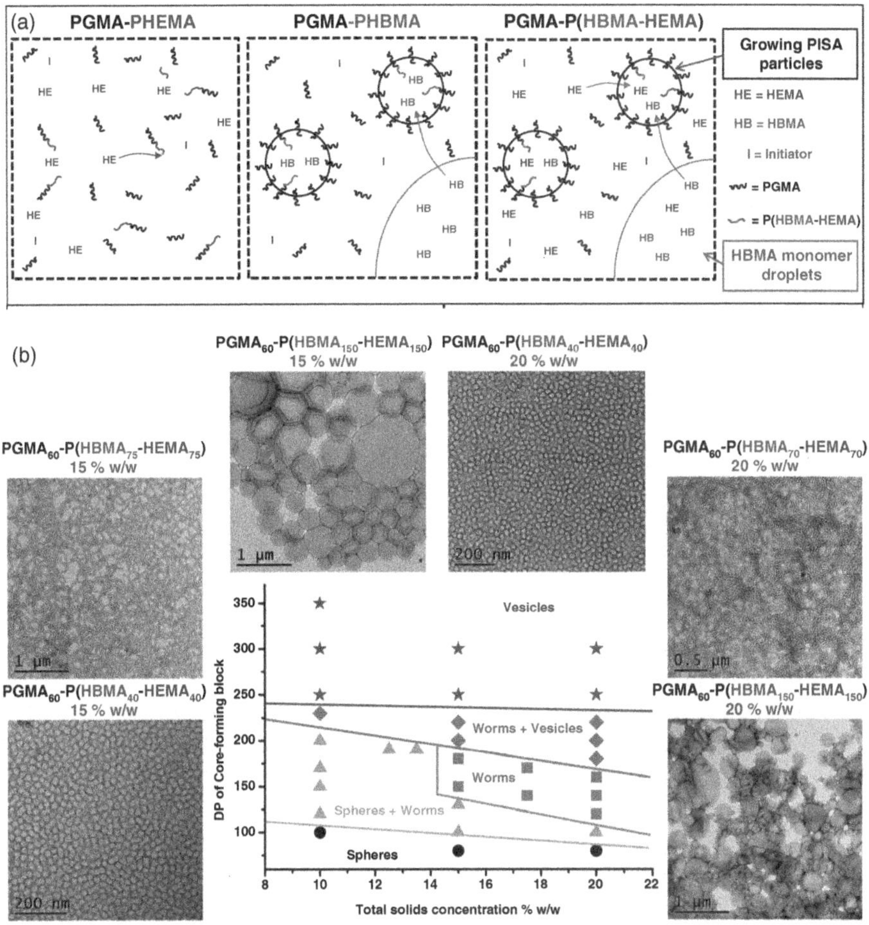

Figure 2.15 (a) Schematic representation of the initial spatial location of the HEMA and HBMA comonomers in the RAFT synthesis of $PGMA_{60}$-*b*-P($HBMA_x$-*co*-$HEMA_y$) diblock copolymer nanoparticles, and (b) Phase diagram constructed for $PGMA_{60}$-*b*-P($HBMA_x$-*co*-$HEMA_y$) statistical diblock copolymers prepared at 70°C by RAFT copolymerization of HBMA and HEMA at a 1:1 molar ratio. Selected TEM images are shown for specific diblock compositions and copolymer concentrations, confirming that pure sphere, worm and vesicle morphologies can be obtained under appropriate conditions. Reprinted and adapted with permission from [105]. Copyright 2014 Royal Society of Chemistry. (*See color insert for color representation of this figure*).

for the core-forming block, leading to a range of morphologies: from spheres, to worms and finally vesicles, with intermediate molar masses usually leading to mixtures of morphologies (Figure 2.15). The influence of the solids content was also studied. In subsequent work, PGMA macroRAFT agent was also used for the emulsion polymerization of BzMA [106]. The formed nanoparticles, invariably spherical (20 to 193 nm in diameter, depending on the final molar mass of the PBzMA block, up to 50 wt% solids content), were stable to both freeze–thaw cycles and salt addition. By taking advantage of the cis-diol functionality on PGMA, they were selectively adsorbed onto a micropatterned phenylboronic acid-functionalized planar surface. The particles also proved to be efficient stabilizers for the formation of Pickering emulsions, actually exhibiting more efficient adsorption on *n*-dodecane droplets than their less hydrophobic core-forming PHPMA homologues. For this application the enhanced hydrophobic interactions between the core-forming chains of PBzMA enabled the maintenance of particle integrity during the high-shear emulsification step without the need for any covalent crosslinking (which was the case of PHPMA-based spheres) [107]. In addition, PGMA-*b*-PBzMA worms adsorbed more strongly than their spherical counterparts and enabled the production of smaller droplets for a given copolymer concentration.

An *et al.* performed several studies dedicated to the synthesis of core-crosslinked star (CCS) polymers. MacroRAFT agents of PDMAAm, PPEOMA, and poly(2-methoxyethyl acrylate-*co*-poly(ethylene oxide) methyl ether acrylate) (P(MEA-*co*-PEOA) were used for the copolymerization of *n*-butyl acrylate and poly(ethylene oxide) diacrylate in water, affording very small core-crosslinked nanoparticles (between 11 and 22 nm) that were subsequently used as efficient emulsifiers [108]. P(MEA-*co*-PEOA)'s dual responsiveness to temperature and salts was then exploited to stabilize high internal phase emulsions and trigger their demulsification [109].

In addition to allowing the synthesis of well-defined block copolymer particles, the PISA concept is well suited to the design of more exotic particles such as Janus anisotropic nanoparticles. Pham *et al.* [110] used diblock copolymers of either PAA-*b*-PSt or PAAm-*b*-PSt to first synthesize small crosslinked PSt particles that, after monomer swelling, were used as seeds in St or MMA/BA emulsion polymerization. When first heated to the polymerization temperature, some of the monomer was expelled and polymerized to form a bulge attached to the original seed and produce ellipsoidal or dumbbell-like particles. The shape of this bulge, as well as the surface properties and polymer composition of the two lobes, could be fine-tuned by adjusting the nature of the macroRAFT agent and initiator and the nature and amount of the swelling monomer(s), leading to Janus nanoparticles, which were finally employed as efficient stabilizers in the emulsion polymerization of St or BA/MMA mixtures.

These last examples nicely illustrate the exciting potential of the relatively new PISA process in generating novel polymer particles for which a variety of applications, including the stabilization of Pickering emulsions, can be envisaged. In the last few years, RAFT *ab initio* batch emulsion polymerization following a PISA approach has been employed in this direction for the synthesis of bio-active glyco-particles bearing glucose units [111], degradable redox-sensitive nanoparticles for controlled release [112], nanoparticles for catalysis 113, 114], covalent core-shell latexes for H-bonded films with enhanced properties [115, 116], fluorinated latexes of either diblock of poly(methacrylic acid)-*b*-poly(dodecafluoroheptyl methacrylate) [117], triblock of poly(acrylic acid)-*b*-poly(hexafluoroheptyl acrylate)-*b*-poly(*n*-butyl acrylate-*co*-methyl methacrylate-*co*-hexafluoroheptyl acrylate) [118] or phenylphosphinic acid-functionalized crosslinked PSt particles [119]. High solids content (37 wt%) poly(vinyl acetate) (PVAc) latexes, which are industrially relevant as adhesives, paints, and coatings, have been synthesized via RAFT-mediated PISA performed using three different hydrophilic xanthate-based macroRAFT agents of poly(*N*-vinyl pyrrolidone), poly(*N*-acryloylmorpholine), or PEO [120]. Finally, various PNaSS-stabilized particles were reported by Yeole *et al.* [121–123], and the small particle size obtained under certain conditions suggests that these were also PISA systems, even though no molar mass analyses were provided (probably due to the lack of common solvent for both blocks).

2.4.1.3 When PISA Is Not Fully Controlled When the PISA process was first developed, the main motivation for using the macroRAFT block copolymers was to introduce an active stabilizer that was covalently anchored to the final latex particles, with less emphasis placed on their ability to control the polymerization of the core-forming block. For this reason PISA was used for the synthesis of PSt particles and included macroRAFT agents of polysoaps [124], protonated PDEAEMA [125], PEO, protonated PDMAEMA and protonated PEO-*b*-PDMAEMA [126], as well as PMAA [127]. PVAc particles using a dextran-based macroxanthate [128] and PMMA latexes using protonated PDMAEMA-*b*-PMMA macroRAFT [129, 130] have also been reported. The PISA process therefore provides a valuable new pathway of performing surfactant-free emulsion polymerizations that expand the range of strategies based on the use reactive surfactants that have been investigated for many years [131].

2.4.2 Using RAFT in Dispersion Polymerization

The first reports on PISA by RAFT-mediated dispersion polymerization appeared in the literature in 2006. These experiments mainly used organic solvents, such as cyclohexane [132], isododecane [133], supercritical carbon dioxide [134], and chloroform [135] as a continuous phase. As stated in the introduction of this chapter, we will focus here on water-based systems but will provide some examples of nonaqueous dispersion polymerizations [18, 24, 136, 137].

2.4.2.1 Use of Water as Solvent The first example of RAFT aqueous dispersion polymerization was published by An *et al.* in 2007 [138]. They polymerized, in a controlled way, NIPAM in the presence of poly(*N,N*-dimethylacrylamide) macroRAFT agents (PDMAAm) using microwave irradiation, at 70°C. At this temperature PNIPAM is not soluble, but rather precipitates, and consequently diblock copolymer nanoparticles were formed that could be disassembled post-polymerization by cooling below the LCST of PNIPAM. In parallel, Rieger *et al.* prepared well-defined PEO-*b*-PDEAAm diblock and PEO-*b*-PDMAAm-*b*-PDEAAm triblock copolymers in aqueous dispersion polymerization, at the same temperature (70°C), in short reaction times (90 min) and using a conventional radical initiator [139]. Both groups incorporated a crosslinker, MBAAm, yielding nanosized thermoresponsive gel particles. The importance of the chemical nature of the macroRAFT for achieving stable

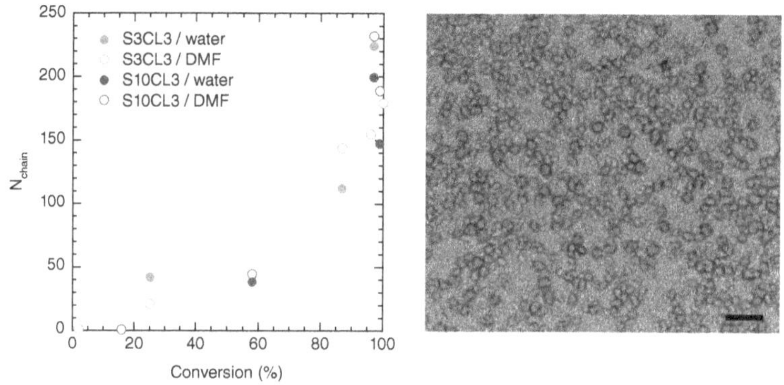

Figure 2.16 Synthesis of PEO-*b*-PDMAAm-*b*-P(DEAAm-*co*-MBAAm) nanogels obtained by RAFT-mediated crosslinking polymerization. *Left*: Monitoring of the number of chains per nanogel by SLS *vs*. monomer conversion. *Right*: TEM micrograph of a final nanogel sample. The scale bar is 200 nm. Reprinted and adapted with permission from [141]. Copyright 2011 Royal Society of Chemistry.

and homogeneously sized nanogels was emphasized in Rieger *et al.*'s work: while a pure PEO_{2K} macroRAFT (M_n(PEO) = 2000 g mol^{-1}) led to large gel particles with broad size distributions, the introduction of a second stabilizing PDMAAm block allowed the formation of narrowly distributed sub-100 nm PEGylated nanogels. The need for introducing a hydrophilic middle block was explained by PEG's inability to sufficiently stabilize the initially formed nuclei due to inadequate hydrophilicity at 70°C, or an insufficient phase separation between PEO and the core-forming PDEAAm polymer. Another reason could be the insufficient molar mass of the stabilizing PEO block, since in later work higher molar mass PEO macroRAFT agents (M_n(PEO) = 5000 g mol^{-1}, PEO_{5K}) were successfully used as stabilizers and control agents in aqueous emulsion polymerization [73], or at lower temperature (50°C) in aqueous dispersion polymerization of HPMA [140].

The formation of PEO-*b*-PDMAAm-*b*-P(DEAAm-*co*-MBAAm) nanogels was monitored by SEC and static light scattering (SLS) in a subsequent mechanistic study [141]. It was found that the nanogels evolved by coagulation and successive reticulation of several initially formed individual nuclei, meaning that the number of particles decreased in the course of the polymerization while the aggregation number (number of chains par particle) increased (Figure 2.16). The nanogels' size was governed mainly by the DEAAm and the crosslinker concentration but depended on the molar mass of the stabilizing macroRAFT agent used in the polymerization. All these parameters should indeed have an impact on the limited and irreversible particle aggregation occurring during polymerization. Such a limited coagulation-based mechanism was later reported by Su *et al.* for the formation of PPEOV-*b*-PSt particles [142].

Since then, other core-shell nanogels have been prepared by RAFT-mediated PISA in water: pegylated cationic PDMAEMA nanogels were synthesized in the presence of PEO-based amphiphilic macroRAFTs [143] using either AIBN or ACPA as an initiator (the choice of initiator had an impact on the nanogel size). PNIPAM [108, 144], P(MEA-*co*-PEOA) [145], and poly[di(ethylene glycol) methyl ether methacrylate-*co*-PEO methyl ether methacrylate)] (P(DEGMA-*co*-PEOMA)) [146] based thermoresponsive nanogels, stabilized by PDMAAm [145], PAA [144], PEO_{5K} [146], or P(PEOMA) [146] coronas have also been reported.

Apart from these core polymers possessing LCSTs above room temperature, only 2-methoxyethyl acrylate (MEA) [147] and HPMA [148] have been polymerized in a controlled manner by RAFT dispersion polymerization in pure water. When a bifunctional monomer was added as a crosslinker from the beginning of the polymerization, spherical core-shell nanogels were generally formed for all types of monomers [149]. Under such conditions, only Sugihara *et al.* observed the formation of nonspherical crosslinked morphologies composed of PMPC-*b*-P(HPMA-*co*-EGDMA) [148]. The observed "lumpy rods" clearly showed the features of individual particles and originated from non-isotropic interparticle crosslinking of individual nanogels. They can thus not be referred to as truly anisotropic worm-like structures as attained in specific PISA systems without crosslinker (see Sections 2.3.1 and 2.4.1). In fact, since 2009, the possibility to obtain all kinds of anisotropic morphologies by PISA through RAFT dispersion polymerization has been demonstrated [18, 24, 71, 137] in both organic solvents [150, 151] and aqueous solutions [152]. Armes's group [152], in particular, has reported various examples of worms, vesicles, and lamellae structures formed by RAFT-mediated dispersion polymerization in water. All systems were based on the same core polymer, PHPMA. They systematically varied the chemical nature of the stabilizing macroRAFT agent (PGMA [152, 153], PEO_{5K} [140]), established detailed and predictive morphological phase diagrams [140], and provided important advances in understanding the formation mechanism of the various morphologies. Since these results have been thoroughly reviewed by Warren *et al.* [154], we will summarize here only the main parameters that determine the resulting morphologies by PISA through RAFT dispersion polymerization: (1) the respective molar masses of the solvophobic and solvophilic block; (2) the chemical nature of the monomer to be polymerized, the macroRAFT, and the solvent (including water); (3) the interaction of the block copolymer segments with each other and the solvent (solvation of the blocks); and (4) the solvophobic monomer concentration. In general, these align with those discussed earlier for NMP and RAFT emulsion (see Sections 2.3.1 and 2.4.1). The last parameter is especially important in dispersion polymerization because—in contrast to emulsion polymerization where the monomer constitutes essentially a separated reservoir phase —the monomer is initially soluble in the continuous phase. For instance, multi-lamellar vesicles were only obtained with initial monomer concentrations higher than 20 wt% (Figure 2.17) [140] or even 70 wt% [155].

The detailed phase diagrams were established by independently varying the monomer concentration, and the respective molar masses of the core and shell polymer segment. The preparation of phase diagrams is laborious and time-consuming, but it allows to a certain extent the prediction of the morphologies. Prediction by theoretical simulations is demanded but remains difficult, as both kinetically trapped and thermodynamically controlled structures are formed.

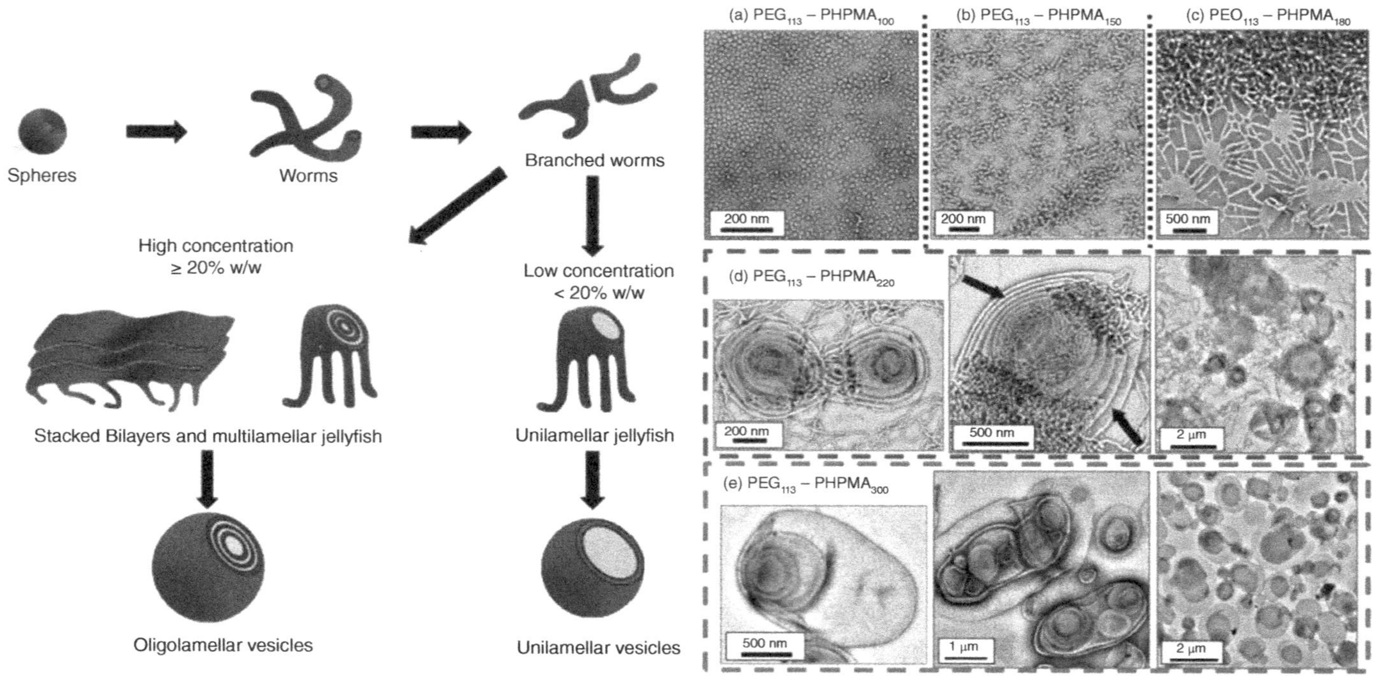

Figure 2.17 *Left*: Schematic representation of the formation of the oligolamellar and unilamellar vesicles obtained from PISA through aqueous RAFT dispersion polymerization of HPMA with PEG_{5K}-macroRAFT agent at 50°C at ≥ 20 wt% or < 20 wt% solids, respectively. *Right*: TEM images of PEO_{5K}-*b*-$PHPMA_x$ nano-objects prepared at ≥ 20% solids. Reprinted with permission from [140]. Copyright 2014 American Chemical Society. (*See color insert for color representation of this figure*).

Continuous monitoring of the polymerizations by ^{1}H-NMR (for determining monomer conversion), SEC, TEM, cryo-TEM, DLS, SLS, and SAXS, provided important insights in the formation mechanism of the different morphologies [140, 142, 153]. It was established that worms are formed from monomer-swollen spherical block copolymer micelles that fuse together. These linear worms then become branched, and octopi-like structures form that reorganize to form vesicles (Figure 2.18).

Recently, Huo *et al.* reported for the first time the possibility to rearrange diblock copolymer vesicles obtained by PISA into spheres through what we could call polymerization-induced self-assembly reorganization or polymerization-induced order-to-order transition [156]. The authors capitalized on "living" chain ends

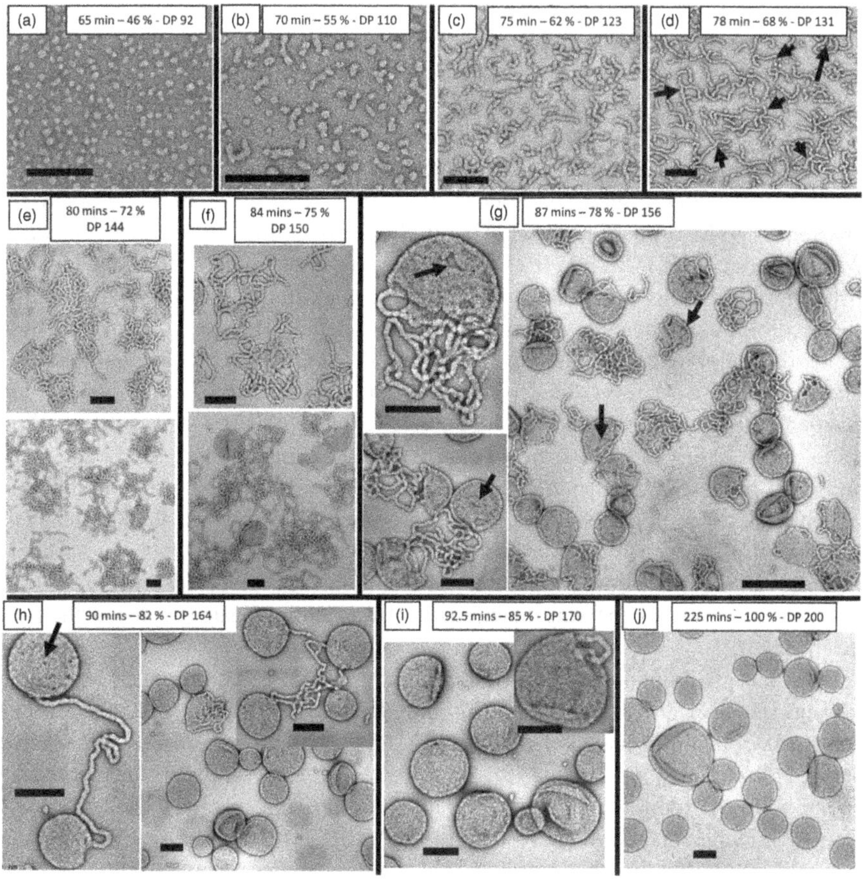

Figure 2.18 TEM monitoring of a typical aqueous dispersion polymerization of HPMA (10 wt% solids) in the presence of a PGMA macroRAFT agent displaying sphere-to-worm and worm-to-vesicle transitions. Scale bars = 200 nm. Reprinted with permission from [153]. Copyright 2011 American Chemical Society.

still available after PISA for chain extension with a third, soluble polymer segment. Following this approach, DMAAm was polymerized in the presence of poly[*N*-(4-vinylbenzyl)-*N,N*-diethylamine]-*b*-polystyrene (PVEA-*b*-PSt) vesicle "seeds" (generated in a first step by RAFT-mediated PISA in ethanol/water (95/5)), whose subsequent morphological reorganizations were monitored by TEM. With growing molar mass of the third PDMAAm block, the initial vesicles transformed to tubules, then jellyfish-like morphologies, followed by worms that divided into spheres. The vesicle "dissociation" mechanism (i.e., the vesicle-to-sphere transition by chain reorganization) was found to be essentially the reverse of the vesicle formation mechanism reported earlier.

It is well-established that morphological transitions of block copolymer nano-objects can be triggered by external stimuli, such as temperature or pH. These external physical parameters impact the interaction of the shell and core polymers with the solvent, namely their solvation and conformation. Known examples are the temperature-dependent swelling and shrinking of nanogels, and the association-dissociation of diblock copolymers possessing a thermo- or pH-responsive segment. It was recently shown that it is also possible to trigger by an external stimulus the morphological transitions of nano-objects obtained by PISA in dispersion (without dissociating them completely). Temperature-induced order-to-order transitions have been observed for PGMA-*b*-PHPMA or PEG-*b*-PHPMA diblock copolymer assemblies in water. Worm-to-sphere (PGMA-*b*-PHPMA) [157] or vesicle-to-sphere (PEG-*b*-PHPMA) [140] reorganizations occurred upon cooling below a certain temperature (depending on the diblock composition, ranging from 7 to 20°C). They were explained by a higher degree of hydration of the core-forming PHPMA block at lower temperature. The transitions were reversible in the sense that both types of morphologies could reversibly be formed, but the PEG-*b*-PHPMA vesicles reformed after a first cooling cycle were smaller and less disperse in size than the original ones. As already mentioned in the case of emulsion polymerization, worm-based dispersions are highly viscous solutions or free-standing gels (depending on the worm length and their concentration), because of entanglements between worms [90]. The microscopic worm-to-sphere transition is therefore macroscopically revealed by a gel-to-sol transition. This paves the way to biomedical applications, such as using them as sterilizable gels.

Such thermally induced gel-to-sol transitions have also been reported in organic solvents [158, 159]. For instance, in *n*-dodecane, poly(lauryl methacrylate)-*b*-poly(benzyl methacrylate) (PLMA-*b*-PBzMA), spherical assemblies are formed during polymerization at 70°C [158]. A sphere-to-worm transition was then observed upon cooling to room temperature, explained by a partial solvation of the PBzMA core at higher temperature (observed by ^{1}H-NMR). The morphological transition was irreversible for diluted dispersions, but a certain degree of reversibility could be maintained at high solids (20 wt%).

Furthermore, it was reported that order-to-order transitions in water can be triggered by a change in pH. In a recent example, a reversible worm-to-sphere morphological transition of HOOC-PGMA-*b*-PHPMA diblock assemblies could be induced by mere ionization of the carboxylic acid α-end-group originating from the RAFT agent through addition of base [160]. Similarly, HOOC-PGMA-*b*-PHPMA vesicles could reorganize to spheres, but in this case the transition was not reversible.

Finally, Shi *et al.* synthesized double responsive (temperature- and pH-sensitive) block copolymer particles by RAFT dispersion polymerization of styrene in ethanol/water (80/20) with a mixture of two different macroRAFT stabilizers based on P4VP and PNIPAM [161]. Spherical particles were obtained that could be transferred to water, in which the mixed stabilizing shell precipitated selectively in response to a change in temperature or pH.

It should be noted that such stimuli-induced morphological transitions have not been described yet for nano-objects obtained by PISA in emulsion polymerization. This is probably due to a too poor solvation of the core polymer by the surrounding solvent (usually water).

Morphological order-to-order transitions or order-to-disorder transitions (disassembly to unimers) triggered by a change in temperature, pH, or in solvent are certainly of interest (especially the gel-to-sol transition corresponding to a microstructural worm-to-sphere transition), but sometimes the opposite, namely stabilization of a fixed morphology, is desired and necessary. Chemical or physical crosslinking of the structures should prevent their dissociation under conditions such as temperature changes (example of PNIPAM-based polymers), the addition of solvents/organic molecules, or transfer to other solvents (through dialysis). Stabilization of the AB diblock nano-objects can be achieved by chain extension of the amphiphilic diblock copolymers to introduce a third polymer segment, in a kind of seeded emulsion polymerization step (the nano-objects being performed, and the monomer swelling the nanostructures). In a third polymerization step or at high monomer conversion during the synthesis of the second block, either solvophobic monomers or bifunctional monomers (crosslinker) can be added, yielding physically crosslinked ABC triblock copolymer structures (as already reported above for NMP-mediated PISA in emulsion [43]) or covalently linked core-crosslinked structures. For the first approach, with the formation of ABC triblock copolymers some unconventional morphologies, such as *framboidal* vesicles [162], have been observed that are driven by microphase separation between polymer segments B and C in the core. In the second strategy, the crosslinker must be added at very high or complete conversion of the monomer forming segment B in order to not hinder morphological transitions during polymerization. Then highly crosslinked and robust vesicles [163, 164] and worms [39] (similar to what has been done for PISA by RAFT *emulsion* polymerization [90]) were obtained that could, for instance, be used as stabilizers for colloidosomes or Pickering emulsions due to their ability to survive the quite harsh shear conditions. It was further shown that crosslinking permitted the transfer of morphologies from one solvent to another without meaningfully altering their structure [90, 164].

Recently, Figg *et al.* reported the first synthesis of nonspherical morphologies possessing a thermoresponsive core. They were composed of PDMAAm-*b*-P(DMAAm-*co*-AA)-*b*-PNIPAM triblock copolymers blocks that assembled during polymerization at 70°C to form vesicles or branched worms. Dissociation of the assembly upon cooling could efficiently be prevented by crosslinking the P(DMAAm-*co*-AA) middle segment with diamines [165].

2.4.2.2 Use of Solvents Other Than Pure Water The PNIPAM [165] and the numerous PHPMA-based structures described above are until now the only anisotropic morphologies synthesized by RAFT-mediated dispersion polymerization performed directly in water, since only a few monomers are water-soluble and thus amenable to aqueous dispersion polymerization. Hence merely around a dozen articles report PISA through RAFT-mediated dispersion polymerization in water, most of them dealing with PHPMA-based polymer assemblies or thermosensitive block polymer assemblies that disassemble after polymerization when not crosslinked. Despite being a "green," clean, and highly efficient process (with conversions close to 100% generally obtained in short reaction times), purely aqueous RAFT dispersion polymerization has thus severe limitations in the chemical composition, functionality, and architecture of the particles. Several research groups therefore have turned their interest to solvents other than water, mostly alcohols or alcohol-water mixtures, even if a final application in water was targeted [19].

A switch from water to organic solvents or their mixtures with water was, for instance, necessary when polyelectrolyte-stabilized morphologies other than spheres were targeted: lateral electrosteric repulsion of the chains was found to hamper PISA and morphological transitions to worms or vesicles in water [166]. This issue can be overcome by the addition of screening salts that assist morphological transitions in aqueous media., as was already observed in RAFT emulsion polymerization [79]. In contrast to pure water, in ethanol it was, for example, possible to synthesize worm-like and vesicular PMAA-*b*-PBzMA diblock copolymer morphologies [19, 164]. Their subsequent transfer to water via dialysis was reported. BzMA is indeed one of the most studied monomers used to form a large variety of morphologies by dispersion polymerization in nonpolar organic solvents (*n*-alkanes) [158, 167], alcohols [164, 168, 169], and solvent/water mixtures [81]. Zhang *et al.* [81] and Jones *et al.* [169] were the first to use BzMA for PISA by RAFT dispersion polymerization. Zhang *et al.* performed the polymerization of BzMA with P(MAA-*co*-PEOMA) macroRAFT agents in various mixtures of ethanol/water and dioxane/water [81], that is, in solvents in which the monomer was either soluble or not but PBzMA was not. Thus the polymerizations proceeded either according to an emulsion or dispersion polymerization mechanism, and a strong influence of the solvent on polymerization kinetics and the morphological phase diagram was highlighted. For a given diblock copolymer composition, an increasing amount of organic cosolvent in the polymerization medium favored the formation of worms instead of spherical objects. Compared to the ethanol-water system, in which nonspherical objects existed only above 80 vol% of ethanol, in dioxane-water mixtures the morphological transition was observed at much lower proportions of dioxane (about 20 vol%). This was explained by a difference in capacity of the organic solvent to swell the PBzMA core.

Similarly, in pure ethanol, Jones *et al.* [169] prepared PDMAEMA-*b*-PBzMA copolymer assemblies of various morphologies. As observed before, the formed morphology was dictated by the targeted molar mass of PBzMA block and the total solids of the reaction solution. It was further shown that it was possible to transfer the nano-objects to water without major alteration of their structure.

Generally, for BzMA very high or complete conversions were reached in dispersion polymerization in relatively short reaction times. This stands in contrast to the RAFT dispersion polymerization of a more conventional monomer, styrene. Whereas styrene has been successfully polymerized to high or complete conversion in aqueous emulsion polymerization conditions following the PISA principle (see Section 2.4.1), only low, or very low, conversions are obtained for styrene in pure organic solvents, such as methanol: monomer conversions between 5% [170] and 10% [155, 171, 172] have been reported even after long polymerization times (40 h) and employing high monomer concentrations (up to 70% [155]). In isopropanol, conversions were higher but still did not exceed 41% [173]. Nevertheless, various morphologies could be prepared. The need to remove the residual monomer (e.g., by centrifugation) is laborious and costly, and diminishes the industrial potential of these systems. In order to reach PSt-based morphologies in high yields and with high monomer conversions, emulsion polymerization of styrene in water (see Section 2.4.1) or water/alcohol mixtures [174, 175] remains the process of choice.

2.4.2.3 Monomers That Influence the Morphological Transition Mechanism

Even if dispersion polymerization has some disadvantages, such as a limited choice of solvophobic monomer–solvent pairs, one advantage over the emulsion process is that monomers that are solid at the reaction temperature can be polymerized, provided that they are soluble in the polymerization medium. This is probably the reason why less-conventional, "exotic" monomers, such as ionic liquid monomers [176], have been polymerized in dispersion-polymerization conditions. Recently, PISA using cholesteryl-based monomers [177] (cholesteryl (meth)acryloyl tetraethylene glycol, Chol-TEG(M)A; structure in Figure 2.19) or nucleobase monomers [178] (2-(2-(adenine-9-yl)acetoxyl) ethyl methacrylate, AMA, and 2-(2-(thymine-1-yl)acetoxyl) ethyl methacrylate, TMA) has been reported. The special feature of the nucleobase monomers is that they are able to interact with each other through hydrogen bonding, and cholesteryl derivatives are known to form liquid-crystalline (i.e., organized) phases. It was conceived that the introduction of an additional level of interaction or ordering in the assemblies' cores might influence the structure of the morphologies formed by PISA. This in turn would affect the reorganization step of chains occurring during PISA (due to a reduced mobility, similar to polymers of high T_g) [179] and also have an effect on the organization of the macromolecular chains in the core (as already observed when they were self-assembled by solvent-displacement techniques) [180]. Aiming at the preparation of biocompatible cholesterol-based fiber supports, Zhang *et al.* [177] polymerized Chol-TEGMA and Chol-TEGA with P(AA-*co*-PEOA) or P(MAA-*co*-PEOMA) macroRAFT agents in a mixture of ethanol and water (95:5 in volume). Worm-like assemblies could be obtained, and cryo-TEM and SAXS indeed revealed an internal liquid-crystalline-type organization of the cholesteryl-monomer units in the fibers' cores. It was probably thanks to this secondary organization that elongated

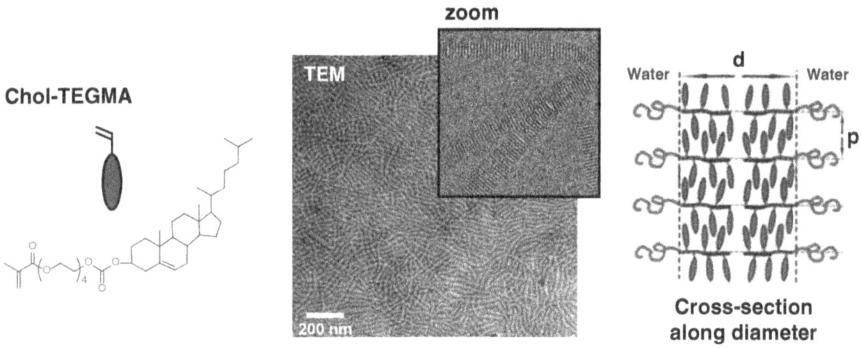

Figure 2.19 Cholesteryl methacrylate monomer (*left*) and the P($MAA_{0.5}$-*co*-$PEOMA_{0.5}$)$_{70}$-*b*-P(Chol-TEGMA)$_{20}$ worms obtained through PISA in ethanol with a P($MAA_{0.5}$-*co*-$PEOMA_{0.5}$)$_{70}$ macroRAFT agent (*middle*). The zoomed inlay shows that the fibers possess an internal nanostructuration due to the formation of a liquid-crystalline order as schematized in the schematic on the right. Adapted with permission from reference [177]. Copyright 2012 Royal Society of Chemistry.

nano-objects were observed over a broad range of molar masses (Figure 2.19). Upon transfer to water, the fibers reorganized to form spheres.

To synthesize nucleobase-containing materials, Kang *et al.* [178] (co)polymerized AMA and TMA in chloroform or 1,4-dioxane (dispersion polymerization conditions) at 60°C in the presence of a PMMA-macroRAFT agent. In chloroform, only popcorn-like particles were observed, attributed by the authors to the frozen state of the core-forming nucleobase polymer that hinders chain reorganization during polymerization. In dioxane, which should be a better solvent for the core-forming polymer, various structures were obtained depending on the polymer's molar mass and composition. The observed structures slightly differed from those obtained using conventional monomer, as described above. For instance, twisted structures instead of spherical vesicles were found.

2.4.3 Using TERP

Emulsion TERP has been investigated in detail by Okubo *et al.* Most of the studies involved the use of a macroTERP agent of poly(methacrylic acid)-methyltellanyl (PMAA-$TeCH_3$) [181–188]. The emulsion polymerizations of BA [181] and St [182, 184, 185] were reported first. Well-defined nanospheres were obtained, but particular attention had to be paid to the stirring rate and to the polymerization temperature in order to favor the desired self-assembly process over homogeneous nucleation: a high stirring rate led to a high monomer concentration inside the particles (via an increased diffusion rate through the aqueous phase), which in turn promoted consumption of the macroTERP agent, while low temperatures caused a slower generation of radicals from the initiator [182, 184]. Building on these optimized conditions, block copolymer particles of PMAA-*b*-PBA-*b*-PSt were prepared [186]. The MMA emulsion polymerization could also be controlled using PMAA-$TeCH_3$ macroTERP agent, provided that free dimethyl telluride was employed as catalyst [187]. The water miscibility of the hydrophobic monomer (MMA versus BMA) again showed the importance of monomer diffusion through the aqueous phase. In the case of the more hydrophilic MMA, the polymerization could be controlled regardless of the stirring rate, whereas the more hydrophobic BMA required high stirring rates to achieve a good control over the reaction [188]. Very recently, PMAA-$TeCH_3$ was successfully substituted by PDMAEMA-TeC_4H_9 to produce PSt particles [189].

2.5 CONCLUDING REMARKS

Polymerization-induced self-assembly using controlled radical polymerization is a young and active field of research. Originally, it was developed by researchers mostly involved in emulsion polymerization [14, 16] who were looking for efficient strategies to conduct controlled radical polymerization in dispersed media [190]. Very quickly, the excellent control PISA offers over the growth of the hydrophobic blocks expanded its appeal as a unique and very simple methodology to design amphiphilic block copolymers in water. These block copolymers self-assemble during polymerization into core-shell structures and, under certain conditions, the formed nano-objects can adopt more elaborate morphologies than the spherical particles typically obtained in conventional aqueous dispersion or emulsion polymerization. Indeed, worms, fibers, and vesicles can be obtained, mirroring the structures formed by the self-assembly of preformed amphiphilic block copolymers in water [6].

The self-assembly of small amphiphilic molecules can be described by the packing parameter p (defined as $v/(al)$, where a is the area of the surfactant head group and v and l are the volume and length of the surfactant tail) [191]. This concept can be extended to the self-assembly of preformed diblock copolymers [53, 54] to qualitatively explain the morphologies obtained experimentally. The self-assembly and the resulting morphology are a balance between (1) the degree of stretching of the core-forming blocks, (2) repulsive interactions among the hydrophilic chains in the shell, and (3) the interfacial tension between the core and the solvent, some of these morphologies being thermodynamically induced, while others are kinetically controlled [6].

This chapter has shown that in addition to the parameters usually affecting the self-assembly of preformed block copolymers (Scheme 2.1), the picture is more complex in PISA as other parameters may also come into play, such as the concentration of the core-forming monomer (solids content), the affinity of each block with the dispersing phase (no common solvent used as for preformed block copolymers), the amount of residual monomer that swells the particles, as well as pH-value and salt effects during the polymerization (in water). Most important, PISA concerns a dynamic system in which the molar mass of the hydrophobic block is constantly growing, while the monomer is being consumed.

For some systems [18, 24, 192], the construction of morphological phase diagrams appears to be a valuable methodology to anticipate the final morphology under

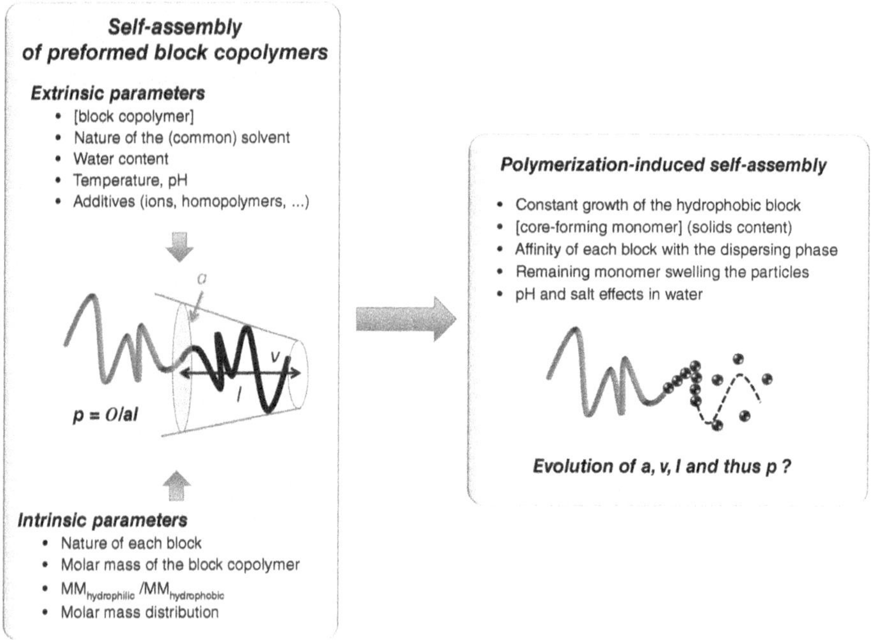

Scheme 2.1 Parameters governing the self-assembly of preformed amphiphilic block copolymers or block copolymers formed during polymerization (PISA).

particular conditions. Pure morphologies could sometimes be obtained by precisely tuning the molar mass ratios of the blocks and the hydrophobic monomer concentrations. In other cases, despite employing a wide range of experimental conditions, morphologies other than spherical particles were not accessible [95, 96, 106].

The control and the simplicity offered by the CRP techniques will facilitate further systematic study of the influence of polymer structure and chemical composition on the resulting self-assembled morphologies. Indeed, it was recently shown that the deprotonation of a carboxylic end-group located at the end of the hydrophilic segment of amphiphilic block copolymer worms was enough to induce a worm-to-sphere transition [160]. This understanding and the corresponding rationalizations will undoubtedly help stimulate widespread use of PISA to capitalize on its best attributes, namely the production of various morphologies with high solids contents under undemanding conditions.

The potential for producing different morphologies should not overshadow the great promise that the PISA process offers for the design of more "conventional" surfactant-free latex particles. In fact, the use of only very low amounts of hydrophilic living chains leads to the *in situ* formation of amphiphilic block copolymers. Here, the hydrophilic blocks act primarily as stabilizers of the polymer particles, with minimal control over the molar masses of the resulting macromolecules. To date, this approach has scarcely been developed [124–130], and it will probably be of real interest in the future for applications in which only a low amount of hydrophilic species can be tolerated.

ACKNOWLEDGMENTS

Drs. Bastian Ebeling and Samuel Pearson (C2P2) are gratefully acknowledged for helpful discussions.

ABBREVIATIONS

AA	Acrylic acid
AAm	Acrylamide
AIBN	2,2′-Azobis(2-methylpropionitrile)
AMA	2-(2-(Adenine-9-yl)acetoxyl) ethyl methacrylate
A(R)GET	Activator (re)generated by electron transfer
ATRP	Atom transfer radical polymerization
BA	*n*-Butyl acrylate
BAI	Bis(alkylphenyl)iodonium hexafluorophosphate
BIEE	1,2-Bis(2-iodo-ethoxy)ethane
BMA	*n*-Butyl methacrylate
Bpy	2,2′-Bipyridine
BzMA	Benzyl methacrylate
CCS	Core-crosslinked star
Chol-TEG(M)A	Cholesteryl (meth)acryloyl tetraethylene glycol carbonate
CMC	Critical micelle concentration
CRP	Controlled radical polymerization
Cryo-TEM	Cryogenic transmission electron microscopy
DEAAm	*N,N*-Diethyl acrylamide
DEAEMA	2-Diethylaminoethyl methacrylate
DEGA	Dimethyleneglycol acrylate
DEGMA	Di(ethyleneglycol) methyl ether methacrylate
DLS	Dynamic light scattering
DMSO	Dimethyl sulfoxide
DT	Degenerative transfer
Đ	Molar mass dispersity
DMAAm	*N,N*-Dimethylacrylamide
DMAEMA	2-Dimethylaminoethyl methacrylate
D_n	Number-average particle diameter
DP_n	Degree of polymerization
D_w	Weight-average particle diameter
EBAAn	*N,N*′-Ethylenebisacrylamide
EGDMA	Ethylene oxide (glycol) dimethacrylate

GMA	Glycerol methacrylate
HBMA	4-Hydroxybutyl methacrylate
HEMA	2-Hydroxyethyl methacrylate
HPMA	2-Hydroxypropyl methacrylate
I	Isoprene
ITP	Iodine transfer polymerization
LCST	Lower critical solution temperature
LMA	Lauryl methacrylate
MA	Methyl acrylate
MAA	Methacrylic acid
MacroRAFT(TERP) agent	Macromolecular RAFT(TERP) agent
MAnA	Mannose-containing acrylate
MBAAm	*N,N'*-Methylenebisacrylamide
MEA	2-Methoxyethyl acrylate
MMA	Methyl methacrylate
M_n	Number-average molar mass
MPC	2-(Methacryloyloxy)ethyl phosphorylcholine
M_6TREN	(2-Dimethylaminoethyl)amine
NaSS	Sodium styrene sulfonate
NIPAM	*N*-Isopropylacrylamide
NMP	Nitroxide mediated polymerization
NMR	Nuclear magnetic resonance
PEO(M)A	Poly(ethylene oxide) methyl ether (meth)acrylate
p	Packing parameter
PEG	Poly(ethylene glycol)
PEG-Br	Bromo end-functionalized poly(ethylene glycol)
PEO	Poly(ethylene oxide)
PEOV	Poly(ethylene oxide) methyl ether vinyl phenyl
PISA	Polymerization-induced self-assembly
PMDETA	*N,N,N',N',N''*-Pentamethyldiethylenetriamine
QCM	Quartz crystal microbalance
RAFT	Reversible addition–fragmentation chain transfer
RDRP	Reversible-activation radical polymerization
RITP	Reverse iodine transfer polymerization
SAXS	Small angle X-ray spectroscopy
SEC	Size exclusion chromatography
SG1	*N-tert*-Butyl-*N*-(1-diethyl phosphono-2,2-dimethyl propyl) Nitroxide
SLS	Static light scattering
SR&NI	Simultaneous reverse and normal initiation
St	Styrene
TEM	Transmission electron microscopy

TEMPO	(2,2,6,6-Tetramethylpiperidin-1-yl)oxy
TEMPO-OMe	(2,2,6,6-Tetramethyl-4-methoxy-piperidin-1-yl)oxy
TERP	Telluride-mediated polymerization
T_g	Glass transition temperature
TIPNO	2,2,5-Trimethyl-4-phenyl-3-azahexane-3-oxyl
TMA	2-(2-Thymine-1-yl)acetoxyl) ethyl methacrylate
VAc	Vinyl acetate
VDC	Vinylidene chloride
VEA	*N*-(4-Vinylbenzyl)-*N,N*-diethylamine
4VP	4-Vinylpyridine

REFERENCES

1. Lodge, T. P. *Macromol. Chem. Phys.* **2003**, *204*, 265–273.
2. Riess, G. *Prog. Polym. Sci.* **2003**, *28*, 1107–1170.
3. Iha, R. K.; Wooley, K. L.; Nyström, A. M.; Burke, D. J.; Kade, M. J.; Hawker, C. J. *Chem. Rev.* **2009**, *109*, 5620–5686.
4. Binder, W. H.; Sachsenhofer, R. *Macromol. Rapid Commun.* **2007**, *28*, 15–54.
5. Zhang, L.; Eisenberg, A. *Science* **1995**, *268*, 1728–1731.
6. Mai, Y.; Eisenberg, A. *Chem. Soc. Rev.* **2012**, *41*, 5969–5985.
7. Cameron, N. S.; Corbierre, M. K.; Eisenberg, A. *Can. J. Chem.* **1999**, *77*, 1311–1326.
8. Chen, Y. *Macromolecules* **2012**, *45*, 2619–2631.
9. Hansen, F. K. Historic overview. In *Chemistry and Technology of Emulsion Polymerisation*. John Wiley & Sons Ltd: **2013**; pp. 1–22.
10. Matyjaszewski, K.; Gnanou, Y.; Leibler, L. Macromolecular engineering. In *Macromolecular Engineering*. Wiley-VCH Verlag GmbH & Co. KGaA, **2007**, pp. 1–6.
11. Charleux, B.; Nicolas, J. *Polymer* **2007**, *48*, 5813–5833.
12. Zetterlund, P. B.; Kagawa, Y.; Okubo, M. *Chem. Rev.* **2008**, *108*, 3747–3794.
13. Monteiro, M. J.;Cunningham, M. F. *Macromolecules* **2012**, *45*, 4939–4957.
14. Ferguson, C. J.; Hughes, R. J.; Pham, B. T. T.; Hawkett, B. S.; Gilbert, R. G.; Serelis, A. K.; Such, C. H. *Macromolecules* **2002**, *35*, 9243–9245.
15. Barrett, K. E. *Dispersion Polymerization in Organic Media*. Wiley: London, **1975**.
16. Ferguson, C. J.; Hughes, R. J.; Nguyen, D.; Pham, B. T. T.; Gilbert, R. G.; Serelis, A. K.; Such, C. H.; Hawkett, B. S. *Macromolecules* **2005**, *38*, 2191–2204.
17. Dire, C.; Nicolas, J.; Brusseau, S.; Charleux, B.; Magnet, S.; Couvreur, L. *Controlled/Living Radical Polymerization: Progress in RAFT, NMP & OMRP*. American Chemical Society: Washington,DC, **2009**, Vol. 1024.
18. Charleux, B.; Delaittre, G.; Rieger, J.; D'Agosto, F. *Macromolecules* **2012**, *45*, 6753–6765.
19. Semsarilar, M.; Jones, E. R.; Blanazs, A.; Armes, S. P. *Adv. Mater.* **2012**, *24*, 3378–3382.
20. Chaduc, I.; Zhang, W.; Rieger, J.; Lansalot, M.; D'Agosto, F.; Charleux, B. *Macromol. Rapid Commun.* **2011**, *32*, 1270–1276.

21. Delaittre, G.; Dire, C.; Rieger, J.; Putaux, J.-L.; Charleux, B. *Chem. Commun.* **2009**, 2887–2889.
22. Okay, O.; Funke, W. *Macromolecules* **1990**, *23*, 2623–2628.
23. Kim, J.; Jeong, S. Y.; Kim, K. U.; Ahn, Y. H.; Quirk, R. P. *J. Polym. Sci., Part A: Polym. Chem.* **1996**, *34*, 3277–3288.
24. Warren, N. J.; Armes, S. P. *J. Am. Chem. Soc.* **2014**, *136*, 10174–10185.
25. *Handbook of Radical Polymerization*. John Wiley & Sons: Hoboken, NJ, **2002**.
26. Nicolas, J.; Guillaneuf, Y.; Lefay, C.; Bertin, D.; Gigmes, D.; Charleux, B. *Prog. Polym. Sci.* **2013**, *38*, 63–235.
27. Matyjaszewski, K. *Macromolecules* **2012**, *45*, 4015–4039.
28. Fischer, H. *Chem. Rev.* **2001**, *101*, 3581–3610.
29. Gromada, J.; Matyjaszewski, K. *Macromolecules* **2001**, *34*, 7664–7671.
30. Jakubowski, W.; Matyjaszewski, K. *Macromolecules* **2005**, *38*, 4139–4146.
31. Magenau, A. J. D.; Strandwitz, N. C.; Gennaro, A.; Matyjaszewski, K. *Science* **2011**, *332*, 81–84.
32. Matyjaszewski, K.; Gaynor, S.; Wang, J. S. *Macromolecules* **1995**, *28*, 2093–2095.
33. Tatemoto, M.; Nakagawa, T. (Daikin Kogyo Co., Ltd.) German Offenlegungsschrift 2729671, **1978**
34. Tonnar, J.; Lacroix-Desmazes, P. *Angew. Chem. Int. Ed.* **2008**, *47*, 1294–1297.
35. Goto, A.; Kwak, Y.; Fukuda, T.; Yamago, S.; Iida, K.; Nakajima, M.; Yoshida, J.-I. *J. Am. Chem. Soc.* **2003**, *125*, 8720–8721.
36. Keddie, D. J.; Moad, G.; Rizzardo, E.; Thang, S. H. *Macromolecules* **2012**, *45*, 5321–5342.
37. Moad, G.; Rizzardo, E.; Thang, S. H. *Polymer* **2008**, *49*, 1079–1131.
38. Couvreur, L.; Lefay, C.; Belleney, J.; Charleux, B.; Guerret, O.; Magnet, S. *Macromolecules* **2003**, *36*, 8260–8267.
39. Delaittre, G.; Nicolas, J.; Lefay, C.; Save, M.; Charleux, B. *Chem. Commun.* **2005**, 614–616.
40. Delaittre, G.; Nicolas, J.; Lefay, C.; Save, M.; Charleux, B. *Soft Matter* **2006**, *2*, 223–231.
41. Delaittre, G.; Charleux, B. *Macromolecules* **2008**, *41*, 2361–2367.
42. Delaittre, G.; Save, M.; Charleux, B. *Macromol. Rapid Commun.* **2007**, *28*, 1528–1533.
43. Delaittre, G.; Save, M.; Gaborieau, M.; Castignolles, P.; Rieger, J.; Charleux, B. *Polym. Chem.* **2012**, *3*, 1526–1538.
44. Nicolas, J.; Charleux, B.; Guerret, O.; Magnet, S. *Macromolecules* **2004**, *37*, 4453–4463.
45. Nicolas, J.; Charleux, B.; Guerret, O.; Magnet, S. *Angew. Chem. Int. Ed.* **2004**, *43*, 6186–6189.
46. Dire, C.; Magnet, S.; Couvreur, L.; Charleux, B. *Macromolecules* **2009**, *42*, 95–103.
47. Qiao, X. G.; Lansalot, M.; Bourgeat-Lami, E.; Charleux, B. *Macromolecules* **2013**, *46*, 4285–4295.
48. Darabi, A.; Shirin-Abadi, A. R.; Pinaud, J.; Jessop, P. G.; Cunningham, M. F. *Polym. Chem.* **2014**, *5*, 6163–6170.
49. Yoshida, E. *Colloid Polym. Sci.* **2011**, *289*, 1625–1630.
50. Yoshida, E. *Open J. Polym. Chem.* **2011**, *3*, 16–22.
51. Yoshida, E. *Colloid Polym. Sci.* **2013**, *291*, 2733–2739.

52. Yoshida, E. *Colloid Polym. Sci.* **2014**, *292*, 763–769.
53. Blanazs, A.; Armes, S. P.; Ryan, A. J. *Macromol. Rapid Commun.* **2009**, *30*, 267–277.
54. Antonietti, M.; Förster, S. *Adv. Mater.* **2003**, *15*, 1323–1333.
55. Yoshida, E. *Colloid Polym. Sci.* **2014**, *292*, 1463–1468.
56. Brusseau, S.; D'Agosto, F.; Magnet, S.; Couvreur, L.; Chamignon, C.; Charleux, B. *Macromolecules* **2011**, *44*, 5590–5598.
57. Groison, E.; Brusseau, S.; D'Agosto, F.; Magnet, S.; Inoubli, R.; Couvreur, L.; Charleux, B. *ACS Macro Lett.* **2011**, *1*, 47–51.
58. Kim, K. H.; Kim, J.; Jo, W. H. *Polymer* **2005**, *46*, 2836–2840.
59. Wan, W.-M.; Pan, C.-Y. *Macromolecules* **2007**, *40*, 8897–8905.
60. Sugihara, S.; Armes, S. P.; Lewis, A. L. *Angew. Chem. Int. Ed.* **2010**, *49*, 3500–3503.
61. Sugihara, S.; Sugihara, K.; Armes, S. P.; Ahmad, H.; Lewis, A. L. *Macromolecules* **2010**, *43*, 6321–6329.
62. Anastasaki, A.; Haddleton, A. J.; Zhang, Q.; Simula, A.; Droesbeke, M.; Wilson, P.; Haddleton, D. M. *Macromol. Rapid Commun.* **2014**, *35*, 965–970.
63. Zhang, Q.; Wilson, P.; Li, Z.; McHale, R.; Godfrey, J.; Anastasaki, A.; Waldron, C.; Haddleton, D. M. J. *Am. Chem. Soc.* **2013**, *135*, 7355–7363.
64. Alsubaie, F.; Anastasaki, A.; Wilson, P.; Haddleton, D. M. *Polym. Chem.* **2015**, *6*, 406–417.
65. Zhang, Q.; Wilson, P.; Anastasaki, A.; McHale, R.; Haddleton, D. M. *ACS Macro Lett.* **2014**, *3*, 491–495.
66. Sprong, E.; Leswin, J. S. K.; Lamb, D. J.; Ferguson, C. J.; Hawkett, B. S.; Pham, B. T. T.; Nguyen, D.; Such, C. H.; Serelis, A. K.; Gilbert, R. G. *Macromol. Symp.* **2006**, *231*, 84–93.
67. Ganeva, D. E.; Sprong, E.; ., de Bruyn, H.; Warr, G. G.; Such, C. H.; Hawkett, B. S. *Macromolecules* **2007**, *40*, 6181–6189.
68. Leswin, J. S. K.; Meuldijk, J.; Gilbert, R. G.; van Herk, A. M. *Macromol. Symp.* **2009**, *275–276*, 24–34.
69. Bar-Nes, G.; Hall, R.; Sharma, V.; Gaborieau, M.; Lucas, D.; Castignolles, P.; Gilbert, R. G. *Eur. Polym. J.* **2009**, *45*, 3149–3163.
70. Božović-Vukić, J.; Manon, H. T.; Meuldijk, J.; Koning, C.; Klumperman, B. *Macromolecules* **2007**, *40*, 7132–7139.
71. Chen, Y.; Luo, W.; Wang, Y.; Sun, C.; Han, M.; Zhang, C. *J. Colloid Interface Sci.* **2012**, *369*, 46–51.
72. Rieger, J.; Stoffelbach, F.; Bui, C.; Alaimo, D.; Jérôme, C.; Charleux, B. *Macromolecules* **2008**, *41*, 4065–4068.
73. Rieger, J.; Osterwinter, G.; Bui, C.; Stoffelbach, F.; Charleux, B. *Macromolecules* **2009**, *42*, 5518–5525.
74. Dos Santos, A. M.; Le Bris, T.; Graillat, C.; D'Agosto, F.; Lansalot, M. *Macromolecules* **2009**, *42*, 946–956.
75. Velasquez, E.; Rieger, J.; Stoffelbach, F.; Charleux, B.; D'Agosto, F.; Lansalot, M.; Dufils, P.-E.; Vinas, J. *Polymer* **2013**, *54*, 6547–6554.
76. Rieger, J.; Zhang, W.; Stoffelbach, F.; Charleux, B. *Macromolecules* **2010**, *43*, 6302–6310.

77. Ji, J.; Yan, L.; Xie, D. *J. Polym. Sci., Part A: Polym. Chem.* **2008**, *46*, 3098–3107.
78. Boissé, S.; Rieger, J.; Belal, K.; Di-Cicco, A.; Beaunier, P.; Li, M.-H.; Charleux, B. *Chem. Commun.* **2010**, *46*, 1950–1952.
79. Boissé, S.; Rieger, J.; Pembouong, G.; Beaunier, P.; Charleux, B. *J. Polym. Sci., Part A: Polym.* Chem. **2011**, *49*, 3346–3354.
80. Zhang, X.; Boissé, S.; Zhang, W.; Beaunier, P.; D'Agosto, F.; Rieger, J.; Charleux, B. *Macromolecules* **2011**, *44*, 4149–4158.
81. Zhang, X.; Rieger, J.; Charleux, B. *Polym. Chem.* **2012**, *3*, 1502–1509.
82. Xu, J.; Xiao, X.; Zhang, Y.; Zhang, W.; Sun, P. *J. Polym. Sci., Part A: Polym. Chem.* **2013**, *51*, 1147–1161.
83. Xu, J.; Wang, X.; Zhang, Y.; Zhang, W.; Sun, P. *J. Polym. Sci., Part A: Polym. Chem.* **2012**, *50*, 2484–2498.
84. Carlsson, L.; Fall, A.; Chaduc, I.; Wagberg, L.; Charleux, B.; Malmstrom, E.; D'Agosto, F.; Lansalot, M.; Carlmark, A. *Polym. Chem.* **2014**, *5*, 6076–6086.
85. Thomas, D. B., Convertine, A. J., Hester, R. D., Lowe, A. B., McCormick, C. L. *Macromolecules* **2004**, *37*, 1735–1741.
86. Zhang, W.; D'Agosto, F.; Boyron, O.; Rieger, J.; Charleux, B. *Macromolecules* **2011**, *44*, 7584–7593.
87. Doussineau, T.; Bao, C. Y.; Antoine, R.; Dugourd, P.; Zhang, W.; D'Agosto, F.; Charleux, B. *ACS Macro Lett.* **2012**, 414–417.
88. Doussineau, T.; Santacreu, M.; Antoine, R.; Dugourd, P.; Zhang, W.; Chaduc, I.; Lansalot, M.; D'Agosto, F.; Charleux, B. *Chem. Phys. Chem.* **2013**, *14*, 603–609.
89. Zhang, W.; D'Agosto, F.; Boyron, O.; Rieger, J.; Charleux, B. *Macromolecules* **2012**, *45*, 4075–4084.
90. Zhang, W.; Charleux, B.; Cassagnau, P. *Soft Matter* **2013**, *9*, 2197–2205.
91. Zhang, W.; D'Agosto, F.; Dugas, P.-Y.; Rieger, J.; Charleux, B. *Polymer* **2013**, *54*, 2011–2019.
92. Gigault, J.; Zhang, W.; Lespes, G.; Charleux, B.; Grassl, B. *Anal. Chim. Acta* **2014**, *819*, 116–121.
93. Zhang, W.; Charleux, B.; Cassagnau, P. *Macromolecules* **2012**, *45*, 5273–5280.
94. Chaduc, I.; Lansalot, M.; D'Agosto, F.; Charleux, B. *Macromolecules* **2012**, *45*, 1241–1247.
95. Chaduc, I.; Girod, M.; Antoine, R.; Charleux, B.; D'Agosto, F.; Lansalot, M. *Macromolecules* **2012**, *45*, 5881–5893.
96. Chaduc, I.; Crepet, A.; Boyron, O.; Charleux, B.; D'Agosto, F.; Lansalot, M. *Macromolecules* **2013**, *46*, 6013–6023.
97. Wang, X.; Luo, Y.; Li, B.; Zhu, S. *Macromolecules* **2009**, *42*, 6414–6421.
98. Luo, Y.; Guo, Y.; Gao, X.; Li, B.-G.; Xie, T. *Adv. Mater.* **2013**, *25*, 743–748.
99. Luo, Y.; Wang, X.; Zhu, Y.; Li, B.-G.; Zhu, S. *Macromolecules* **2010**, *43*, 7472–7481.
100. Wei, R.; Luo, Y.; Xu, P. *J. Polym. Sci., Part A: Polym. Chem.* **2011**, *49*, 2980–2989.
101. Zhou, J.; Wang, H.; Zhang, L.; Ma, J. *Polym. Int.* **2014**, *63*, 2098–2104.
102. Xu, S.; Huang, J.; Xu, S.; Luo, Y. *Polymer* **2013**, *54*, 1779–1785.
103. Huang, J.; Zhao, S.; Gao, X.; Luo, Y.; Li, B. *Ind. Eng. Chem. Res.* **2014**, *53*, 7688–7695.
104. Yang, J.; Bao, Y.; Pan, P. *J. Appl. Polym. Sci.* **2014**, *131*, DOI: 10.1002/APP.40391.

105. Ratcliffe, L.; Blanazs, A.; Williams, C.; Brown, S.; Armes, S. P. *Polym. Chem.* **2014**, *5*, 3643–3655.

106. Cunningham, V. J.; Alswieleh, A. M.; Thompson, K. L.; Williams, M.; Leggett, G. J.; Armes, S. P.; Musa, O. M. *Macromolecules* **2014**, *47*, 5613–5623.

107. Thompson, K. L.; Mable, C. J.; Cockram, A.; Warren, N. J.; Cunningham, V. J.; Jones, E. R.; Verber, R.; Armes, S. P. *Soft Matter* **2014**, *10*, 8615–8626.

108. Qiu, Q.; Liu, G.; An, Z. *Chem. Commun.* **2011**, *47*, 12685–12687.

109. Chen, Q.; Xu, Y.; Cao, X.; Qin, L.; An, Z. *Polym. Chem.* **2014**, *5*, 175–185.

110. Pham, B. T. T.; Such, C.; Hawkett, B. *Polym. Chem.* **2015**, *6*, 426–435.

111. Ting, S. R. S.; Min, E. H.; Zetterlund, P. B.; Stenzel, M. H. *Macromolecules* **2010**, *43*, 5211–5221.

112. Jiang, G.; Wang, Y.; Zhang, R.; Wang, R.; Wang, X.; Zhang, M.; Sun, X.; Bao, S.; Wang, T.; Wang, S. *ACS Macro Lett.* **2012**, *1*, 489–493.

113. Zhang, X.; Cardozo , A. F.; Chen , S.; Zhang, W.; Julcour , C.; Lansalot, M.; Blanco , J.-F.; Gayet , F.; Delmas , H.; Charleux , B.; Manoury , E.; D'Agosto, F.; Poli, R. *Chem. Eur. J.* **2014**, *20*, 15505–15517.

114. Chen , S.; Cardozo, A. F.; Julcour, C.; Blanco, J.-F.; Barthe, L.; Gayet, F.; Lansalot, M.; D'Agosto, F.; Delmas , H.; Manoury , E.; Poli, R. *Polymer* **2015**, *72*, 327–335.

115. Chenal, M.; Bouteiller, L.; Rieger, J. *Polym. Chem.* **2013**, *4*, 752–762.

116. Chenal, M.; Rieger, J.; Véchambre, C.; Chenal, J.-M.; Chazeau, L.; Creton, C.; Bouteiller, L. *Macromol. Rapid Commun.* **2013**, *34*, 1524–1529.

117. Xu, S.; Liu, W. *J. Fluorine Chem.* **2008**, *129*, 125–130.

118. Zhou, J.; Zhang, L.; Ma, J. *Chem. Eng. J.* **2013**, *223*, 8–17.

119. Hua, D.; Tang, J.; Jiang, J.; Zhu, X.; Bai, R. *Macromolecules* **2009**, *42*, 8697–8701.

120. Binauld, S.; Delafresnaye, L.; Charleux, B.; D'Agosto, F.; Lansalot, M. *Macromolecules* **2014**, *47*, 3461–3472.

121. Yeole, N.; Hundiwale, D.; Jana, T. *J. Colloid Interface Sci.* **2011**, *354*, 506–510.

122. Yeole, N.; Hundiwale, D. *Coll. Surf. A: Phys. Eng. Asp.* **2011**, *392*, 329–334.

123. Yeole, N.; Hundiwale, D. *RSC Adv.* **2013**, *3*, 22213–22218.

124. Tichagwa, L.; Götz, C.; Tonge, M.; Sanderson, R.; Pasch, H. *Macromol. Symp.* **2003**, *193*, 251–260.

125. Manguian, M.; Save, M.; Charleux, B. *Macromol. Rapid Commun.* **2006**, *27*, 399–404.

126. Dos Santos, A. M.; Pohn, J.; Lansalot, M.; D'Agosto, F. *Macromol. Rapid Commun.* **2007**, *28*, 1325–1332.

127. Wi, Y.; Lee, K.; Lee, B. H.; Choe, S. *Polymer* **2008**, *49*, 5626–5635.

128. Bernard, J.; Save, M.; Arathoon, B.; Charleux, B. *J. Polym. Sci., Part A: Polym. Chem.* **2008**, *46*, 2845–2857.

129. Zhang, Q.; Yu, G.; Wang, W.-J.; Li, B.-G.; Zhu, S. *Macromol. Rapid Commun.* **2012**, *33*, 916–921.

130. Zhang, Q.; Yu, G.;Wang, W.-J.; Yuan, H.; Li, B.-G.; Zhu, S. *Macromolecules* **2013**, *46*, 1261–1267.

131. Guyot, A. *Adv. Colloid Interface Sci.* **2004**, *108–109*, 3–22.

132. Zheng, G.; Pan, C. *Macromolecules* **2006**, *39*, 95–102.

133. Houillot, L.; Bui, C.; Save, M.; Charleux, B.; Farcet, C.; Moire, C.; Raust, J.-A.; Rodriguez, I. *Macromolecules* **2007**, *40*, 6500–6509.

134. Zong, M.; Thurecht, K. J.; Howdle, S. M. *Chem. Commun.* **2008**, 5942–5944.

135. Ji, W.; Yan, J.; Chen, E.; Li, Z.; Liang, D. *Macromolecules* **2008**, *41*, 4914–4919.

136. Sun, J.-T.; Hong, C.-Y.; Pan, C.-Y. *Soft Matter* **2012**, *8*, 7753–7767.

137. Sun, J.-T.; Hong, C.-Y.; Pan, C.-Y. *Polym. Chem.* **2013**, *4*, 873–881.

138. An, Z.; ., Shi, Q.; ., Tang, W.; Tsung, C.-K.; Hawker, C. J.; Stucky, G. D. *J. Am. Chem. Soc.* **2007**, *129*, 14493–14499.

139. Rieger, J.; Grazon, C.; Charleux, B.; Alaimo, D.; Jérôme, C. *J. Polym. Sci., Part A: Polym. Chem.* **2009**, *47*, 2373–2390.

140. Warren, N. J.; Mykhaylyk, O. O.; Mahmood, D.; Ryan, A. J.; Armes, S. P. *J. Am. Chem. Soc.* **2014**, *136*, 1023–1033.

141. Grazon, C.; Rieger, J.; Sanson, N.; Charleux, B. *Soft Matter* **2011**, *7*, 3482–3490.

142. Su, Y.; Xiao, X.; Li, S.; Dan, M.; Wang, X.; Zhang, W. *Polym. Chem.* **2014**, *5*, 578–587.

143. Yan, L.; Tao, W. *Polymer* **2010**, *51*, 2161–2167.

144. Liu, X.-B.; Zhou, J.-F.; Ye, X.-D. *Chin. J. Chem. Phys.* **2012**, *25*, 463–468.

145. Liu, G.; Qiu, Q.; An, Z. *Polym. Chem.* **2012**, *3*, 504–513.

146. Shen, W.; Chang, Y.; Liu, G.; Wang, H.; Cao, A.; An, Z. *Macromolecules* **2011**, *44*, 2524–2530.

147. Liu, G.; Qiu, Q.; Shen, W.; An, Z. *Macromolecules* **2011**, *44*, 5237–5245.

148. Sugihara, S.; Armes, S. P.; Blanazs, A.; Lewis, A. L. *Soft Matter* **2011**, *7*, 10787–10793.

149. Sanson, N.; Rieger, J. *Polym. Chem.* **2010**, *1*, 965–977.

150. Wan, W.-M.; Sun, X.-L.; Pan, C.-Y. *Macromolecules* **2009**, *42*, 4950–4952.

151. Wan, W. M.; Hong, C. Y.; Pan, C. Y. *Chem. Commun.* **2009**, 5883–5885.

152. Li, Y.; Armes, S. P. *Angew. Chem. Int. Ed.* **2010**, *49*, 4042–4046.

153. Blanazs, A.; Madsen, J.; Battaglia, G.; Ryan, A. J.; Armes, S. P. *J. Am. Chem. Soc.* **2011**, *133*, 16581–16587.

154. Sugihara, S.; Blanazs, A.; Armes, S. P.; Ryan, A. J.; Lewis, A. L. *J. Am. Chem. Soc.* **2011**, *133*, 15707–15713.

155. Zhang, W.-J.; Hong, C.-Y.; Pan, C.-Y. *Macromolecules* **2014**, *47*, 1664–1671.

156. Huo, F.; Li, S.; He, X.; Shah, S. A.; Li, Q.; Zhang, W. *Macromolecules* **2014**, *47*, 8262–8269.

157. Blanazs, A.; Verber, R.; Mykhaylyk, O. O.; Ryan, A. J.; Heath, J. Z.; Douglas, C. W., Armes, S. P. *J. Am. Chem. Soc.* **2012**, *134*, 9741–9748.

158. Fielding, L. A.; Lane, J. A.; Derry, M. J.; Mykhaylyk, O. O.; Armes, S. P. *J. Am. Chem. Soc.* **2014**, *136*, 5790–5798.

159. Pei, Y.; Dharsana, N. C.; van Hensbergen, J. A.; Burford, R. P.; Roth, P. J.; Lowe, A. B. *Soft Matter* **2014**, *10*, 5787–5796.

160. Lovett, J. R.; Warren, N. J.; Ratcliffe, L. P. D.; Kocik, M. K.; Armes, S. P. *Angew. Chem. Int. Ed.* **2015**, *54*, 1279–1283.

161. Shi, P.; Li, Q.; He, X.; Li, S.; Sun, P.; Zhang, W. *Macromolecules* **2014**, *47*, 7442–7452.

162. Chambon, P.; Blanazs, A.; Battaglia, G.; Armes, S. P. *Macromolecules* **2012**, *45*, 5081–5090.

163. Thompson, K. L.; Chambon, P.; Verber, R.; Armes, S. P. *J. Am. Chem. Soc.* **2012**, *134*, 12450–12453.

164. Semsarilar, M.; Ladmiral, V.; Blanazs, A.; Armes, S. P. *Polym. Chem.* **2014**, *5*, 3466–3475.

165. Figg, C. A.; Simula, A.; Gebre, K. A.; Tucker, B. S.; Haddleton, D. M.; Sumerlin, B. S. *Chem. Sci.* **2015**, *6*, 1230–1236.

166. Semsarilar, M.; Ladmiral, V.; Blanazs, A.; Armes, S. P. *Langmuir* **2012**, *28*, 914–922.

167. Fielding, L. A.; Derry, M. J.; Ladmiral, V.; Rosselgong, J.; Rodrigues, A. M.; Ratcliffe, L. P. D.; Sugihara, S.; Armes, S. P. *Chem. Sci.* **2013**, *4*, 2081–2087.

168. Zehm, D.; Ratcliffe, L. P. D.; Armes, S. P. *Macromolecules* **2013**, *46*, 128–139.

169. Jones, E. R.; Semsarilar, M.; Blanazs, A.; Armes, S. P. *Macromolecules* **2012**, *45*, 5091–5098.

170. Karagoz, B.; Yeow, J.; Esser, L.; Prakash, S. M.; Kuchel, R. P.; Davis, T. P.; Boyer, C. *Langmuir* **2014**, *30*, 10493–10502.

171. Bleach, R.; Karagoz, B.; Prakash, S. M.; Davis, T. P.; Boyer, C. *ACS Macro Lett.* **2014**, *3*, 591–596.

172. Karagoz, B.; Boyer, C.; Davis, T. P. *Macromol. Rapid Commun.* **2014**, *35*, 417–421.

173. Zhao, W.; Gody, G.; Dong, S.; Zetterlund, P. B.; Perrier, S. *Polym. Chem.* **2014**, *5*, 6990–7003.

174. Huo, F.; Wang, X.; Zhang, Y.; Zhang, X.; Xu, J.; Zhang, W. *Macromol. Chem. Phys.* **2013**, *214*, 902–911.

175. Wang, X.; Xu, J.; Zhang, Y.; Zhang, W. *J. Polym. Sci., Part A: Polym. Chem.* **2012**, *50*, 2452–2462.

176. Yuan, J.; Soll, S.; Drechsler, M.; Muller, A. H.; Antonietti, M. *J. Am. Chem. Soc.* **2011**, *133*, 17556–17559.

177. Zhang, X.; Boissé, S.; Bui, C.; Albouy, P.-A.; Brûlet, A.; Li, M.-H.; Rieger, J.; Charleux, B. *Soft Matter* **2012**, *8*, 1130–1141.

178. Kang, Y.; Pitto-Barry, A.; Willcock, H.; Quan, W.-D.; Kirby, N.; Sanchez, A. M.; O'Reilly, R. K. *Polym. Chem.* **2014**, *6*, 106–117.

179. Yang, P.; Ratcliffe, L. P. D.; Armes, S. P. *Macromolecules* **2013**, *46*, 8545–8556.

180. Boissé, S.; Rieger, J.; Di-Cicco, A. l.; Albouy, P.-A.; Bui, C.; Li, M.-H.; Charleux, B. *Macromolecules* **2009**, *42*, 8688–8696.

181. Okubo, M.; Sugihara, Y.; Kitayama, Y.; Kagawa, Y.; Minami, H. *Macromolecules* **2009**, *42*, 1979–1984.

182. Kitayama, Y.; Chaiyasat, A.; Minami, H.; Okubo, M. *Macromolecules* **2010**, *43*, 7465–7471.

183. Kitayama, Y.; Chaiyasat, A.; Okubo, M. *Macromol. Symp.* **2010**, *288*, 25–32.

184. Kitayama, Y.; Moribe, H.; Minami, H.; Okubo, M. *Polymer* **2011**, *52*, 2729–2734.

185. Moribe, H.; Kitayama, Y.; Suzuki, T.; Okubo, M. *Macromolecules* **2011**, *44*, 263–268.

186. Kitayama, Y.; Kishida, K.; Minami, H.; Okubo, M. *J. Polym. Sci., Part A: Polym. Chem.* **2012**, *50*, 1991–1996.

187. Kitayama, Y.; Moribe, H.; Kishida, K.; Okubo, M. *Polym. Chem.* **2012**, *3*, 1555–1559.

188. Kitayama, Y.; Kishida, K.; Okubo, M. *J. Polym. Sci., Part A: Polym. Chem.* **2013**, *51*, 716–723.

189. Kitayama, Y.; Okubo, M. *Polym. Chem.* **2014**, *5*, 2784–2792.
190. Cunningham, M. F. *Prog. Polym. Sci.* **2008**, *33*, 365–398.
191. Israelachvili, J. *Intermolecular and Surface Forces*, 3rd ed. Academic Press, London, **2011**.
192. Charleux, B.; D'Agosto, F.; Delaittre, G. *Adv. Polym. Sci.* **2010**, *233*, 125–183.

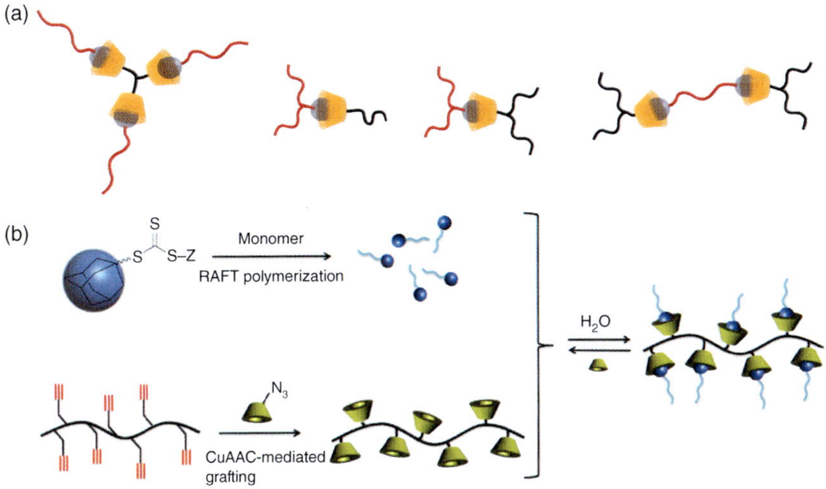

Figure 1.6 (a) Various supramolecular star architectures from left to right: Three arm stars [78], A_2B miktoarm stars [79], A_2B_2 miktoarm stars [80], and H-shape stars [80]. (b) Supramolecular brush formation via adamantyl end-functionalized PAA and CD conjugated poly(acrylate) [40a]. Reproduced from [39a] with permission of The Royal Society of Chemistry.

Macromolecular Self-Assembly, First Edition. Edited by Laurent Billon and Oleg Borisov.

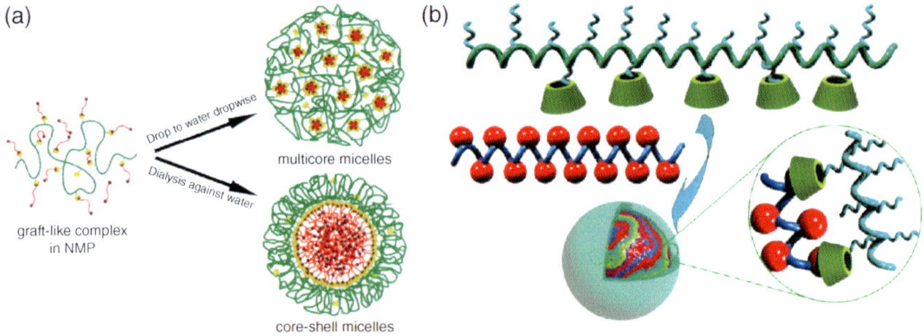

Figure 1.7 (a) Formation of micelles based on poly(*N*-vinyl-2-pyrrolidone) with β-CD side chains and adamantyl end-functionalized PCL leading to different micelle architectures depending on the preparation technique [97] (Reproduced from [96] with permission of John Wiley and Sons) and (b) formation of micelles based on β-CD conjugated poly(ethylene imine) and poly(β-benzyl-L-aspartate) [98] (Reproduced with permission from [97]. Copyright 2010 American Chemical Society).

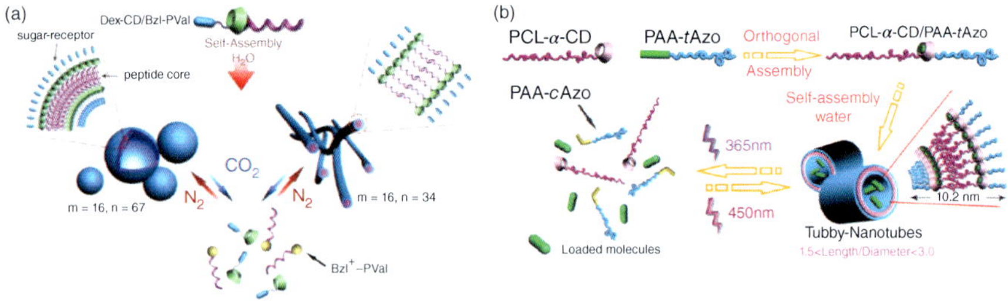

Figure 1.8 (a) Formation of pH/CO_2 responsive vesicles or fibers based on β-CD/benzimidazole interactions [47] (Reprinted with permission from [46]. Copyright 2014 American Chemical Society) and (b) formation of supramolecular light-responsive nanotubes based on azobenzene/α-CD interactions [53] (Reproduced from [52] with permission of The Royal Society of Chemistry).

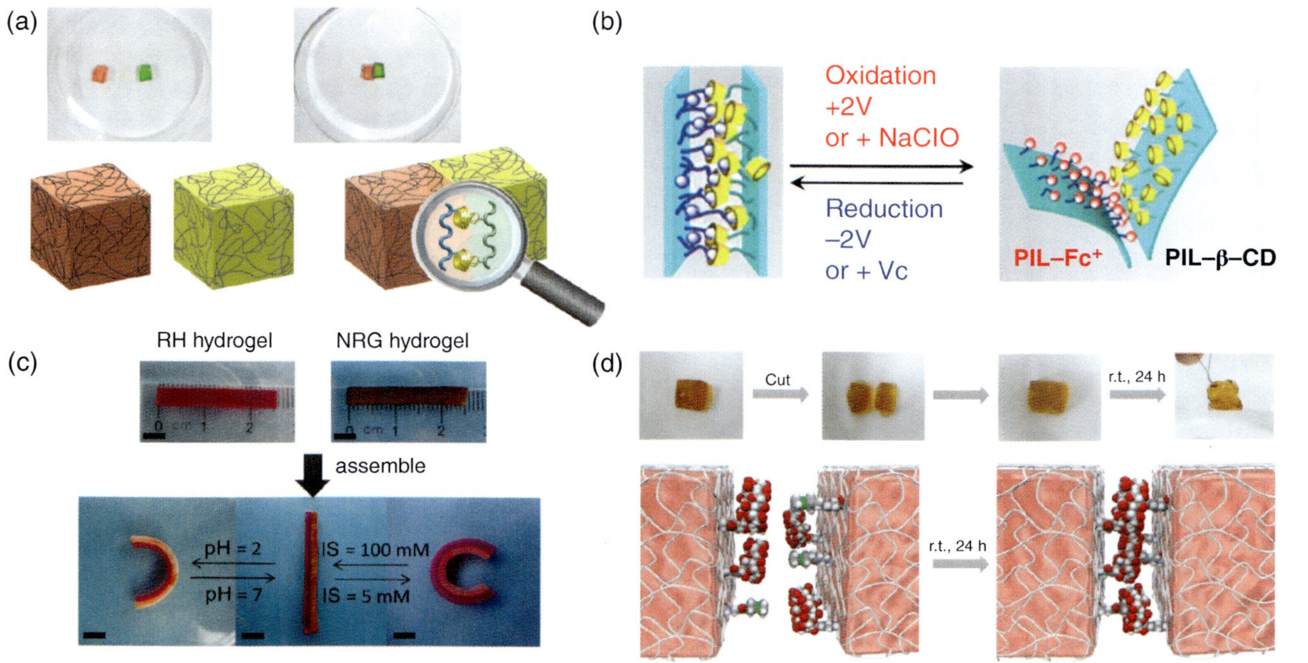

Figure 1.10 (a) Macroscopic self-assembly of an adamantyl containing hydrogel cube (green) and a β-CD containing hydrogel cube (red) [128] (Reprinted by permission from Macmillan Publishers Ltd: *Nat. Chem*. [126], copyright 2011), (b) β-CD/ferrocene based supramolecular velcro with reversible zipping/unzipping depending on the ferrocene oxidation state [129] (Reproduced from [127] with permission of The Royal Society of Chemistry), (c) combination of a β-CD and carboxylic acid containing gel (RH hydrogel) and a ferrocene containing gel (NRG hydrogel) with pH or ionic strength depending bending [130] (Adapted from [128]with permission of John Wiley and Sons), and (d) β-CD/ferrocene based self-healing hydrogel [22a] (Reprinted by permission from Macmillan Publishers Ltd: *Nat. Commun*. [21a], copyright 2011).

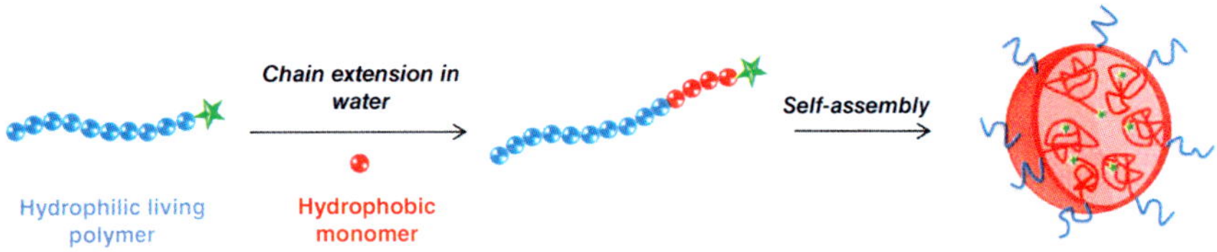

Figure 2.2 Polymerization-induced self-assembly (PISA) using CRP in water.

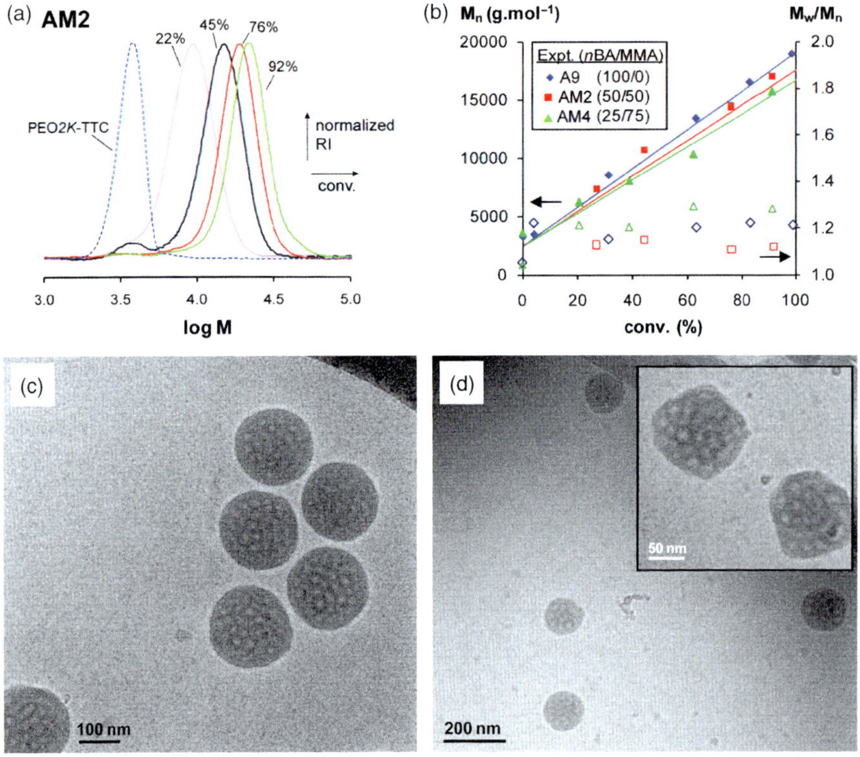

Figure 2.10 Surfactant-free, batch emulsion homopolymerization of BA (A9) and copolymerizations of BA with MMA (BA:MMA molar ratio 50:50, AM2 and 25:75, AM4) in water at 70°C: (a) evolution of the size exclusion chromatograms with conversion for experiment AM2; (b) number-average molar mass, M_n, and dispersity, $Đ = M_w/M_n$, determined by SEC as a function of the conversion. The straight lines correspond to the theoretically expected M_n versus conversion functions. Cryo-TEM micrographs of particles obtained in (c) experiment A9 and (d) experiment AM4. Reproduced with permission from reference [73]. Copyright 2009 American Chemical Society.

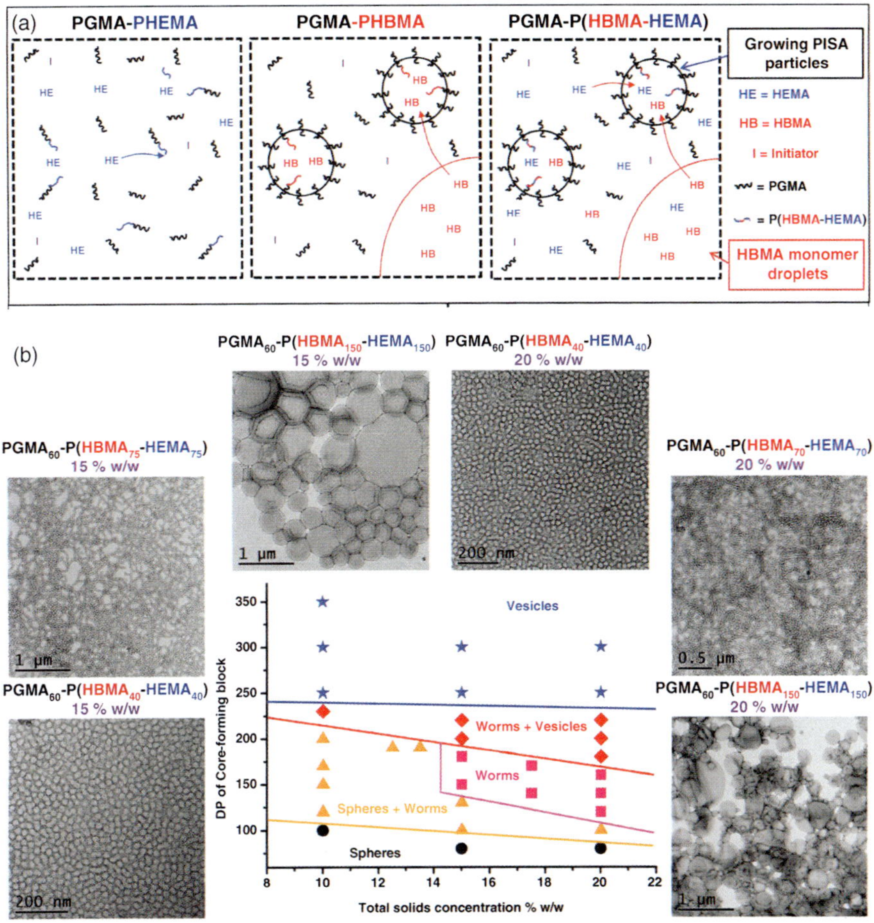

Figure 2.15 (a) Schematic representation of the initial spatial location of the HEMA and HBMA comonomers in the RAFT synthesis of $PGMA_{60}$-*b*-P($HBMA_x$ – *co*-$HEMA_y$) diblock copolymer nanoparticles, and (b) Phase diagram constructed for $PGMA_{60}$-*b*-P($HBMA_x$-co-$HEMA_y$) statistical diblock copolymers prepared at 70°C by RAFT copolymerization of HBMA and HEMA at a 1:1 molar ratio. Selected TEM images are shown for specific diblock compositions and copolymer concentrations, confirming that pure sphere, worm and vesicle morphologies can be obtained under appropriate conditions. Reprinted and adapted with permission from [15]. Copyright 2014 Royal Society of Chemistry.

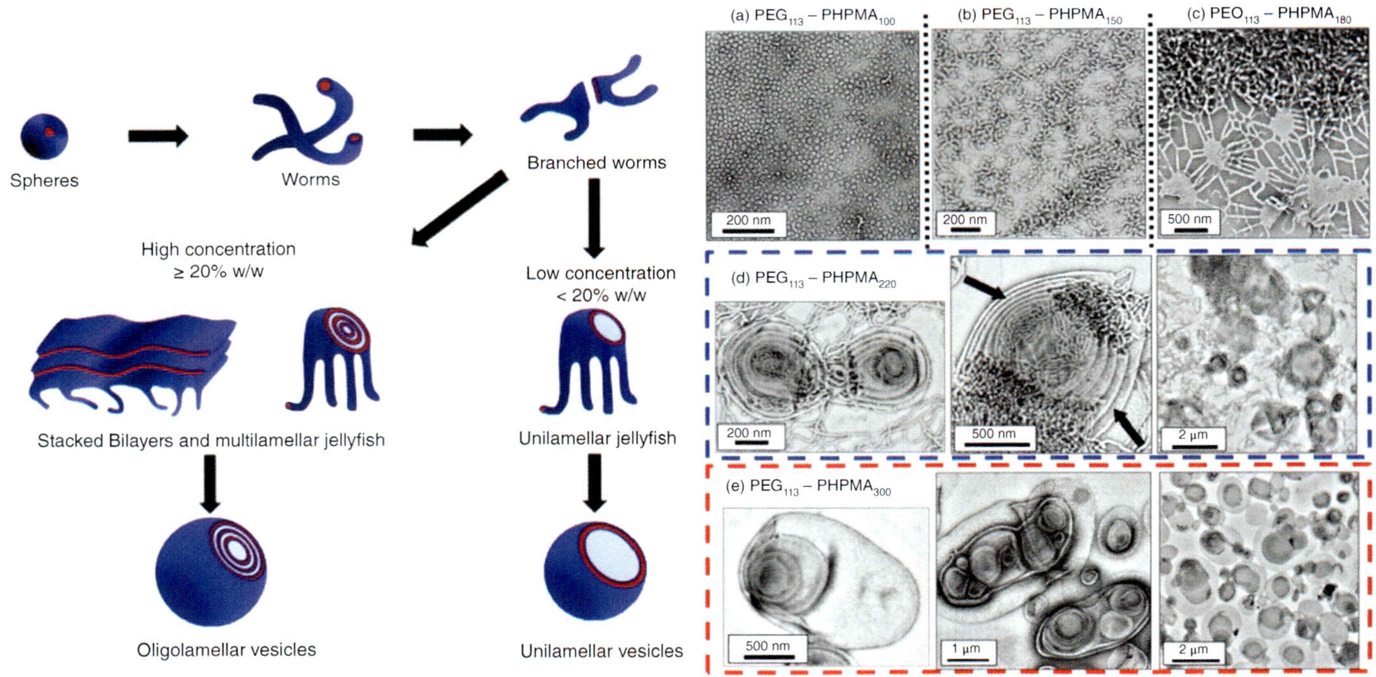

Figure 2.17 *Left*: Schematic representation of the formation of the oligolamellar and unilamellar vesicles obtained from PISA through RAFT aqueous dispersion polymerization of HPMA with PEG_{5K}-macroRAFT agent at 50°C at ≥ 20 wt% or <20 wt% solids, respectively. *Right*: TEM images of PEO_{5K}-b-$PHPMA_x$ nano-objects prepared at ≥ 20% solids. Reprinted with permission from [140]. Copyright 2014 American Chemical Society.

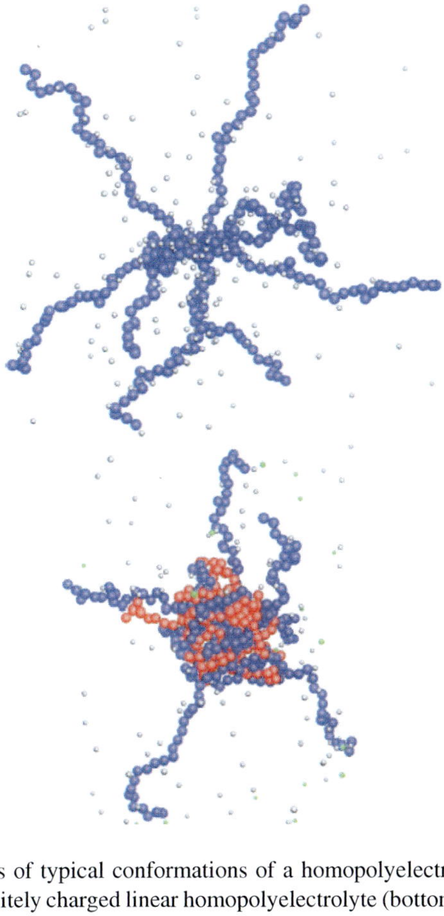

Figure 4.4 Snapshots of typical conformations of a homopolyelectrolyte star (top) and its complex with an oppositely charged linear homopolyelectrolyte (bottom). Reprinted from [18] with permission from the American Chemical Society.

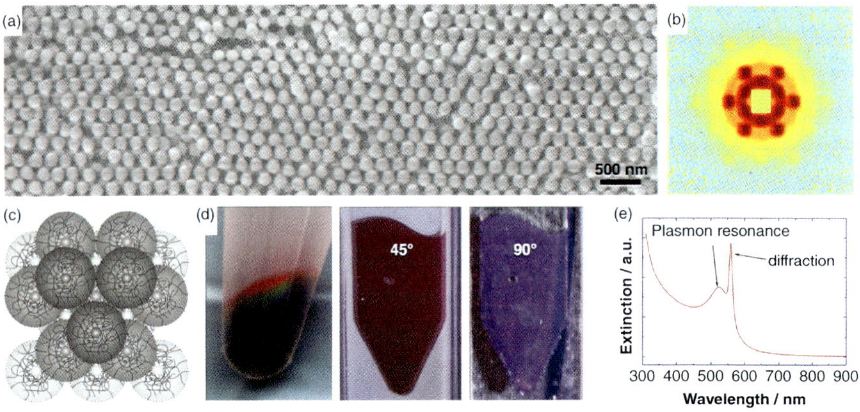

Figure 6.4 Distance control in 3D assemblies: (A) SEM image of a colloidal crystal of gold nanoparticles with shells of crosslinked PNIPAm prepared by precipitation polymerization [Karg *et al.*, unpublished data]. (B) 2D detector image from small-angle neutron scattering of a crystalline Au/PNIPAm core/shell sample measured on the D11 instrument of the Institute Laue–Langevin at a sample-to-detector distance of 28 m [Karg *et al.*, unpublished data]. (C) The pronounced Bragg peaks can be attributed to a face-centered cubic structure [86]. (D) Photographs of the crystalline samples under different angles of illumination [86]. The strong Bragg diffraction can be seen with the naked eye. (E) UV/vis extinction spectrum of a photonic crystal of gold/PNIPAm core/shell particles showing the typical localized surface plasmon resonance of spherical gold nanoparticles and a diffraction peak [86]. (C–E) Adapted with permission from [86]. Copyright 2011 Wiley.

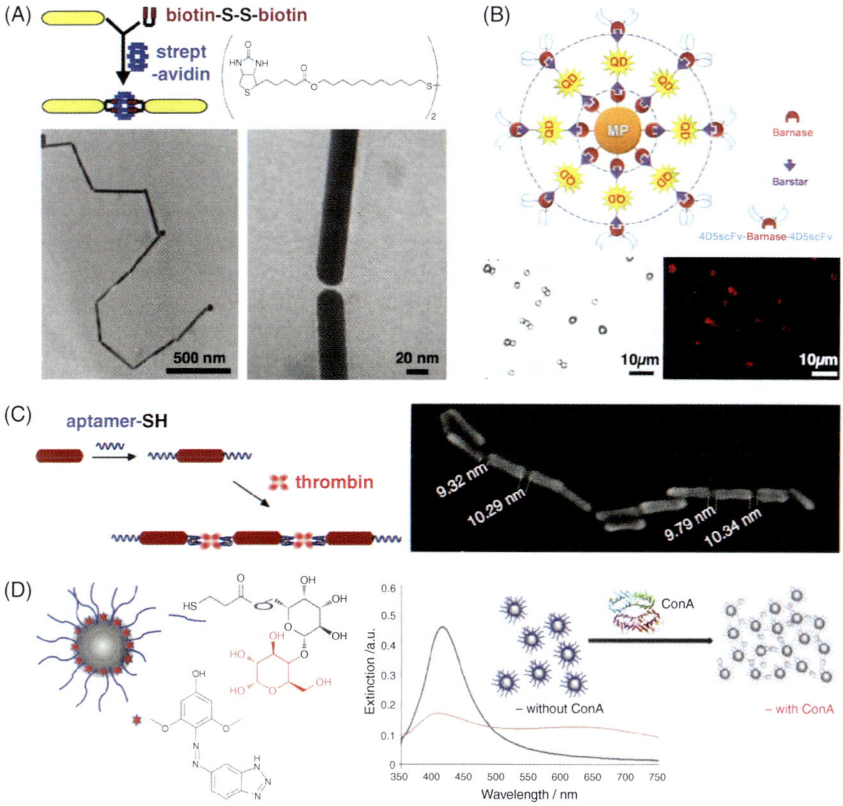

Figure 6.9 Examples of protein-mediated assembly of nanoparticles: (A) End-to-end assembly (linear aggregation) of gold nanorods tip-functionalized by biotin disulfide with streptavidin as linker. TEM images showing tip-to-tip aggregation of nanorod chains. Adapted with permission from [129]. Copyright 2003 ACS. (B) Superstructure of a magnetic core particle (MP), quantum dots (QD), and superfacial tags (4D5scFv-Bn) utilizing the key–lock type bio-recognition of two proteins (Barstar/Barnase). Bright-field (*left*) and fluorescence microscopy images (*right*) showing trifunctional nanoparticle assemblies. Adapted with permission from [130]. Copyright 2010 PNAS. (C) End-to-end assembly of gold nanorods tip-functionalized with a thiolatedaptamer with thrombin as linker. TEM image with selected tip-to-tip distances Adapted with permission from [131]. Copyright 2009 ACS. (D) Mixed monolayer assembly by aggregation of glyco-decorated nanoparticles using a protein (ConA) as linker. UV/vis extinction before and after the addition of ConA. Adapted with permission from [132]. Copyright 2013 RSC.

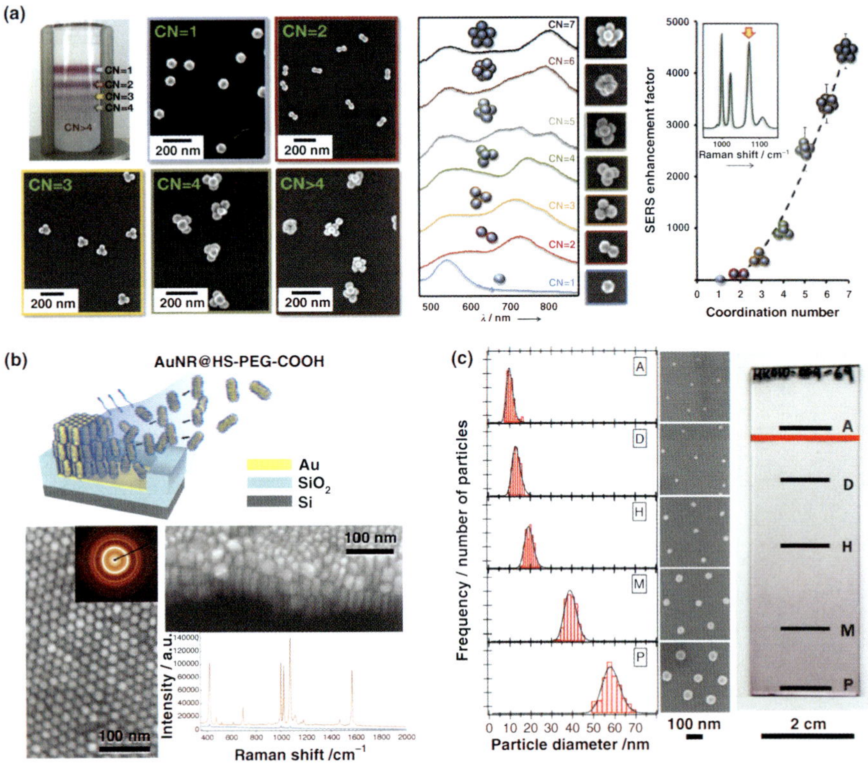

Figure 6.10 Application examples of assemblies of macromolecule-decorated nanoparticles: (A) Organized plasmonic nanoclusters assembled from gold nanoparticles with block copolymer shell. SEM images show clusters of discrete coordination numbers (CN), separated by density gradient centrifugation. Dark-field single-particle optical spectroscopy next to the corresponding clusters; SERS enhancement as a function of CNs normalize to the enhancement of a single particle. (Adapted with permission from [136]. Copyright 2012 Wiley.) (B) 3D arrays of standing gold nanorods, modified with thiol-PEO-carboxyl ligands assembled into micropatterned substrates. SEM images show top and side view of the hexagonally close packing of nanorods with interparticle gaps of 5 nm as evaluated from the transmission SAXS pattern (*see inset*). SERS enhancement was studied for benzenethiol (*see spectrum*). (Adapted with permission from [79]. Copyright 2012 Wiley.) (C) Plasmonic library based on substrate-supported gradiential plasmonic arrays with well-defined interparticle spacing: size distributions and SEM images of gold nanoparticle arrays on a macroscopic glass slide (*see photograph*). The gradual color change from nearly transparent to almost purple indicates the increasing particle sizes from top (9 nm) to bottom (57 nm) of the substrate. The monolayer assembly was achieved by spin-coating Au/PNIPAm core/shell colloids onto the substrate and subsequent overgrowth of the gold cores. (Adapted with permission from [137]. Copyright 2014 ACS.)

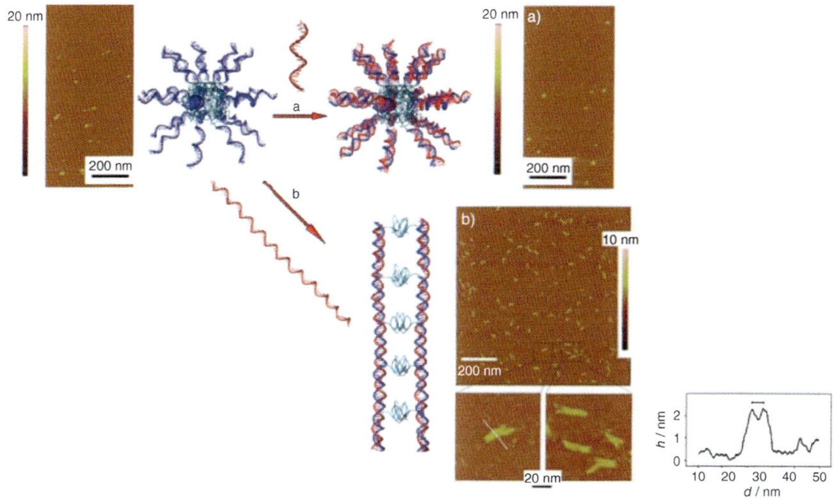

Figure 7.4 (a) Hybridization of complementary sequences does not affect the shape of the DNA-copolymer micelle. (b) Hybridization with an elongated complementary sequence results in the formation of rod-like structures [38]. Adapted with permission of Wiley-VCH.

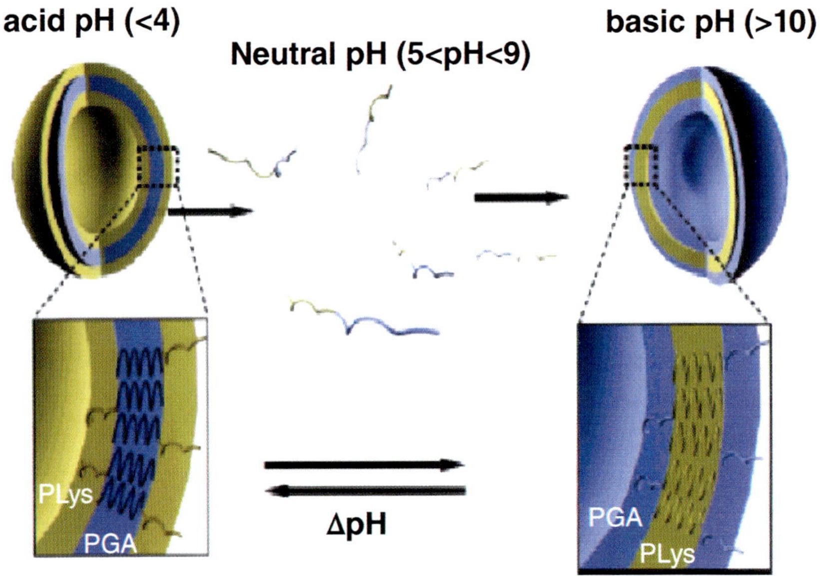

Figure 7.8 Schematic representation of the schizophrenic self-assembly into vesicular structures of PGA_{15}-b-$PLys_{15}$. Adapted with permission from American Chemical Society.

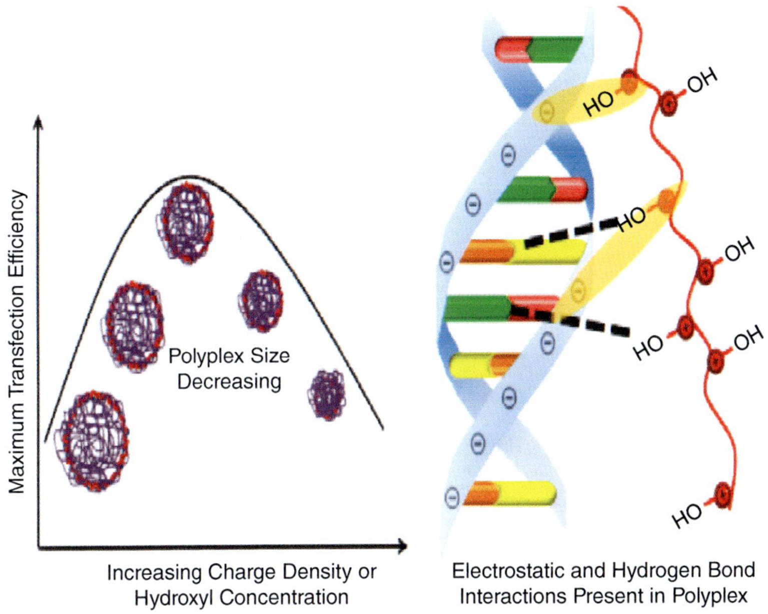

Figure 7.14 IPEC formed between a cationic polymer and condensed anionic plasmid DNA. Various factors including hydrogen bonding, impact the IPEC stability. Reprinted with the permission from [134].

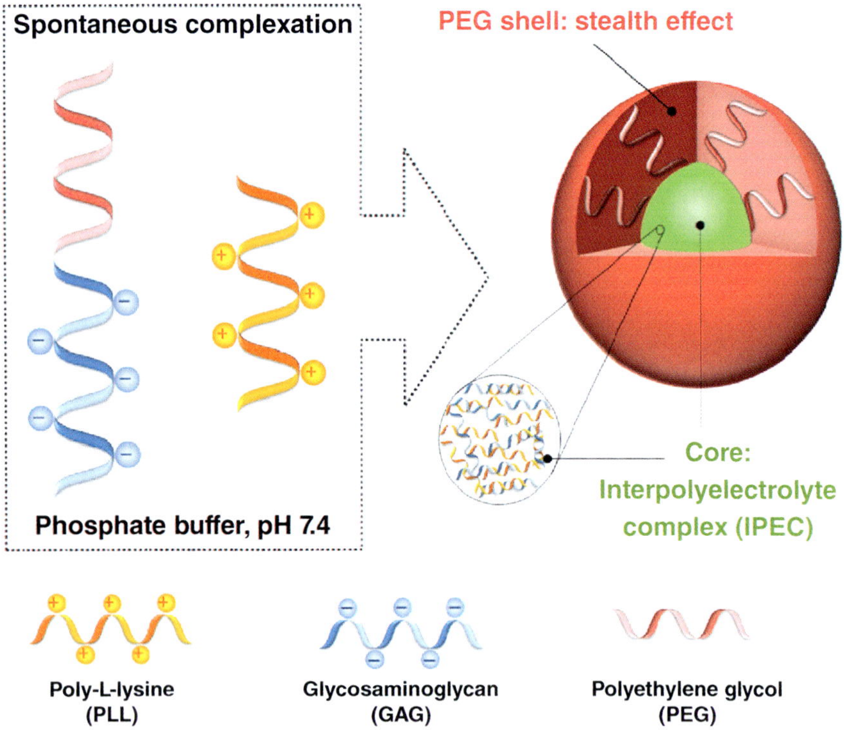

Figure 7.15 Schematic presentation of the complexation process between GAG-b-PEG and polycationic protein (poly-L-lysine (PLL)). Reprinted with the permission from [139].

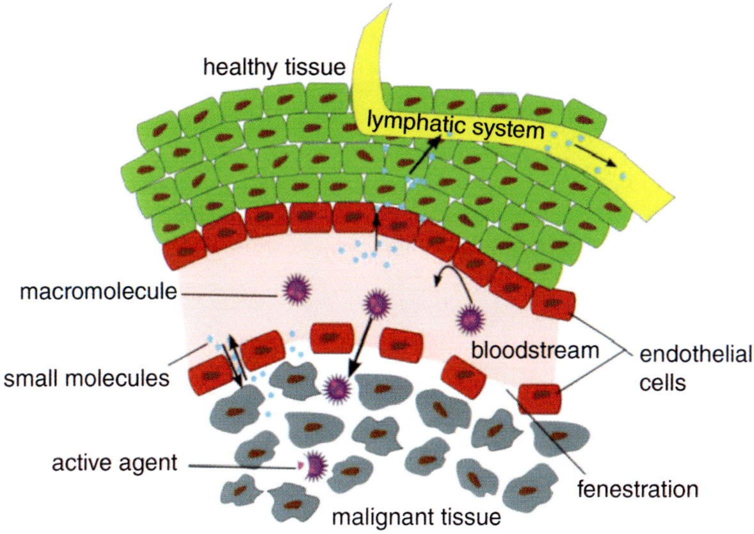

Figure 8.3 EPR effect [13]. Macromolecules and nanoparticles do not penetrate the compact endothelium but can enter the defective malignant tissue.

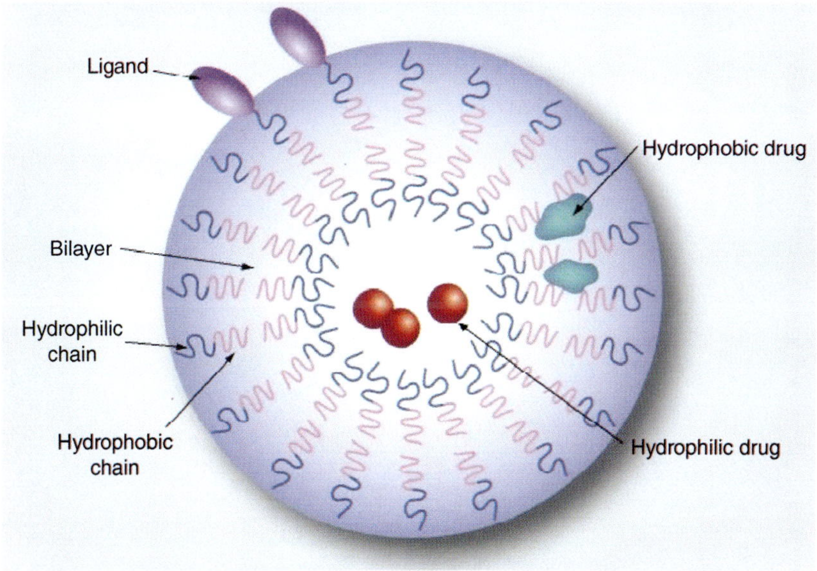

Figure 8.6 Structure of a polymersome with both a hydrophobic and hydrophilic drug incorporated [59].

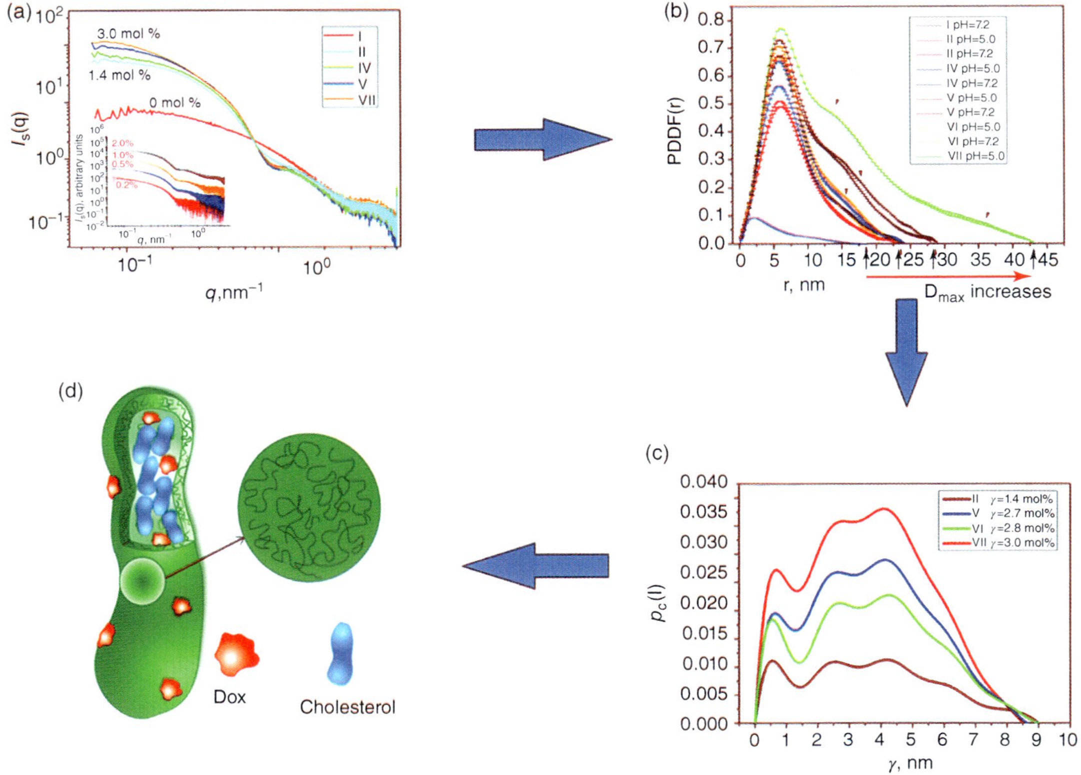

Figure 8.7 (a) SAXS intensity profiles for conjugates with different amounts of cholesterol: I (0 mol%)—red, II (1.4 mol%)—cyan; IV (2.3 mol%)—green, V (2.7 mol%)—blue, and VII (3.0 mol%)—black. (b) Pair-distance-distribution function for some conjugates. (c) Cross-sectional pair-distance distribution functions, $p_c(r)$, for some conjugates. (d) 3D structure of HPMA-Dox-cholesterol NP with bound Dox [69].

3

AMPHIPHILIC GRADIENT COPOLYMERS: SYNTHESIS AND SELF-ASSEMBLY IN AQUEOUS SOLUTION

ELISE DENIAU-LEJEUNE, OLGA BORISOVA, PETR ŠTĚPÁNEK, LAURENT BILLON, AND OLEG BORISOV

Institut des Sciences Analytiques et de Physico-Chimie pour l'Environnement et les Matériaux IPREM, Université de Pau & Pays de l'Adour, France; Department of Polymer Science, Moscow State University, Russia, and Institute of Macromolecular Chemistry AS CR, Praha, Czech Republic

3.1 INTRODUCTION

Although the term "gradient copolymer" was first proposed in 1977 [1], it was only first used in the literature in 1996 by Pakula and Matyjaszewski [2]. Since then, the study of synthesis and properties of these polymers and the exploration of their application have attracted significant attention [3–6].

Gradient copolymers are characterized by a gradual variation in local chemical composition (instantaneous fractions of A and B comonomers) along the polymer chain, from mostly A at one chain end to mostly B at the other end. They can be considered the intermediate copolymer architecture between block and random copolymers (Figure 3.1) [2, 3, 7–10]. Typical distributions of the comonomers along the chains for these three types of polymers are illustrated in Figure 3.1.

In a block copolymer the instantaneous local fraction of each type of monomers jumps stepwise from zero to unity at the point where the two homopolymer blocks are

Macromolecular Self-Assembly, First Edition. Edited by Laurent Billon and Oleg Borisov.

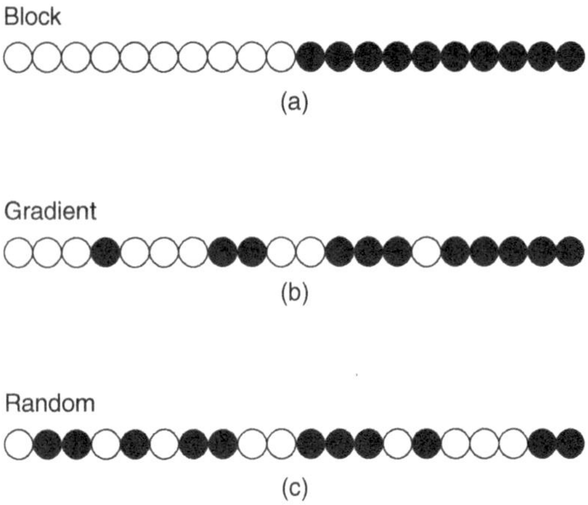

Figure 3.1 Schematic representation of the composition in (a) block, (b) gradient, and (c) statistical copolymers, in which the open circles denote monomer A and the closed circles monomer B.

bonded (Figure 3.2a). In a random copolymer the average fraction of each component can be regarded as uniform along the chain (Figure 3.2c). In a gradient copolymer, the local fraction of one of the monomer units increases continuously along the polymer chain. The shape of the gradient of local composition, that is, the probability to find comonomer unit of type A at a certain position along the chain, depends on the reactivity of the comonomers and on the copolymerization reaction conditions. The average local composition as a function of the monomer unit ranking number for the ideal case of constant gradient is shown in Figure 3.2b by a solid line. However, in the general case, various local composition profiles can be obtained, while random and block copolymers can be considered extreme cases of gradient copolymers (Figure 3.2).

Gradient copolymers with sufficiently sharp gradient profiles can be considered "fuzzy" block copolymers, and they may exhibit self-organization and nanostructuration both in bulk and in solution. The phase boundary in nanostructured morphologies is, however, more diffuse than that found in pure block copolymers [11]. Compared to block copolymers, the gradual variation in the instantaneous comonomer fraction along the polymer chain leads to weaker intra- and interchain repulsions [12–16] and thus accounts for its unique properties and interfacial behavior, especially the unique thermal properties of gradient copolymers in bulk. Gradient copolymers exhibit an unusually broad glass transition, whereas statistical copolymers exhibit one narrow glass transition at some well-defined Tg due to the absence of microphase separation. Block copolymers, however, present two discernable narrow glass transitions due to the presence of well-defined domains and thus the incompatibility of the two blocks [13, 17, 18]. Gradient copolymers can be used potentially in many fields; for example, as compatibilizers they help blend immiscible polymers by reducing the interfacial tension of each homopolymer domain [19, 20], reinforcing agents

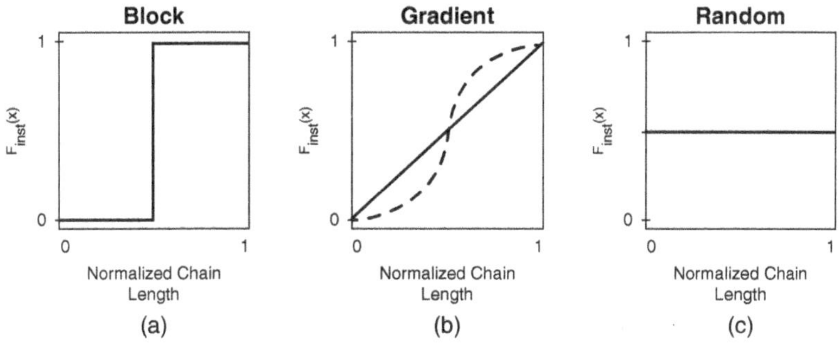

Figure 3.2 Schematic illustration of the instantaneous composition (F_{inst}) as a function of the normalized chain length for (a) block, (b) gradient (the solid line corresponds to the ideal case, the dotted line gives an example of a gradient characterized by an S-shape), and (c) statistical copolymers.

[21], damping materials [20, 22], thermoplastic elastomers [10], and biocompatible materials [23]. Moreover, as compared to the homologous block copolymer [24–26], gradient copolymers exhibit higher surface activity, which can be a useful property in many applications. The solution properties and self-assembling behavior of gradient copolymers in selective solvents are also markedly different from those of diblock copolymers and exhibit some special features [27, 28].

The self-assembly of block copolymers in selective solvents has been actively studied in the past decades because they can form diverse nanostructures (spherical or cylindrical micelles, polymersomes, etc.) [29–32]. The assembly of amphiphilic diblock copolymers in aqueous environment is driven by the hydrophobic attraction of the insoluble (associating) blocks and is counterbalanced by repulsions between the soluble (hydrophilic) blocks. The soluble forms hydrate (swollen) coronal domains, and the repulsive interactions between these domains provide aggregative stability to micellar solutions. The well-defined shape and narrow distributions with respect to the aggregation number in such assemblies are thus assured by the large surface tension at the sharp interface between the collapsed hydrophobic core and the highly hydrated coronae. The specific morphology of the copolymer nano-aggregates is determined by the balance of repulsive interactions in the corona and the conformational entropy losses of the core-forming insoluble blocks.

The ability of gradient copolymers to form well-defined nanostructures in selective solvents has remained controversial in the literature. A comprehensive study of their association in a solution of poly(acrylic acid-*grad*-styrene) (P(AA-*grad*-S)) copolymers [33] has demonstrated that, instead of micelles, aggregates of irregular shapes and wide-ranging sizes are distributed in the solution. This finding is in line with results of computer simulations [34], confirming that copolymers with smooth gradients cannot form well-defined nanostructures due to the absence of sharp interfaces between collapsed and solvated domains (absence of the mechanism of close association). Lefebvre *et al.* [13] found that the sizes of aggregates of the gradient copolymers are smaller than those of the corresponding block copolymers. In addition, they determined that a microphase separation cannot occur in bulk, though a cooperative intermolecular association can lead to the formation of well-defined nano-assemblies by copolymers whose compositions have a sharp intramolecular gradient resembling that of the block-copolymers.

This raises another possibility, that gradient copolymers can be upended by a soluble homopolymer block and result in so-called block-gradient copolymers. It has been recently demonstrated that block-gradient copolymers are capable of micellization under conditions where the gradient block is the associating block of the copolymer [35–37]. Moreover, in contrast to micelles of homologous diblock copolymers (with the same net composition), the block-gradient copolymers exhibit pronounced stimuli-responsive properties, which are attributed to the presence of a fraction of soluble monomer units in the associating gradient blocks. In this respect, the self-assembly behavior of block-gradient copolymers is remarkably similar to that of block-statistical copolymers [38–46]. This property in combination with the robustness of the one-pot synthesis makes block-gradient copolymers highly attractive for the development of smart stimuli-responsive materials.

3.2 SYNTHETIC STRATEGIES FOR THE PREPARATION OF GRADIENT COPOLYMERS

The existing synthesis methods for gradient copolymers were developed based on the criterion that their instantaneous composition, F_{inst}, or cumulative composition, F_{cum}, varies continuously as a function of the apparent normalized chain length (Figure 3.2b) [2, 7, 47, 48].

Cumulative and instantaneous compositions are calculated as [49]

$$F_{cum,1} = \frac{(\%conv)_1[M_1]_0}{(\%conv)_1[M_1]_0 + (\%conv)_2[M_2]_0} \tag{3.1}$$

$$F_{inst,1} = F_{cum,1} + (\%conv)\frac{\Delta F_{cum,1}}{\Delta(\%conv)} \tag{3.2}$$

Here (%conv) denotes the conversion of monomer and $[M]_0$ denotes the initial concentration of monomer (subscripts 1 and 2 denote monomer 1 and 2, respectively). In general, plotting composition versus chain length helps to visualize the compositional changes occurring over the length of the chain (Figure 3.2).

In order to achieve this continuous change in instantaneous composition, all chains must be initiated simultaneously, and must survive until the end of the polymerization. Therefore gradient copolymers cannot be prepared by conventional radical polymerization or by polycondensation. Living (ionic) or controlled radical polymerization (CRP) techniques, currently named reversible–deactivation radical polymerization (RDRP) [50], must be employed because the significant presence of chain-breaking reactions can lead to heterogeneity in both the chains' composition and molecular weight. In addition to these mechanistic requirements, synthesizing well-defined gradient copolymers also requires facile cross-propagation. This is very hard to fulfill in ionic copolymerizations because reactivities of monomers in ionic systems do not favor cross-propagation [51]. Only few examples of anionic polymerization [10, 52–55], cationic polymerization [56–64], and ring-opening metathesis polymerizations [12, 65, 66] have been employed to synthesize gradient copolymers.

In contrast, free radical polymerization abounds with examples of monomers that easily cross-propagate [67]. Therefore, for synthesizing a well-defined gradient copolymer, controlled radical copolymerizations (CRcoP) offer an attractive and powerful option. It is also worth noting that using CRcoP in a "one-step strategy" represents a significant advance. The following section will only focus on the preparation of gradient copolymer using CRcoP methods.

3.2.1 Preparation of Gradient Copolymers by Controlled Radical Copolymerization

Two experimental approaches have been reported to control the compositional gradient, namely the "spontaneous"—and the "forced"—gradient methods. The first is the use of batch copolymerizations in which a spontaneous gradient in instantaneous composition is formed based on the differences in the reactivity ratios of the comonomers and the concentrations of comonomers in the monomer feed. The second method is the use of semi-batch copolymerizations to form "forced" controlled gradients in instantaneous composition.

3.2.1.1 Batch Process The simplest approach to prepare gradient copolymers, called spontaneous gradient polymerization, is a batch copolymerization in which the comonomer composition change is promoted by the difference in reactivity of the comonomers [68]. In the absence of either azeotropic points or ideal random copolymerizations the monomer composition will change gradually with the monomer conversion because of the different reactivities of the monomers.

In a copolymerization of two monomers, M_1 and M_2, with reactivity ratios, r_1 and r_2, the relative rate of consumption of each monomer (df_1/df_2) is given by the Mayo–Lewis equation [69]:

$$\frac{df_1}{df_2} = \frac{(r_1f_1^2 + f_1f_2)}{(r_2f_2^2 + f_1f_2)} = \frac{F_1}{F_2} \tag{3.4}$$

in which f_1, f_2 are the mole fraction of M_1, M_2 in the feed solution and F_1/F_2 represents the instantaneous copolymer composition. In most cases, F_1/F_2 is different to f_1/f_2, resulting in composition drift given by the integrated copolymerization composition equation (3.5) (Skeist's equation [70]):

$$X = 1 - \left(\frac{f_1}{f_{1,0}}\right)^{\alpha} \left(\frac{f_2}{f_{2,0}}\right)^{\beta} \left(\frac{[f_0 - f_{az}]}{[f_1 - f_{az}]}\right)^{\gamma} \tag{3.5}$$

where

$$\alpha = \frac{r_2}{1-r_2},\quad \beta = \frac{r_1}{1-r_1},\quad \gamma = \frac{1-r_1r_2}{(1-r_1)(1-r_2)},\quad f_{az} = \frac{1-r_2}{2-r_1-r_2},$$

X represents the conversion, and $f_{1,0}$ the initial mole fraction of M_1 in the feed solution. As determined by this Skeist's equation [70] the composition along the polymer chain varies as a function of chain length.

To prepare a copolymer with a high gradient by batch polymerization, it is necessary to select monomer pairs for which the reactivity ratios differ, at least, by several times. The second condition for formation of gradient macromolecules under this regime is that high conversions close to ultimate values should be achieved, since only in this case will the head and tail of a molecule have appreciably different compositions.

In this subsection, we describe the synthesis of gradient copolymers using batch methods by the most relevant CRcoP processes used up to date, which are Atom-Transfer Radical Polymerization (ATRP), Nitroxide-Mediated (radical) Polymerization (NMP, currently named Aminoxyl-Mediated Radical Polymerization, AMRP [71]), and Reversible-Addition-Fragmentation chain-Transfer polymerization (RAFT).

3.2.1.1.1 ATRP Batch copolymerization using ATRP to prepare gradient copolymers has been shown to be successful for a number of monomer pairs [3, 4, 49, 72–96]. Matyjaszewski *et al.* mainly contributed to the ATRP synthesis of gradient copolymers using the batch process [3, 4, 11, 49, 72, 80, 83, 84, 86, 97]. Batch copolymerization was performed and focused on styrene-acrylates and methyl methacrylate-acrylates gradient copolymers using catalytic systems based on copper complexes with various multifunctional nitrogen-containing ligands. Spontaneous gradient copolymers with a continuous tapering along the copolymer chain could be obtained for these systems. The gradient structure of these copolymers was confirmed by an increase in the instantaneous composition of the less active monomer (acrylates) with conversion and a decrease of the one of the more active monomers (styrene, methyl methacrylate) while maintaining a constant number of polymer chains.

Gu *et al.* [87] synthesized gradient poly(methyl methacrylate-*grad-n*-butyl acrylate) (P(MMA-*grad-n*BA) copolymer chains from the surface of aluminum oxide particles by ATRP using the CuBr/PMDETA catalyst system. Here again, in analyzing the kinetics of copolymerization, it was observed that the controlled gradient copolymers obtained are MMA-rich near the surface and become progressively richer in *n*BA content on going further out. This observation has suggested more ways to synthesize novel ceramic/polymer architectures.

Indeed, the gradient copolymers of *n*BA/MMA were also synthesized by batch copolymerization in ATRP by Madruga *et al.* both in bulk and in benzonitrile solution [88, 98]. These studies were carried out on two copolymer series using CuCl/bpy as the catalyst systems with two different initiators at 100°C. A Nuclear Magnetic Resonance (NMR) technique was used to determine the copolymer microstructure, particularly the stereochemical arrangement of monomers in poly(MMA-*grad-n*BA) copolymers. The relative intensity dependence of the monomer triad fractions as a

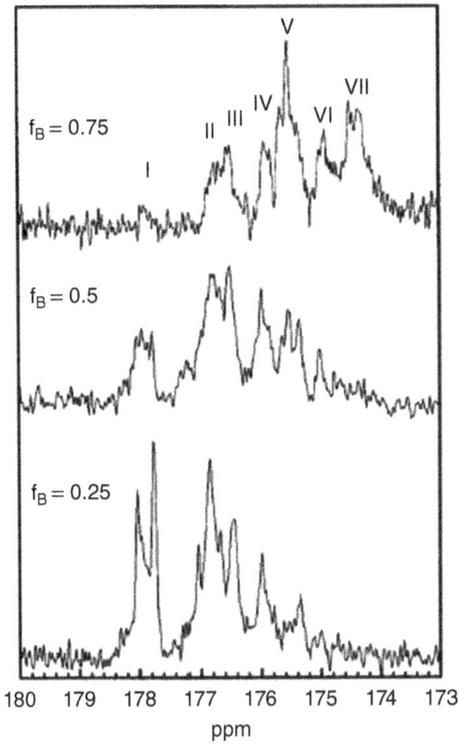

Figure 3.3 Expanded ^{13}C NMR spectra showing the carbonyl region of P(MMA-*grad*-*n*BA) copolymers prepared by ATRP at different feed compositions. Reprinted from [88].

function of conversion was studied by analysis of the carbonyl region of these copolymers, and seven distinguishable signals in the carbon carbonyl region were observed in the ^{13}C NMR spectra (Figure 3.3). Their relative intensities changed with the composition of the monomer in the copolymer and depended on both the diad tacticity (racemic (*r*) or meso (*m*)) and the proportion of the different diads and triads sequences. According to Aerdts's method [99], four of these signals (I, II, IV, and VI; see Figure 3.3) were assigned to sequences and configurations of MMA-centered triads. The three others (III, V, and VII) corresponded to *n*BA-centered triads. Later, Brar *et al.* [89, 90] reported the complete spectral assignments for the microstructure of spontaneous poly(MMA-*grad*-*n*BA) copolymers produced by ATRP with a complete assignment of the peaks by 2D NMR spectroscopy.

Because of the differences in reactivity of acrylates and methacrylates, many methacrylates-acrylates-based spontaneous gradient copolymer systems were synthesized by ATRP using the batch process. The ATRP copolymerization of two (meth)acrylate-based PEO monomers—that is, a methacrylate PEO macromonomer with a methyl end-group (PEOMeMA, DPPEO = 23, *r*(PEOMeMA) = 1.51 +/− 0.02) and a PEO acrylate with a phenyl end-group (PEOPhA, DPPEO = 4,

TABLE 3.1 Reactivity Ratios of Comonomers Reported

M_1	M_2	Process	Solvent	Initiator T, °C (control agent)	r_1	r_2	Ref.
			Methacrylate-acrylate				
AMA	*n*BA	ATRP	Toluene	MBrP, 100°C (PMDETA/CuBr)	2.58	0.51	[93]
HEMA	*t*BA		Bulk	EB*i*B, 50°C (PMDETA/CuCl)	4.49	0.1	[94]
PEOMeMA	PEOPhA		Anisole	EtBr*i*Bu, 80°C (Me6TREN/CuCl)	1.51	0.67	[92]
MMA	*n*BA	RAFT	Bulk	AIBN, 60°C (CDB)	1.7	0.2	[124]
MMA	MA	FRP	Bulk	AIBN, 50°C	2	0.40	[129]
AAEM	FDA	RAFT	TFT	AIBN, 65°C (ECEDB)	1.41	0.56	[126]
			Methacrylate-methacrylate				
BMA	DMAEMA	ATRP	Water/2-propanol	MPEG, 25°C (Bpy/CuBr)	1.24	1.07	[23]
	MMA		Bulk	EB*i*B, 80°C (PCPI/CuBr)	1.26	0.98	[140]
			Methacrylate—other				
MMA	DMA	NMP	1,4-dioxane	BB, 100°C (SG1)	2.36	0.330	[116]
MMA	VC2	FRP	Bulk	AIBN, 50°C	2	0.36	[129]
TFEMA	AA	RAFT	1,4-dioxane	AIBN, 70°C (DSCSP)	0.85	0.8	[159]

				Acrylate—other			
VBPDE	FDA	RAFT	TFT	AIBN, 65°C (ECEDB)	1.41	0.24	[126]
*n*BA	VAc	CMRP	Toluene	V-70, 30°C ($Co(acac)_2$,)	15	0.02	[167]
MA	VC2	FRP	water	IPP, 45°C	0.95	0.90	[127]
				Styrene-acrylate			
S	*n*BA	FRP	bulk	AIBN, 120°C	0.81	0.23	[105]
		FRP	benzene	AIBN, 50°C	0.698	0.164	[141]
	*t*BA	NMP	bulk	Benzoyl peroxide, 120°C (TEMPO)	0.9	0.3	[6]
	MA	NMP	bulk	MONAMS, 120°C (SG1)	0.89	0.22	[107]
	AcGalEA	RAFT	DMAc	AIBN, 90°C (MCPDT)	0.7	0.07	[160]
				Styrene-methacrylate			
S	MMA	FRP	bulk	Benzoyl peroxide, 60°C	0.52	0.46	[150]
				Styrene-acylic acid			
S	AA	RAFT	bulk	AIBN, 60°C (BPATC)	0.21	0.08	[68]
		NMP	1,4-dioxane	MONAMS, 120°C (SG1)	0.72	0.27	[77]
				BB, 120°C (SG1)	0.93	0.25	[171]
				PAA-SG1, 120°C (SG1)	0.22	0.88	[36]
				Styrene—other			
AS	S	FRP	bulk	AIBN, 60°C	1.218	0.887	[150]

r(PEOPhA) = 0.67 +/− 0.02) (all reactivity ratios reviewed in this chapter are summarized in Table 3.1)—was shown to lead to well-defined spontaneous gradient copolymers with narrow molecular weight distribution (1.2 < *Mw/Mn* < 1.4) [92]. Functional spontaneous gradient copolymers having a pendant, crosslinkable functional group of allyl methacrylate (AMA) and *n*-butyl acrylate (*n*BA) of P(AMA-*grad*-*n*BA), were also synthesized via ATRP polymerization [93]. The copolymer composition and microstructure, which demonstrated the gradient character, were analyzed. The experimental data agreed well with data calculated using the Mayo–Lewis terminal model and Bernoullian statistics, with monomer reactivity ratios of 2.58 +/− 0.37 and 0.51 +/− 0.05 for AMA and *n*BA, respectively (calculated by the Kelen–Tüdos method [100]). Functional spontaneous gradient copolymers of 2-hydroxyethylmethacrylate (HEMA) and *tert*-butyl acrylate (*t*BA) were also synthesized by ATRP in cyclohexanone at 50°C [94]. The monomer reactivity ratios *r*(HEMA) = 4.50 and *r*(*t*BA) = 0.21 were estimated by an extended Kelen–Tüdos method using copolymer composition data [101].

Acrylates derivatives often present similar reactivities, and therefore batch copolymerization often results in a statistical distribution. However, blocky-gradient copolymers of *t*BA and 2-ethylhexyl acrylate (2EHA) were prepared by Vidts *et al.* [91] by polymerizing first *t*BA alone until conversion occurred and then adding the 2EHA. This procedure obtained copolymers that consist of a P*t*BA segment, followed by a random structure of both monomers with a high 2EHA content.

Different system based on styrene-(meth)acrylates copolymerization were also widely studied by ATRP batch copolymerization. For example, well-defined amphiphilic gradient copolymers of poly(ethylene glycol) methyl ether methacrylate (PEGMA) and styrene (S) were successfully synthesized by ATRP [95]. The authors reported that the initial feed ratio of the two monomers had a significant effect on the copolymer's gradient composition. A gradient structure of the copolymer was seen when equimolar amounts of the comonomers were used, and the instantaneous composition changed continuously along the chain contour. The main reason for the formation of gradient copolymers might be considered to be the difference in relative reactivity between styrene and PEGMA due to their incompatibility.

The development of gradient copolymers is inhibited by the absence of a proper experimental technique to quantify the gradient quality, namely the gradient profile in the copolymer and the dispersity of this profile in the chains. However, the control over chain length and end-group functionality should be related to the control over gradient quality, since a perfect gradient implies a uniform distribution of the comonomers. In other words, the gradient quality could be measured indirectly by a measurement of the PDI and/or end-group functionality. Recently, Van Stanbeerge *et al.* have verified this hypothesis by an evaluation of the gradient quality using computational simulation [96]. The linear gradient quality and the control over chain length and end-group functionality for the copolymerization of acrylates, methacrylates, and styrenes via ATRP was evaluated by detailed kinetic Monte Carlo (kMC) simulations with explicit tracking of macromolecules. The kinetic Monte Carlo simulations confirmed that batch ATRP conditions allow those copolymers with good linear gradient qualities to be synthesized, and this was quantified by the value of a parameter ⟨GD⟩ that ranges between 0 and 1. This parameter denotes the average deviation for each polymer chain from the corresponding theoretically perfect linear gradient of the same chain length (e.g., for a uniform distribution of comonomers, the ⟨GD⟩ value is close to 0) and hence was reported to be a new copolymer property besides the average chain length, composition, and Đ. Moreover, for sufficiently high conversions and well-chosen comonomers, it was demonstrated that a strong correlation exists between ⟨GD⟩ and Đ. Thus the measured Đ was used as a first assessment of the linear gradient quality. The physical meaning of ⟨GD⟩ was illustrated by discussing copolymer compositions with varying linear gradient qualities. Under batch ATRP conditions, copolymerization of *n*BA and MMA could lead to a copolymer with good linear gradient quality if high conversion is reached and an appropriate catalytic system is selected, and thus lead to a low Đ. Further it was shown that this correlation between ⟨GD⟩ and Đ depends on the monomer reactivity ratios.

3.2.1.1.2 NMP Nitroxide-mediated polymerization (NMP) is a rather widespread method for preparing gradient copolymers but mostly using the semi-batch process [6], as will be discussed later in this chapter. Charleux *et al.* have mostly contributed to the NMP synthesis using the batch process [33, 76, 77, 102–104]. These authors reported on the batch copolymerization of styrene with *n*-butyl acrylate [76] and acrylic acid [33, 77, 102] using an alkoxyamine initiator based on the *N*-tertbutyl-*N*-(1-diethylphosphono-2,2-dimethylpropyl)nitroxide, SG1, in mini-emulsion and in 1,4-dioxane, respectively. As in these cases the reactivity differences of the comonomers were not so strong (r(S) = 0.81 and r(*n*BA)) = 0.23 at 120°C in bulk [105]; r(S) = 0.72 for styrene and r(AA) = 0.27 for acrylic acid at 120°C in 1,4-dioxane [77]) (Figure 3.4), the chains exhibited no or small composition drift depending on the initial comonomer composition, and hence a slightly pronounced gradient structure was obtained. The gradient microstructure was confirmed experimentally by liquid adsorption chromatography (LAC) analyses, performed at different conversions showing a net composition drift of the copolymer chains toward higher F_{AA} values with the increase in comonomer conversion, on the one hand, and a narrow composition distribution, on the other hand. A comprehensive characterization of the controlled feature of NMP copolymerization of styrene and *n*-butyl acrylate comonomers in bulk and in butyl acetate solution, using SG1 as the control agent and methyl propionate-SG1 (MONAMS) as the initiator, which led to a gradient microstructure, was obtained by automatic continuous online monitoring of polymerization reactions (ACOMP) [106].

NMP copolymerization of styrene and methylacrylate in bulk, with SG1 used as the control agent and MONAMS as the initiator, was also found to lead to a gradient distribution along the polymer chain [107]. Here too, a slightly pronounced gradient structure was obtained due to the value of the reactivity ratios of the comonomers (r(S) = 0.89 and r(MA) = 0.22) and was confirmed by a ^{1}H NMR study that showed the triad MA-MA-MA to increase with the time of polymerization.

The NMP of methacrylates does not exhibit any features of a controlled system, typically reaching a low conversion plateau in a short time period. This is due to an extensive irreversible termination of the propagating radicals caused by

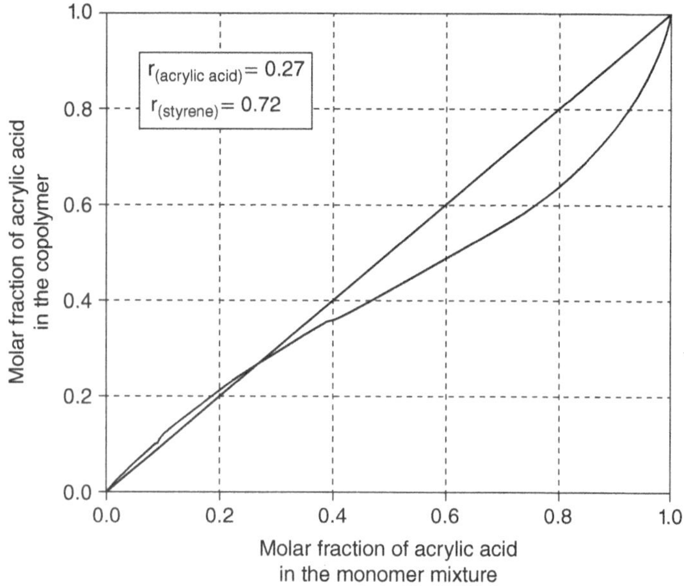

Figure 3.4 Reactivity ratios for acrylic acid and styrene determined in 1,4-dioxane at 120°C using MONAMS/SG1 system. Reprinted from [77].

the large activation–deactivation equilibrium constant and β-hydrogen transfer from the propagating radical to the nitroxide no matter which nitroxide is used (TEMPO [108, 109] or SG1 [110–114]). However, a strongly improved, controlled nitroxide-mediated polymerization of methyl methacrylate was attempted using a small amount of styrene [103, 115]. The controlled feature of this copolymerization was demonstrated both experimentally [103, 115] and theoretically [104] and was explained by the fact that due to the low value of the activation–deactivation equilibrium constant K of styrene ($K_S = 4 \times 10^{-10}$ mol.L^{-1} for SG1-mediated polymerization at 90°C), the average activation–deactivation equilibrium constant $<K>$ of the copolymerization of S with a large K value such as MMA ($K_{MMA} = 1 \times 10^{-7}$ mol.L^{-1} for SG1-mediated polymerization at 90°C) remains close to the lowest K value, and thus the presence of styrene promotes the reversible deactivation of the propagating radicals by SG1 and the impact of the side reactions is decreased. The controlled feature of the NMP of MMA was also observed by adding a small proportion of dimethyl acrylamide (DMA −10 mol%), which resulted in well-defined P(MMA-*grad*-DMA) gradient copolymers [116].

More recently, Kalugin *et al.* studied the optimum condition for the NMP copolymerization of styrene and *t*BA mediated by TEMPO [117]. It is known that, in contrast to styrene, *t*BA is little involved in the controlled radical polymerization mediated by TEMPO [118]. This result may be due to the high bond strength in the P*t*BA–TEMPO polymer adduct, that is, its inability to reinitiate polymerization at temperatures usual for this process (110–130°C). However, the copolymerization of *t*BA with styrene can proceed in a controlled manner to high conversions because the probability of occurrence of the styrene unit at the end of the propagating radical is much higher than that of the *tert*-butyl acrylate unit (*r*(S) = 0.9 and *r*(*t*BA) = 0.3) [6]. Therefore, at a sufficient content of styrene in the monomer mixture, the adducts with styrene–TEMPO end-groups would be the main products. An initial content of styrene of 30 and 15 mol% was found to lead to a controlled feature of the copolymerization as well as to a gradient structure. Indeed, at initial contents of styrene of 30 and 15 mol%, the changes in instantaneous compositions F_{inst} of the copolymer were 45 and 30%, respectively. These values correspond to a change in cumulative composition F_{cum} of the copolymer equal to 15% in both cases. However, the use of an additional source of initiation, obtained via the addition of cumenehydroperoxide (CHP), was necessary to increase the rate of polymerization in order to maintain the control feature of the copolymerization.

3.2.1.1.3 RAFT Another way of preparing gradient copolymers is by controlled radical polymerization with a Reversible Addition–Fragmentation chain Transfer (RAFT). Copolymerization conducted under RAFT conditions can generate copolymers with targeted compositions and controlled architectures [119–122]. Since RAFT is classed as a controlled radical polymerization [123], it is possible to obtain gradient copolymers, depending on the reactivity ratios. For instance, Rizzardo *et al.* studied the copolymerization of an equimolar mixture of MMA and *n*BA (*r*(MMA) = 1.7, *r*(*n*BA) = 0.2) in the presence of a dithio-compound as chain transfer agent (CTA) [124]. The values of the reactivity ratios clearly indicated that this copolymerization provides copolymer chains that should be rich in MMA at one end and rich in *n*BA at the other. The authors reported that when an initial mixture of MMA and *n*BA in the molar ratio of 1:0.91 was polymerized with AIBN in the presence of CTA at 60°C, the ratio of MMA to *n*BA in the polymer was 1:0.45 at 22% conversion and reached 1:0.80 at 93% conversion.

Harrison *et al.* [68] attempted RAFT copolymerizations using styrene (S) and acrylic acid (AA) to prepare block-like copolymers. The reactivity ratios of the S–AA system are unusually sensitive to temperature and solvent (due to the formation of hydrogen-bonded AA dimers). So under NMP conditions (120°C, 1,4-dioxane solution, *r*(S) = 0.72, *r*(AA) = 0.27) [77], reported above, relatively weak gradients were formed. The use of RAFT allowed Harrison *et al.* to conduct the polymerization at lower temperatures and in bulk, and the disparate reactivity that resulted in the comonomers (*r*(S) = 0.21 and *r*(AA) = 0.082) created block-like structures comprising a S-rich segment, a relatively short transitional segment, and a segment of AA homopolymer. The copolymers displayed properties characteristic of block copolymers, such as microphase separation (observed by AFM) and self-assembly in a selective solvent (measured by DLS).

The group of Lacroix-Desmazes widely used batch RAFT polymerization for the preparation of new functional copolymers [125, 126]. Rixens *et al.* studied the preparation of terpolymers containing vinylidenechloride (VC2), methyl acrylate (MA), and phosponated groups by RAFT polymerization using 1-(ethoxycarbonyl)-ethyl dithiobenzoateas CTA [125]. Such copolymer combining barrier properties (VC2), solubility in organic solvents (MA), and good adhesion and anticorrosion properties (phosphonated moiety) could be useful as additives in many applications, including paints and surface treatments. Two routes were tried: a direct RAFT copolymerization of a phosphonated methacrylate with VC2 and MA and the preparation of functional polymers by RAFT copolymerization of hydroxyethylacrylate (HEA) with VC2 and MA, followed by a chemical modification of the hydroxyl pendant groups of the HEA units with a phosphonatedepoxide. Two architectures were obtained that were predicted from the reactivity ratios of the monomers involved in RAFT polymerization. The gradient terpolymer was prepared in one step by RAFT terpolymerization of VC2 (1), MA (2), and a phosphonated methacrylate monomer (3) (reactivity ratios are $r_{1-2} = 0.90$ and $r_{2-1} = 0.95$ [127], $r_{1-3} = 0.36$ and $r_{3-1} = 2$ [128, 129], $r_{2-3} = 0.40$ and $r_{3-2} = 2$ [129, 130]), whereas the terpolymer poly(VC2-*co*-MA-*co*-HEA) was expected to be almost ideal statistical because the reactivity ratios of VC2/MA and MA/HEA and VC2/HEA are very similar. This was confirmed experimentally by a composition drift in the case of the gradient terpolymers, whereas no composition drift was observed for the statistical case.

Fluorinated gradient copolymers with complexing groups, Poly(1,1,2,2-tetrahydroperfluorodecyl acrylate-*co*-acetoacetoxyethyl methacrylate) (poly(FDA-*co*-AAEM)) and poly(1,1,2,2-tetrahydroperfluorodecyl acrylate-*co*-vinylbenzylphosphonic acid diethylester) (poly(FDA-*co*-VBPDE)), were prepared by Ribaut *et al.* using batch RAFT polymerization [126]. The RAFT polymerization was performed in α,α,α-trifluorotoluene using 1-(ethoxycarbonyl)-ethyl dithiobenzoate as CTA. Because the reactivity ratio of an acrylate monomer (r_1) is less than the reactivity ratio of the methacrylate and styrenic monomers (r_2), the AAEM (M_2) thus reacted faster than FDA (M_1) and the obtained copolymer was a gradient copolymer (calculated reactivity ratios were $r_2 = 1.61$, $r_1 = 0.56$). The same held for the copolymerization of FDA with the styrenic monomer VBPDE (M_2) (calculated reactivity ratios: $r_2 = 1.41$, $r_1 = 0.24$). The faster consumption of AAEM and VBPDE in comparison with FDA was confirmed experimentally by performing a kinetic analysis of the copolymerization. Such gradient copolymers were soluble in supercritical carbon dioxide ($scCO_2$) and thus could be advantageously synthesized directly in CO_2 medium, which allowed the use of organic solvents to be avoided, promoting a green chemistry. Moreover, the authors reported that these gradient copolymers exhibit remarkable CO_2 solubility that may allow them to be used as complexing surfactants in many applications, particularly for extraction of metal derivatives in $scCO_2$ in the context of the nuclear decontamination.

Other functional gradient copolymers have been prepared by batch RAFT polymerization. Gody *et al.* reported the first synthesis of glycopolymer architectures by the RAFT process [131]. In this work, biotin end-functionalized hydrophilic glycopolymers were synthesized using copolymerization of an acrylamide galactose derivative (GalAm) with *N*-acryloylmorpholine (NAM) in the presence of a biotin CTA. The individual monomer conversions indicated that the conversion of NAM is slightly higher than that of the acrylamide galactose derivative, so that a slight gradient microstructure can be obtained. Subsequently, this glycopolymer has been successfully used to prepare gold nanoparticles [132].

This reviewed literature shows that, under the correct conditions, gradient copolymers can be made in batch copolymerizations. Although the batch technique is quite simple, it suffers from four drawbacks [5]: (1) it is only applicable to strictly nonrandom copolymerization systems and will fail in close vicinity to azeotropic points; (2) in order to prepare a copolymer with a high gradient by batch polymerization, it is necessary to select monomer pairs for which the reactivity ratios differ, at a minimum, by several times and so limits the comonomers choice; (3) it cannot be used to cross the full compositional space from $F_1 = 0$ to $F_1 = 1$; and (4) it is virtually impossible to generate constant compositional gradients over the full range of the monomer's composition.

Thus more often it is better to use forced gradient techniques.

3.2.1.2 ***Semi-batch Process*** Semi-batch copolymerization, called "forced gradient copolymerization," is employed to tweak the gradient copolymers by continuously introducing one of the comonomers to the polymerization mixture during the reaction, often using syringe pump (Figure 3.5) [17, 97]. For comonomers having close reactivity ratios, the semi-batch process presents an interesting way to form a more significant compositional gradient along the chain.

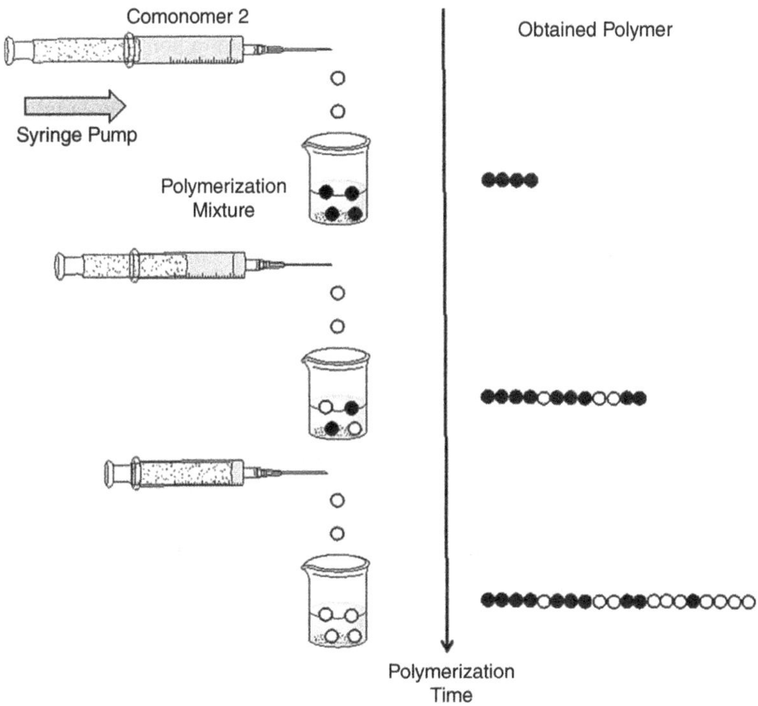

Figure 3.5 Schematic illustration of the synthetic approach used for gradient copolymers.

With the semi-batch process, the obtained gradient distribution is strictly related to the addition rate; in other words, it is related to the change of the molar fractions of the monomeric units in the mixture feed. Initially, the system is enriched in monomer A (this could be even pure monomer A). Then, over the entire process, monomer B is steadily added to the reaction system. Thus the composition of the monomer mixture changes during the growth of macromolecules. Under these conditions, an increase in the rate of addition of the second monomer is equivalent to an increase in a difference in reactivity ratios [65, 84, 133, 134], and thus the higher the rate, the sharper is the gradient of composition. The addition rate as well as the composition of the added mixture can be varied to allow for full control on the composition of the generated copolymer chain.

Two important parameters influence this technique of polymerization [135]. First, the choice of added monomer, as it is essential to add the most reactive monomer to force the incorporation of this monomer in the macromolecular chains to form a specific gradient molecular structure. Second, conversion in the gradient copolymerization system depends on the monomer addition rate. Indeed, if monomer addition is faster than monomer consumption, it will result in decreasing conversion.

As for the batch process, this section will focus on the synthesis of gradient copolymers using semi-batch methods by ATRP, NMP, and RAFT.

3.2.1.2.1 ATRP A semi-batch process using ATRP for the synthesis of gradient copolymers was widely studied by the group of Matyjaszewski [3, 23, 136–139]. These authors demonstrated the control of compositional gradient through the use of controlled monomer addition in the case of ATRP copolymerization of styrene and *n*-butyl acrylate [3, 136, 139]. They reported that, by slowly adding *n*-butyl acrylate to a batch polymerization of styrene, a much more significant gradient in composition was obtained, as opposed to the behavior of this monomer pair in a batch copolymerization. They also used semi-batch copolymerization for the formation of a gradient copolymer based on styrene and acrylonitrile that tend to polymerize in an alternating fashion when using batch process [3, 137].

Min *et al.* reported that Activator Generated by Electron Transfer (AGET) ATRP in mini-emulsion is a suitable process for the preparation of forced gradient copolymers [139]. In this case, all polymer chains were initiated within the mini-emulsion droplets exclusively and the mini-emulsion remained stable for the entire polymerization even when additional monomers fed to the reaction diffused to the active droplets, which is crucial for a successful forced gradient copolymerization in mini-emulsion. Several monomer pairs with a range of reactivity ratios were copolymerized: *n*BA/*t*BA, BMA/MMA, and *n*BA/S. The shape of the gradient along the polymer sequence was affected by the reactivity ratios of the comonomers, the molar ratios of the monomers, as well as the feed rate of the monomers. Furthermore, the gradient shape was influenced by the hydrophobicities of the comonomers. For example, in the case of *n*BA/*t*BA copolymerization, where reactivity ratios of both monomer were very close, the feeding process of *t*BA could be adjusted for the entire polymerization to generate a gradually decrease rate of addition. Indeed, for a rate of addition of 0.01 mL/min, *t*BA was consumed rather slowly at the beginning of the reaction and thus the cumulative composition of *t*BA didn't show evident increase until the normalized chain length reached 0.2 (Figure 3.6a). However, when the feeding rate changed from 0.015 to 0.005 mL/min, *t*BA started incorporating into the polymer chain almost immediately after the feeding process. The cumulative composition of *t*BA was already detectable when the normalized chain length reached 0.1 (Figure 3.6b). In the case of BMA/MMA copolymerization (reactivity ratios of BMA and MMA in ATRP at 80°C were reported to be *r*(BMA) = 1.26 and *r*(MMA) = 0.98 [140]), it was reported that when the feeding rate of BMA increased the cumulative composition of BMA in the polymer increased quickly, indicating that BMA rapidly diffused to mini-emulsion droplets through the aqueous phase and efficiently incorporated into the polymer chains (Figure 3.7). The authors reported that this observation could be due to the slight higher reactivity of BMA in the monomer pair but could be also due to the solubility of MMA

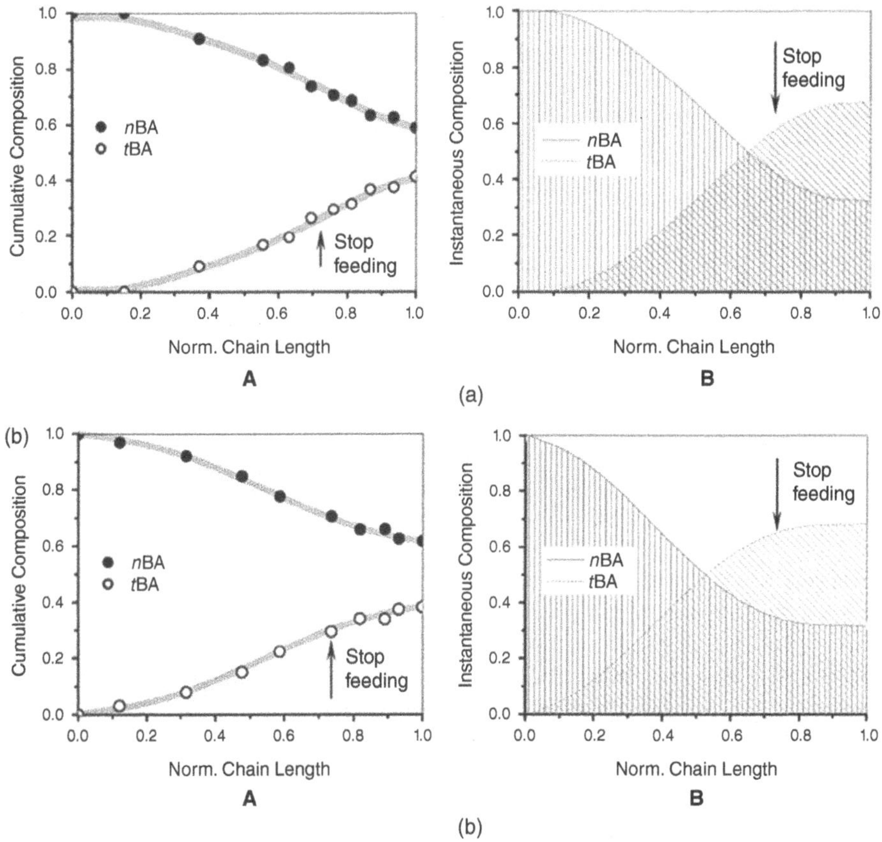

Figure 3.6 (A) Cumulative compositions and (B) instantaneous compositions of *n*BA and *t*BA in the forced gradient copolymer resulting from AGET ATRP in mini-emulsion with *t*BA as a fed monomer and a feeding rate of (a) 0.01 mL/min or (b) 0.015 to 0.005 mL/min for 200 min. Reprinted from *Min et al.* [139].

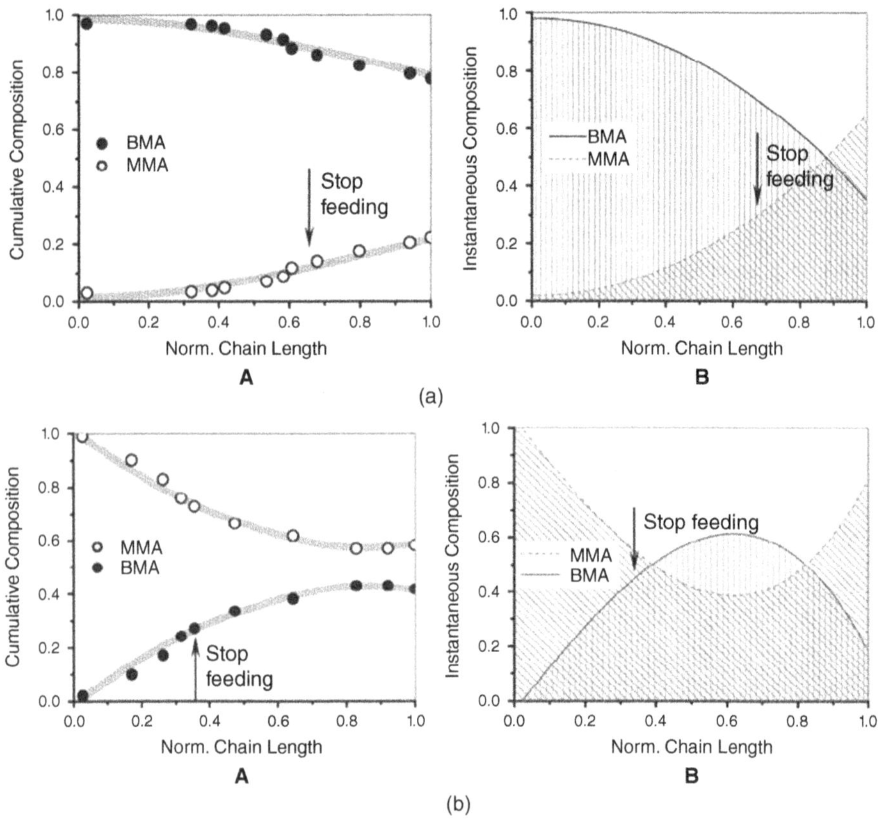

Figure 3.7 (A) Cumulative compositions and (B) instantaneous compositions of BMA and MMA in the forced gradient copolymer resulting from AGET ATRP in mini-emulsion with BMA as a fed monomer and the feeding rate for (a) 3 h or (b) 2 h. Reprinted from *Min et al.* [139].

in water, which could explain the slow diffusion of MMA out of the monomer droplet, while at the same time highly hydrophobic BMA was faster incorporated into the polymer chain. However, toward the end of the copolymerization, because a large fraction of the BMA had been consumed, the incorporation rate of BMA started to decrease. This reaction resulted in a pronounced gradient shape for the MMA/BMA copolymer with a MMA-rich head and tail and a BMA-rich midsection. The copolymerization of monomer pairs *n*BA/S could lead to spontaneous gradient copolymer using the batch process, as described previously, because the values of reactivity ratios are sufficiently different ($r(\text{S}) = 0.698$ and $r(n\text{BA}) = 0.164$ [141]). However, forced gradient copolymerization can still be used for this monomer pair to further adjust the shape of the gradient. The incorporation of S into the polymer chain was initially rapid, because S has a higher reactivity ratio than *n*BA in this monomer pair. When the normalized chain length increased to 0.25, the

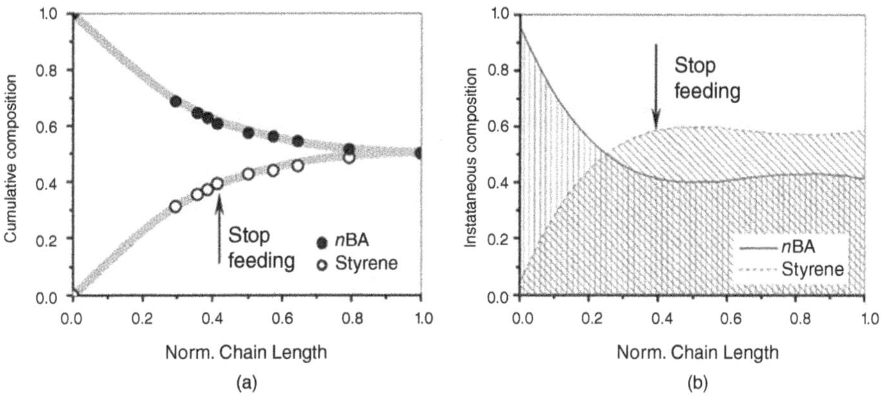

Figure 3.8 (A) Cumulative compositions and (B) instantaneous compositions of *n*BA and S in the forced gradient copolymer resulting from AGET ATRP in mini-emulsion with S as a fed monomer. Reprinted from *Min et al.* [139].

incorporation rate of S and *n*BA to the polymer had already reached the same level (Figure 3.8).

The semi-batch process for the ATRP copolymerization of S and *n*BA was also studied in toluene at 110°C [142]. In this study, random, gradient, and block copolymers were obtained depending on the additional strategy applied. Simultaneous adding of both monomers produced random copolymers of various compositions. A gradient copolymer was obtained by feeding *n*BA over 3 hours in the reaction mixture with pre-charged styrene. The opposite procedure of adding styrene to a *n*BA pre-charge resulted in a block rather than a gradient copolymer.

Wang *et al.* developed a mathematical model based on a semi-batch atom-transfer radical copolymerization (ATRcoP) [143]. Three copolymerization model systems with representative reactivity ratios ($r_1 = 0.8$, $r_2 = 0.2$; $r_1 = 2$, $r_2 = 0.5$; $r_1 = 10$, $r_2 = 0.1$) were theoretically investigated. The models predicted the polymerization rate, composition distribution, molecular weight, dispersity, and both number and weight fractions of dead chains. These models were used to design monomer-feeding profiles with constraint conditions derived from the targeted gradient structures. The strategy was experimentally confirmed via semi-batch ATRP copolymerization of MMA and *t*BMA [144], which demonstrated that the model-based computer-programmed semi-batch ATRP process can provide unprecedented control over the composition profile along copolymer chains. Later Wang *et al.* used the same strategy for synthesizing amphiphilic gradient copolymers of methacrylic acid (MAA) and methyl methacrylate (MMA) [28]. Semi-batch ATRcoP of *tert*-butyl methacrylate (*t*BMA) and MMA was used in the design of uniform, linear, triblock and diblock copolymer. Hydrolysis of *t*BMA units leaded to amphiphilic MAA/MMA copolymers. The experimental composition profiles were in an excellent agreement with the theoretical designs (Figure 3.9). The effect of the compositional profile (uniform, linear, triblock, and diblock) on the pH responsiveness of the copolymers in

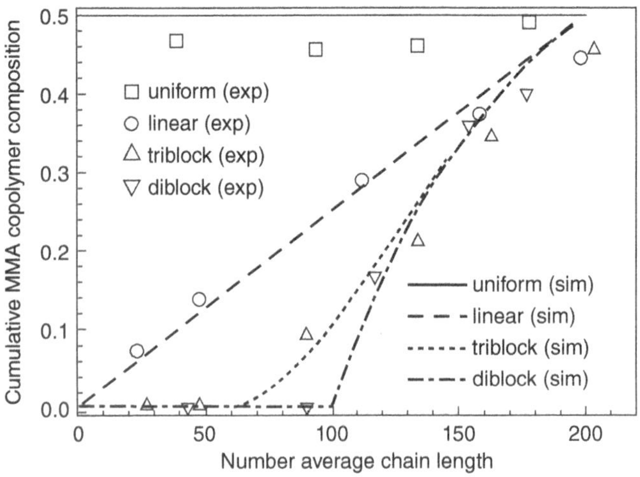

Figure 3.9 Cumulative MMA composition in the gradient *t*BMA/MMA copolymers as a function of the number-average chain length: the points show experimental data, and the lines show the theoretical designs of cumulative compositions profiles corresponding to the instantaneous composition profiles. Reprinted from [28].

aqueous solution was studied and demonstrated that the compositional profile could be a new parameter for fine-tuning the physicochemical properties of (meth)acrylic acid copolymers. However, the profile of the gradient was sharp (blocky structure) as it self-assembled in aqueous solution but still provided a useful parameter for tuning the micelle structure compared to uniform and pure block copolymers.

Recently, methyl methacrylate (MMA) and *tert*-butyle acrylate (*t*BA) gradient copolymers were synthesized by facile semi-batch copper(0)-mediated radical copolymerization technique in DMF ($Đ >$ 1.38) and DMSO ($Đ = 2.08$) at 25°C and in toluene ($Đ = 1.29$) at 120°C [145]. The addition of a small amount of reducing agent into the reaction system allowed the reaction to proceed in the oxygen tolerant system without loss of control and a decrease in total conversion.

To obtain gradient monomer sequences of methyl methacrylate (MMA) and *n*-butyl acrylate (*n*BA) in controlled fashion, an initiator for continuous activator regeneration (ICAR) ATRP was recently used [146]. The authors reported that the highest gradient quality for the dormant polymer molecules was obtained if a multi-component fed-batch procedure was selected. In this procedure, not only the conventional radical initiator but also the comonomer and deactivator are added in a fed-batch manner under starved-feed conditions.

3.2.1.2.2 NMP Nitroxide-mediated radical polymerization is widely used for synthesizing gradient copolymers by the semi-batch process. Frequently this method was used in preparing gradient copolymers of styrene and its derivatives, since they are involved in controlled radical polymerization under the action of common nitroxides.

The group of Torkelson have mostly contributed to the elaboration of gradient copolymers by NMP using the semi-batch process [8, 9, 15, 17, 19, 22, 78, 79, 147–149]. They synthesized first the high molecular weight (MW) gradient copolymers using styrene (S) and 4-acetoxystyrene (AS) using *N*-(R-methylbenzyloxy)-di-*tert*-butylamine) (A-T, first-generation nitroxides) [19, 78, 149] and (2,2,5-trimethyl-3-(1-phenylethoxy)-4-phenyl-3-azahexane) (second-generation nitroxides) [9, 17, 22] as unimolecular initiator. Because the reactivity ratios are close to the unity ($r(\text{S}) = 0.887$ and $r(\text{AS}) = 1.218$ [150]), such gradient copolymers can be obtained only by the semi-batch method. High MW can be obtained by working at a moderate temperature (around 100°C). Hydrolysis of S/AS gradient copolymers produced styrene/4-hydroxystyrene (S/HS) gradient copolymers. Compared to their homologous block and random copolymers, these S/HS gradient copolymers showed the unusual broad glass transition temperatures *Tg* [17] that were predicted by Lebfevre *et al.* [13] and Buzin *et al.* [151]. These authors later showed that, to control the strength of the composition gradient along the chain length, these glass transition breadths could be fine-tuned by the choice of monomer pairs with different degrees of incompatibility and by an adjustment of the semi-batch process [148]. These S/HS gradient copolymers showed very interesting physical properties. They were used as a compatibilizer of immiscible polystyrene (PS)/polycaprolactone (PCL) blends [19]. This blend system was reported to be special because of the ability of the hydroxyl functionalities on the HS units to form hydrogen bonds with the ester functionalities in the PCL. A similar NMP semi-batch procedure was also used for the synthesis of P(S-*grad*-MMA) and P(S-*grad*-*n*BMA) gradient copolymers [147]. In all these cases, the methacrylate comonomers were added to the styrene at a constant flow rate. The P(S-*grad*-MMA) and P(S-*grad*-*n*BMA) gradient copolymers were then used for compatibilization of the melt-processed polymer blends of PS/PMMA and PS/P*n*BMA. When the P(S-*grad*-MMA) gradient copolymers were later used for compatibilizing PS/PMMA nanoblends processed by SSSP [20], they showed, for the first time, that the industrially scalable approaches could yield compatibilized, thermoplastic nanoblends that do not rely on *in situ* polymerization or reactive processing. Moreover, these gradient copolymers could be modified with the organic dye Disperse Red 1 in order to form new functional microspheres that exhibit an enhancement of shear viscosity under application of an electric field [149].

Similar work was done for styrene (S) and 4-methylstyrene (MS) system [79]. At moderate-temperature NMP, S/MS random copolymers were synthesized using low conversion-batch reactions while S/MS gradient copolymers were synthesized using semi-batch reactions. The resulting gradient copolymers exhibited broad enthalpy recovery peaks (*Tg*), whereas the random copolymers exhibited narrower enthalpy recovery peaks.

More recently, P(S-*grad*-*n*BA) gradient copolymers were synthesized by the NMP semi-batch process [8, 9, 15]. Several P(S-*grad*-*n*BA) gradient copolymers were obtained by adding *n*BA to styrene at different rates (either at constant or increasing flow rates) using (2,2,5-trimethyl-3-(1-phenylethoxy)-4-phenyl-3-azahexane) as the initiator in order to study the influence of the obtained gradient sequences on their physical properties.

The group of Billon has widely contributed to the elaboration of gradient copolymers using semi-batch NMP. They synthesized styrene/*n*-butylacrylate (S/*n*BA) [135], *N,N*-dimethylacrylamide/*n*-butyl acrylate(DMA/*n*BA) [152], octadecyl acrylate/methyl acrylate(ODA/MA) [153], *n*-butyl acrylate/methyl methacrylate (*n*BA/MMA) [154], and, more recently, styrene/acrylic acid (S/AA) [35, 36] gradient copolymers. *N-tert*-butyl-*N*-(1-diethylphosphono-2,2-dimethylpropyl) nitroxide (SG1) was always used as the control agent and 2-methylaminoxypropionic-SG1 (MAMA, BlocBuilder®) as the initiator, and the NMP was performed either in bulk or in solution at temperatures varying from 112°C to 120°C depending on the studied system. Batch and semi-batch copolymerization methods were used to form controlled gradient copolymers, and the obtained structures were compared in order to understand the effect of the synthetic methodology on the molecular architecture.

In all cases, it was indicated that comparing to the batch process the semi-batch process was an excellent way to obtain gradient copolymers for monomers that do not present a strong difference between their reactivity ratios and that are difficult to obtain using the batch process. Moreover, the influence of the addition rate of monomer 2 (the most reactive of the two comonomers) on the incorporation of this monomer in the macromolecular chains, and consequently on the total molecular structure of copolymers, was clearly demonstrated. The authors noticed that an increase of the addition rate leads to an increase in the incorporation of the comonomer 2. The elaboration of gradient copolymers was confirmed by the variation of the instantaneous fraction of the comonomer in copolymers and also by the variation of the *Tg* measured by DSC and dynamic mechanical experiments. The physical properties of the obtained nanostructured polymeric materials indicated that they present new properties that are different from those of block or statistical copolymers. These results will be discussed in more details later in this chapter (see Section 3.3.3).

Semi-batch NMP process was also theoretically studied by Wang *et al.* using S/MMA system [155, 156]. A simulation framework based on kinetic Monte Carlo (kMC), which can predict the explicit sequence formed along each chain by tracking the growth of each individual chain, was reported [155]. The copolymerization of MMA/S gradient copolymers prepared by NMP was simulated using this kMC framework as an example. It was shown that kMC simulations are a powerful tool that can be used to predict the explicit sequence of copolymers and can support experimental efforts to precisely design the sequence lengths along copolymer chains. In another work, these authors used kMC simulations to investigate the influence of different synthesis conditions on the formation of monomer sequence along the chain in S/MMA forced gradient copolymerization with BlocBuilder® as the unimolecular initiator using NMP [156]. The synthesis conditions studied included the feed profile of the second monomer, the initiator concentration, and the addition of free nixtroxyl radicals. It was demonstrated that the monomer sequences along the copolymer chain were mainly determined by the feed profile of a monomer as well as that of nitroxyl radicals during the reaction. It was also found that the constant feed profile of a monomer, which is widely used in current syntheses of gradient copolymers, was not suitable for creating gradient copolymers. Through these kMC simulations, it was suggested that a variable feed profile of the monomer may be a better choice for creating gradient copolymers. Furthermore, it was found that only short segments could be formed near the head of the chains if free nitroxyl radicals were added to the system all at once at the beginning of the reaction. On the basis of these findings, a new feed profile of free nitroxyl radicals was proposed so as to generate a more pronounced structural gradient along the chains.

3.2.1.2.3 RAFT As already mentioned, RAFT polymerization has proved to be a reliable method to use in preparing functional gradient copolymers because of the versatility of the monomers and the wide tolerance of the reaction conditions [119, 122]. New functional gradient copolymers have been synthesized by the semi-batch method, such as fluorinated polymers [157–159] and glycopolymers [160]. For fluorinated copolymers, diverse monomers have been used. First, Inoue *et al.* [157] copolymerized 2-Methacryloyloxyethyl phosphorylcholine (MPC) and 2,2,2-trifluoroethyl methacrylate (TFEMA) using 4-Cyanopentanoic acid dithiobenzoate (CPD) as the chain transfer agent (CTA) and AIBN as the initiator. The reaction was performed at 70°C in a mixture of ethanol/1,4-dioxane, and the gradient structure was obtained by the addition of TFEMA at a controlled feed rate. These comonomers were selected because they have disparate polarity that could lead to a stable and distinct phase separation in solution or on a solid surface. Later Chen *et al.* [159] used the same method to synthesize an amphiphilic copolymer of acrylic acid (AA) and TFEMA using 2-{[(dodecylsulfanyl) carbonothioyl] sulfanyl} propanoic acid as CTA. More recently, Wang *et al.* [158] synthesized gradient copolymers of styrene (S) and fluorinated acrylate (F_6BA) by a RAFT emulsifier-free emulsion polymerization under the control of an amphiphilic RAFT agent. The gradient copolymer was used as a compatibilizer to improve the compatibility of the polystyrene/fluoropolymer blend. They employed the semi-batch process to form a forced gradient chain structure. Because the RAFT homopolymerization rate of F_6BA is faster than that of S, the strategy of adding F_6BA was chosen. The instantaneous composition of the S unit increased with the increase of polymeric chain length, while that of F_6BA units deceased, indicating that a gradient chain structure was formed.

It is worth noting that RAFT polymerizations have been reported as the first experimental examples of control over the copolymer composition distribution through semi-batch operations using styrene (S)/*n*-butyl acrylate (*n*BA) as a model system [133, 161]. The comonomer feeding rate profiles for the targeted distributions were designed from a previously developed computer model that was tailored for batch RAFT copolymerizations of S and *n*BA at different monomer compositions [143]. The semi-batch copolymerizations yielded precise copolymer products, and with their composition distributions exactly as targeted and the polymerization rate and molecular weight profiles as predicted by the model.

Later, a chain model and a sequence model were developed to calculate the sequence length and distributions for controlled radical copolymerization via RAFT chemistry in a semi-batch reactor using two sets of reactivity ratios that cover a wide range of monomer reactivities [162]. The authors reported that with a sequence model it was possible to control an additional level of molecular architecture: the monomer sequence, which provides additional insights into understanding the effect of sequence and distribution on the copolymer properties. More important, this model allowed the sequence distribution to be predicted and related to the polymer properties. This model, as developed using RAFT, could be used to simulate free radical copolymerization and could be easily extended to other CRP system such as ATRP and NMP.

In summary, by using semi-batch process, one can synthesize a larger range of possible composition gradients for a given monomer pair, since effects caused by differences in reactivity ratios can be compensated by appropriate rates and order of addition of one monomer to the other. That is, strong gradients of the monomer sequence can be formed by choosing appropriate experimental conditions, even though these monomers form small gradient or even random structures in batch processes.

3.2.2 Preparation of Block-Gradient Copolymers Using Controlled Radical Polymerization

Block copolymers can be efficiently prepared if the first block is synthesized with high end-group functionality, followed by an efficient cross-propagation after the addition of the second monomer. Thus the ability to prepare block copolymers is one of the most useful attributes of controlled radical polymerization.

Block copolymers can be prepared by two general routes: (1) generation of reactive sites on the chain end of the polymer A, followed by polymerization of monomer B (C, D, etc.) using the activated polymer as (macro) initiator, or (2) by coupling of preformed polymers A and B (C, D, etc.) having at one chain end functional groups able to undergo covalent bond formation.

In the case of block-gradient copolymers, a spontaneous or forced gradient microstructure in the gradient block could be obtained depending on the reactivity ratios of the comonomers, as explained previously.

Amphiphilic block-gradient copolymers of 2-dimethylaminoethyl methacrylate (DMAEMA) with *n*-butyl methacrylate (*n*BMA) were synthesized by ATRP in aqueous media using a MPEG macro initiator previously synthesized by ATRP [23]. The gradient block formed by DMAEMA and BMA was obtained by semi-batch ATRP copolymerization, although these monomers have very similar reactivities (r(DMAEMA) = 1.07, r(BMA) = 1.24). Because of these similar values of reactivity ratios, the synthesis required slow feeding of one monomer to the ATRP of the second one. The molecular weight distributions of the resulting block copolymers were fairly narrow. Later, Phan *et al.* [116] prepared diblock and triblock copolymers made of poly(*n*-butyl acrylate) (P*n*BA) as the central block and PMMA/PDMA gradient copolymer as the external block using NMP method. Mono- and difunctional P*n*BA blocks were first synthesized using bulk polymerization of *n*BA at 120°C mediated by the alkoxyamines MAMA-SG1 or DIAMS (SG1-based dialkoxyamine), respectively. A small excess of free SG1 was used to improve the control of the polymerization, leading to polymers with narrow molecular weight distributions. After purification steps, the resulting SG1-P*n*BA and SG1-P*n*BA-SG1 were used as macro initiators for the copolymerization of MMA and DMA in 1,4-dioxane at 100°C. As was shown above, in this copolymerization system, the experimental obtained reactivity ratios (*r*(MMA) = 2.36 and *r*(DMA) = 0.33), which were in good agreement with the literature [163], led us to think that a spontaneous gradient copolymer could be formed, with densely MMA segments incorporated at the beginning of the copolymerization. The copolymerization kinetics of MMA/DMA system initiated by the macro initiators exhibited more similar data than that initiated by MAMA-SG1. Recently, Steinhauer *et al.* [164] synthesized a triblock based on poly(hydroxyethyl acrylate) (P(HEA)) and poly(methoxyethylacrylate) (P(MEA)) blocks, and also P(HEA)-*b*-P(MEA)-*b*-P(HEA) and P(HEA)-*b*-P(HEA-*grad*-MEA)-*b*-P(HEA) via a RAFT process, using previously prepared HEA-based homopolymers with degrees of polymerization of DP*n* ~ 20–70 as the macro-chain transfer agents (MCTAs). After adjusting the addition rate of MEA to the preset HEA/MCTA ratio during the polymerization reaction, good control was attained over the microstructure of the middle block of the P(HEA)-*b*-P(HEA-*grad*-MEA)-*b*-P(HEA)copolymers. Such copolymers are interesting due to their thermos-responsive properties in aqueous solutions, revealing adjustable cloud point (CP) temperatures between 0°C and 80°C for copolymers with a pure block microstructure and between 0°C and 60°C for copolymers with gradient microstructures.

There remain two major drawbacks to the current block copolymer preparation techniques via controlled radical polymerization. The first is that two separate polymerization reactions are required to produce a diblock copolymer. These reactions, with their preparation and workup stages, are time-consuming, and it typically takes several days to produce a single block copolymer. Second, the nature of controlled radical polymerization, with low but unavoidable levels of termination, requires that polymerizations be stopped before reaching full conversion in order to limit the formation of dead polymer chains. This leads to lower reproducibility in their synthesis, as it is difficult to stop a polymerization at a predetermined target conversion. As a result, besides being time-consuming, the preparation of block copolymers by controlled radical polymerization can be expensive and poorly reproducible. Thus the development of a general, one-step strategy for the preparation of block-gradient copolymers presents a significant advance. Indeed, block-gradient copolymers can be prepared by a one-step strategy in two stages: (1) the polymerization of one monomer until a certain conversion and then (2) the addition of the second monomer in the reaction mixture (either straight or at a continuous rate) to obtain the spontaneous or forced gradient, respectively. The advantage of this method is that, depending on the reactivity ratio and on the order of the two stages (either pure block first and then gradient block or first gradient and then pure block), two different architectures can be obtained (Figure 3.10).

Amphiphilic block-gradient copolymers of poly(maleic anhydride-*grad*-styrene)-*b*-polystyrene were obtained for the first time using a one-step strategy by an NMP

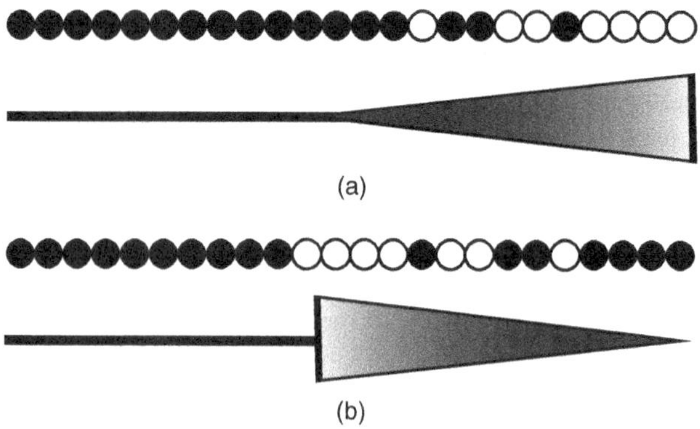

Figure 3.10 Schematic representation of the architectures of block-gradient copolymers in which the open circles denote monomer A and the closed circles monomer B.

copolymerization of a 9:1 mixture of styrene and maleic anhydride [165]. Thanks to the dramatically different reactivity ratios of the comonomers, a preferential incorporation of maleic anhydride occurred, and upon depletion of the maleic anhydride, the monomer feed consisted of essentially pure styrene. The authors reported that, in contrast to the normal free radical case in which pure homopolystyrene chains are produced at this stage, chain growth of a polystyrene block occurred in the controlled free radical case yielding poly(maleic anhydride-*grad*-styrene)-*b*-polystyrene block-gradient copolymers.

Later, Luo *et al.* [166] reported a RAFT copolymerization of MMA and S in mini-emulsion. Well-defined P(S-*co*-MMA) gradient copolymers and P(S-*co*-MMA)-*b*-PS block gradient copolymers were proved to be very efficiently synthesized by a RAFT mini-emulsion polymerization using 1-phenylethyl phenyldithioacetate (PEPDTA) as CTA. Because the reactivity ratios of MMA/S are r(MMA) = 0.46 and r(S) = 0.52 [150], an azeotropic point at f_1 = 0.471 could be predicted, and the copolymerization occurred without a drift in the monomer composition. When the molar fraction of MMA in the feed monomers was lowered below 0.471, MMA was consumed faster during the copolymerization, resulting in well-controlled P(MMA-*grad*-S) gradient copolymers. P(MMA-*grad*-S)-*b*-PS block-gradient copolymers were then obtained by the addition of pure styrene using a syringe pump at 8 mL/h. The comparison of the SEC chromatograme before and after the S addition demonstrated that the molecular weight rose after the addition, indicating the formation of a well-defined block-gradient copolymers.

Poly(n-butyl acrylate-*grad*-vinyl acetate)-*block*-P(vinyl acetate) (P(nBA-*grad*-VAc)-*b*-P(VAc) were also obtained using a one-step synthetic strategy involving Cobalt-Mediated Radical Polymerization (CMRP) [167]. Here again, the reactivity ratios of the comonomers were dramatically different (rnBA = 15 and rVAc = 0.02), and the copolymerization of nBA and VAc conducted at VAc 77% in the initial feed resulted in a block-gradient copolymer composed of a poly(*n*BA-grad-VAc) segment and a pure poly(VAc) segment. By using another comonomer with a higher reactivity ratio than VAc—that is, Poly(ethyleneglycol) acrylate (PEGA)—P(PEGA-*grad*-VAc) gradient copolymers and P(PEGA-*grad*-VAc)-*b*-PVAc block-gradient copolymers were also obtained by the same strategy [168]. The kinetic reveals that that both radicals add PEGA around 30 times faster than VAc. After PEGA was completely consumed, the CMR polymerization of the remaining VAc led to a diblock graft copolymer P(PEGA-*grad*-VAc)-*b*-PVAc in one step. The amphiphilic nature of these copolymers and their (quasi-)diblock structure led to the formation of well-defined micelles in water.

More recently, the synthesis of amphiphilic block-gradient copolymers containing 2-(2′, 3′, 4′, 6′-tetra-*O*-acetyl-*β*-D-galactosyloxy)ethylacrylate (AcGalEA) glycomonomer and styrene (S) by RAFT was reported for the first time [160]. The copolymerizations were performed in dimethylacetamide (DMAc) using S-methoxycarbonylphenylmethyldodecyltrithiocarbonate (MCPDT) as the control agent. The reactivity ratios of both monomers were estimated by *in situ* NMR experiments using Skeist's equation [70] along with the Fineman–Ross [169] and Kelen–Tudos [100] methods: r(AcGalEA) = 0.07 +/−0.01 and r(S) = 0.7 +/−0.1. Three different kinds of amphiphilic copolymers—namely statistical P(S-*co*-GalEA), block PS-*b*-PGalEA, and block-gradient PS-*b*-P(S-*grad*-GalEA)—were prepared by deacetylation of the AcGalEA moiety in order to study their ability to form honeycomb porous films by the breath figure process. The block copolymer was synthesized by a two-stage experiment, whereas the statistical and gradient copolymers were obtained in one-pot synthesis using batch and semi-batch processes, respectively. For an initial monomer fraction of 80% styrene, required for the formation of honeycomb film, the batch process led to statistical copolymers based on the azeotropic curve. Gradient copolymers were obtained using the semi-batch process. By this method, S was polymerized first and then the AcGalEA monomer was slowly added. The molar fraction of AcGalEA in the copolymer for each sample versus the normalized chain length evolved in an "S-shaped" curve, and this can be associated with the slight increase in the instantaneous AcGalEA molar fraction in the copolymer up to 12%.

Block copolymers of styrene (S) and acrylic acid (AA) present the most typical examples of amphiphilic ionic/hydrophobic block copolymers capable of self-assembly in aqueous solutions. The synthesis of di- and triblock gradient copolymers of PAA-*b*-P(AA-*grad*-S) and P(AA-*grad*-S)-*b*-PAA-*b*-P(AA-*grad*-S) was developed by Borisova *et al.* using NMP in a one-step strategy [35, 36]. The reaction was performed in dioxane at 120°C in the presence of alkoxyamine, 2-methylaminoxypropionic-SG1 (BlocBuilder® or MAMA) as the initiator and a slight excess of *N-tert*-butyl-(1-diethyl-phosphono-2,2-dimethylpropyl) nitroxide SG1 as the counter radical in order to avoid side reactions and to ensure a good control of the chains' growth. The procedure consisted of two stages: in the first stage the AA was polymerized for 4 hours, to obtain a mono- or bifunctional macro initiator to use for the second reaction stage. At the second stage S was added stepwise to the reaction mixture in the amount equal to the residual amount of

AA, and the reaction was continued. The gradient sequence of the P(AA-*grad*-S) block(s) with increasing instantaneous fraction of styrene monomer units toward the end(s) was confirmed by ^{1}H NMR. This gradient type was not expected from the reactivity ratios determined earlier for this system (rS = 0.93 and rAA = 0.25; BlocBuilder/SG1 at 120°C in 1,4-dioxane) [170]. This gradient sequence arose because of the increase in the apparent reactivity ratio of AA in the polymerization process due to the PAA macro-initiator (rS = 0.22, rAA = 0.88). The increase in the apparent activity of AA monomers was explained by the "boot-strap" effect [37, 171–177], that is, by enrichment of the vicinity of the reaction center, the AA monomers were preferentially adsorbed by the PAA chain. This effect was reported to be of practical importance for the synthesis of block-gradient copolymers. For example, if the controlled radical copolymerization of styrene with AA is performed in 1,4-dioxane in the presence of a low molecular mass initiator, copolymerization will produce a copolymer in which the head is enriched with styrene units and the tail is enriched with AA units. However, if the functionalized PAA is taken as an initiator, the reaction gives rise to the block-gradient copolymer, in which the content of styrene gradually increases to the chain's end. The self-assembly of the obtained amphiphilic ionic gradient copolymers in water was further characterized, as will be discussed later in this chapter.

3.3 SELF-ASSEMBLY

3.3.1 Gradient Copolymers

The association of poly(acrylic acid-*grad*-styrene) (P(AA-*grad*-S)) gradient copolymers with ~70% of AA in aqueous solution has been studied by combination of static and dynamic light scattering methods [36]. It was found that, in contrast to homologous diblock PAA-*b*-PS copolymers, the gradient copolymers do not undergo cooperative association into micelles and that the solution contains unimers coexisting with large and loose aggregates. At the same time it was demonstrated that the gradient copolymers exhibit sufficiently high surface activity.

A comprehensive theoretical study of self-assembly of symmetric gradient copolymers with variable shape of compositional gradient was performed using Monte Carlo simulations [37]. It was demonstrated that selective solvent copolymers with sharp gradient undergo self-assembly following the closed association mechanism. The micelle-like core-corona structures with a narrow size distribution that result are similar to conventional diblock copolymer micelles that arise in a solution of gradient copolymers beyond the CMC threshold. However, copolymers with smooth gradients associate and form aggregates of ill-defined shape and wide size distribution, which is a monotonically decreasing function of the number of chains in the aggregate. This feature is typical for the open (noncooperative) association mechanism. At high copolymer concentrations, ordered arrays of micelles are observed in a solution of copolymers with sharp gradients of composition whereas copolymers with smooth gradients form gel-like structures.

3.3.2 Diblock-Gradient Copolymers

In contrast to symmetric gradient copolymers, the diblock-gradient copolymers form a gradient block related to a pure solvophilic block and exhibit cooperative associations in selective solvents that give rise to well-defined nano-aggregates. In this case the poor solubility of the gradient block is the driving force behind the association of the block-gradient copolymers. Moreover, the presence of a (small) fraction of solvophilic monomer units in the gradient block brings to the self-assembly process a feature that ensures both its reversibility and stability.

The pH-controlled self-assembly of amphiphilic ionic block-gradient copolymers of styrene and acrylic acid, P(AA-*grad*-S)-*b*-AA, in dilute aqueous solutions was studied in reference [36] by combination of DLS, SANS, and TEM. The molar fraction of hydrophobic S monomers in the gradient blocks was varied in the range 40–50%, whereas the molar fraction of hydrophilic AA monomers in the whole block-gradient copolymer was ~70%.

The DLS titration experiments indicated that the apparent hydrodynamic radius of the solute particles remains virtually constant (~2 nm) in the alkaline conditions ($pH > 7–7.5$), which corresponds to molecular solution of block-gradient copolymers with mostly deprotonated (ionized) AA functions. A decrease in pH below $pH = 7$ provokes a monotonous increase in the hydrodynamic radius of the particles that can be interpreted to result from the formation of finite size (7–8 nm) aggregates followed by their coalescence into larger aggregates and eventually, below $pH = 4$, by macrophase separation. Interestingly, the onset of association caused by partial protonation (deionization) of AA functions in both P(AA-*grad*-S) and PAA blocks occurs at pH higher than pK_a for the AA monomers. This is a consequence of shift of local pH inside the aggregates toward lower values compared to the pH in the surrounding solution. The coupling between association and ionization is an inherent feature for amphiphilic ionic pH-sensitive block copolymers [178, 179].

The most remarkable feature of the pH-controlled assembly of P(AA-*grad*-S)-*b*-PAA copolymers is its perfect reversibility in cycles of pH variation (with the accuracy of effects caused by increasing ionic strength of the medium). This behavior is strikingly different from irreversible association of PAA-*b*-PS block copolymers with the same net composition: once formed (e.g., upon dialysis from the common for both blocks solvent against water), the micelles of the block-copolymers cannot be re-arranged upon action of such external stimuli as pH or ionic strength due to the glassy state of the solvent-free PS core. On the contrary, presence of 50–60% of AA units in the gradient blocks does not prevent their association at neutral or slightly acidic pH but ensures the persistence of the dynamic character of this association, possibly due to larger residual amount of water in the styrene-rich core domains formed by the gradient blocks.

The SANS experiments performed in solutions of P(AA-*grad*-S)-*b*-PAA in D_2O confirm the effect of pH-triggered reversible association. The typical SANS spectra for solution of P(AA-*grad*-S)-*b*-PAA copolymer obtained at different pH are presented in Figure 3.11.

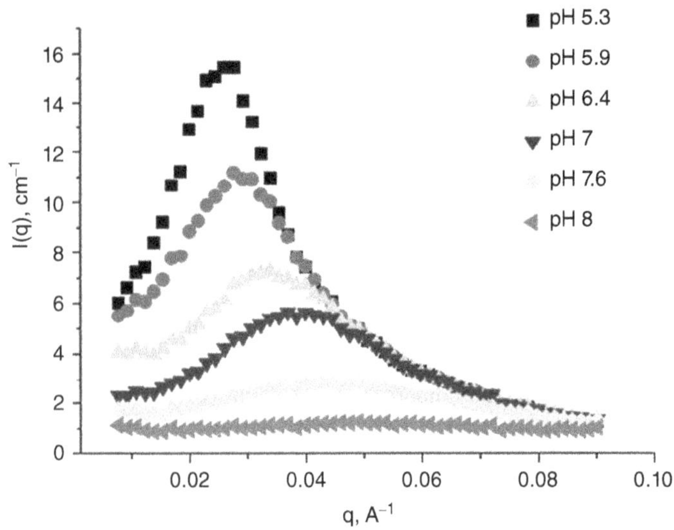

Figure 3.11 SANS spectra from 10 w% solution of P(AA_{24}-grad-S_{46})-*b*-PAA_{100} copolymer at varied pH.

As can be seen in Figure 3.11, the scattering curves exhibit a pronounced correlation peak at pH < 7.5. Upon lowering of the pH the magnitude of the peak increases, whereas the peak position is displaced to lower q. This behavior should be interpreted in terms of increasing repulsion between aggregates of growing size (increasing aggregation number) and concomitant increase of the average interparticle distance, which is in a line with the DLS results on pH-controlled association.

Another remarkable feature of the SANS spectra presented in Figure 3.11 is the presence of only single (correlation) peak, whereas no signatures of the form-factor can be observed. This points to a wide size and/or shape distribution of the scattering particles. In reference [36], the SANS curves were successfully fitted by using superposition of the form-factor of polydisperse spheres and Percus–Yevick structure factor.

Similar to DLS-titration curves, the SANS spectra exhibit complete reversibility in cycles of pH variation, thus confirming the dynamic (equilibrium) nature of the block-gradient copolymers assembly.

Finally, the TEM images (Figure 3.12) provide an unambiguous evidence of self-assembly of the block gradient copolymers P(AA-*grad*-S)-*b*-PAA into spherical micelles with a wide size distribution, and this explains both relative smoothness of the DLS titration curves and absence of any characteristic features in the SANS spectra in the $q > q_{max}$ range.

This broad size distribution of the aggregates points to relatively low cooperativity of the association of block-gradient copolymers that may be explained by low surface tension at the diffuse interface between the P(AA-*grad*-S) core and PAA corona domains.

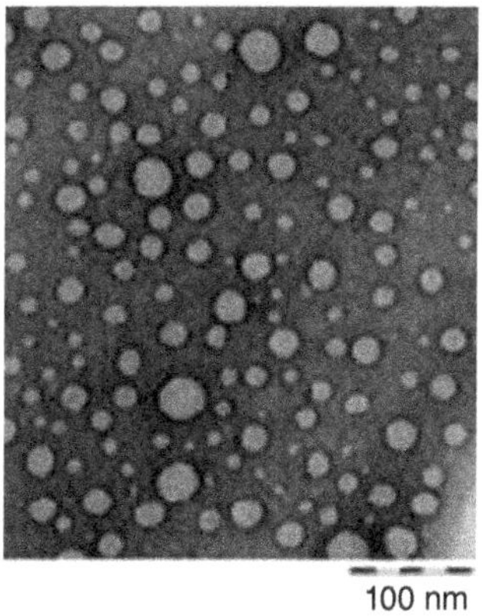

Figure 3.12 TEM image of micelles formed P(AA_{24}-grad-S_{46})-*b*-PAA_{100} at pH = 6. The light gray spots correspond to contrasted styrene-rich domains.

3.3.3 Triblock-Gradient Copolymers

Triblock gradient copolymers P(AA-*grad*-S)-*b*-PAA-*b*-P(AA-*grad*-S) comprise a central hydrophilic PAA block terminated at both ends with amphiphilic gradient P(AA-*grad*-S) blocks. The pH-triggered association of the terminal blocks governs the assembly of the triblock-gradient copolymers in aqueous solutions that was studied in reference [35]. At low polymer concentration, this assembly leads to formation of flower-like spherical micelles. In such micelles both gradient blocks are incorporated in the styrene-rich core domains, whereas the central PAA blocks form loops that constitute the hydrated corona of the micelles. These flower-like micelles resemble the star-like micelles formed by the P(AA-*grad*-S)-*b*-PAA diblock gradient copolymers. However, at the higher copolymer concentration, the second association mechanism comes into play: The bridging attraction emerges due to the possibility of exchange of terminal association blocks between neighboring hydrophobic domains, namely the conversion of some loops into bridges that is accompanied by the entropy gain on the order of thermal energy per bridge [180].

Therefore concentrated solution of triblock-gradient copolymers can be envisioned as an array of partially interpenetrating flower-like micells connected by bridges, namely representing a physical gel. Hence one can expect that, upon variation in pH, a reversible sol-to-gel transition can occur in semi-dilute solution of triblock gradient P(AA-*grad*-S)-*b*-PAA-*b*-P(AA-*grad*-S) copolymers. This transition is manifested in specific pH-dependent rheological properties of solution of the triblock-gradient copolymers as compared to those of homologous diblock-gradient copolymers. Indeed, it was demonstrated [35] that at high pH both behave as Newtonian liquids with zero-shear viscosity, which is (at 10 w% copolymer concentration) 5 to 7 larger than that of pure water. However, a decrease in pH leads to a sharp increase of viscosity (by more than 4 orders of magnitude) in a narrow range of the pH-variation in the case of triblock-gradient copolymer. Remarkably, only a

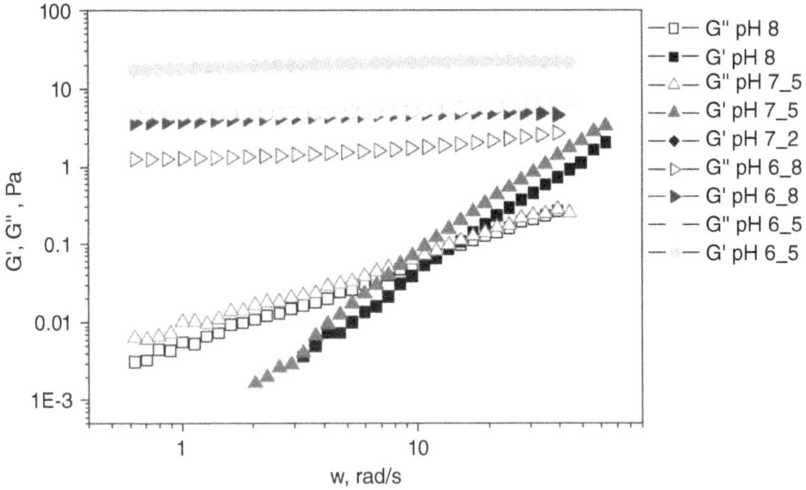

Figure 3.13 The frequency dependences of the storage and loss moduli in aqueous solution of P(AA_4-*grad*-S_{24})-*b*-PAA_{140}-*b*-P(AA_4-*grad*-S_{24}) copolymer at different pH-values.

moderate increase in viscosity was observed in the same pH-range in the solution of a homologous diblock gradient copolymer. This finding points unambiguously that the much stronger increase in viscosity in the solution of triblock-gradient copolymers is due to formation of bridges interconnecting the hydrophobic domain. Clearly, such bridging does not occur in the solution of homologous diblock-gradient copolymers.

The reversible pH-triggered sol-to-gel transition is also manifested in the viscoelastic properties of the solution of triblock-gradient copolymers. The analysis of the frequency dependences of the storage and loss moduli in Figure 3.13 shows that at high pH both G' and G'' are increasing functions of frequency, $G'' > G'$ and $G'' < G'$ at low and high frequency, respectively. This kind of the frequency dependence of the dynamic moduli is typical for viscous liquids. A decrease in pH results in a pronounced increase in the magnitude of both moduli, and they become independent of frequency, G' is larger than G'' by the order of magnitude. This behavior is characteristic for strong gels.

Hence solutions of triblock-gradient copolymer undergo a reversible pH-triggered sol-to-gel transition, which is manifested in dramatic changes in their viscoelastic behavior.

3.4 CONCLUSION AND OUTLOOK

There are many synthetic strategies that give rise to gradient or block-gradient copolymers. In this review, we have reported the methods employed to form well-defined gradient or block-gradient copolymers. Using controlled radical copolymerizations, such as ATRP, NMP, and RAFT, offers a very attractive and powerful option, especially given the one-step strategy used to form block-gradient copolymers. Two experimental approaches have been reported that prove to control the compositional gradient, namely these are the spontaneous and the forced gradient methods. Although the batch technique is quite simple, it is limited by the selected comonomers and by the intrinsic reactivity ratio values. In fact the forced gradient techniques are in many cases more appropriate in fine-tuning first the shape of the gradient profile in the copolymers and then their self-assemblies in solutions.

Amphiphilic ionic/hydrophobic block-gradient copolymers demonstrate a stimuli-induced reversibility of self-assembly in aqueous solution. In this respect their behavior is strikingly different from that of homologous diblock copolymers: the latter associate under the same conditions into micelles with “frozen” hydrophobic cores that cannot change their aggregation state in response to external stimuli. In contrast, gradient copolymer blocks “alone” form polydisperse aggregates of ill-defined shape or do not associate at all. The dynamic nature of the nano-assemblies of block-gradient copolymers is inherently related to a relatively weak cooperativity of association and, as a result, a wide micelle size distribution. Remarkably, up to now there have been no literature reports on other morphologies of nano-aggregates (spherical versus cylindrical micelles or polymersomes) observed as a result of self-assembly of amphiphilic block-gradient copolymers.

Also remarkably, the ability of block-gradient copolymers to reversible self-assembly can be compared to that of block-statistical copolymers [38–46]. The stimuli-responsiveness of nano-structures formed by amphiphilic block-gradient copolymers is a highly valuable property for their potential applications as drug delivery systems, smart nano-containers, and so forth. Reversible stimuli-controlled gelation in solutions of triblock-gradient copolymers enables these polymers to be used for rheology control or as medical adhesives. Altogether, the ability to dynamic stimuli-controlled assembly in combination with robustness of preparation (one-pot synthesis) of the block-gradient copolymers makes them highly attractive for the development of novel stimuli-responsive smart materials.

ABBREVIATIONS

2EHA	2-Ethylhexyl acrylate
AA	Acrylic acid
AAEM	Acetoacetoxyethyl methacrylate
AcGalEA	2-(20,30,40,60-Tetra-O-acetyl-β-D-galactosyloxy)ethylacrylate
ACOMP	Automaticcontinuous online monitoring of polymerization
AFM	Atomic force microscopy
AGET	Activator generated by electron transfer
AIBN	Azobisisobutyronitrile
AMA	Allyl methacrylate

AS	4-Acetoxystyrene
ATRcoP	Atom-transfer radical copolymerization
ATRP	Atom-transfer radical polymerization
***n*BA**	*n*-Butyl acrylate
***t*BA**	*tert*-Butyl acrylate
BB	Bloc Builder® or MAMA
***t*BMA**	*tert*-Butyl methacrylate
BPATC	*S*-Butyl-*S*-2-propionic acid trithiocarbonate
CDB	Cumyldithiobenzoate
CMRP	Cobalt-mediated radical polymerization
CP	Cloud point
CPD	4-Cyanopentanoic acid dithiobenzoate
CRcoP	Controlled radical copolymerizations
CRP	Controlled radical polymerization
CTA	Chain transfer agent
DIAMS	SG1-based dialkoxyamine
DLS	Dynamic light scattering
DMA	Dimethyl acrylamide
DMAc	Dimethylacetamide
DMAEMA	2-dimethylaminoethyl methacrylate
DMSO	Dimethylsulfoxide
DSC	Differential scanning calorimetry
DSCSP	2-{[(Dodecylsulfanyl) carbonothioyl] sulfanyl} propanoic acid
EB*i*B	Ethyl 2-bromoisobutyrate
ECEDB	1-(Ethoxycarbonyl)ethyldithiobenzoate
EtBr*i*Bu	Ethyl 2-bromoisobutyrate
F_6BA	2,2,3,4,4,4-Hexafluorobutyl acrylate
FDA	1,1,2,2-Tetrahydroperfluorodecyl acrylate
FRP	Free radical polymerization
GalAm	Acrylamide galactose
HEA	Hydroxyethylacrylate
HEMA	Hydroxyethylmethacrylate
HS	4-Hydroxystyrene
ICAR	Initiator for continous activator regenearation
IPP	di-Isopropyl peroxydicarbonate
kMC	Kinetic Monte Carlo
LAC	Liquid Adsorption Chromatography
MA	Methylacrylate
MAA	Methacrylic acid
MAMA	2-Methylaminoxypropionic-SG1 (BlocBuilder®)
MBrP	Methyl bromopropionate
MCPDT	S-Methoxycarbonylphenylmethyldodecyltrithiocarbonate
MEA	Methoxyethylacrylate
MMA	Methyl methacrylate

MONAMS	Alkoxyamine derived from 1-(methoxycarbonyl)eth-1-yl initiating radical and SG1 nitroxide
MPC	2-Methacryloyloxyethyl phosphorylcholine
MS	4-Methylstyrene
NAM	*N*-Acryloylmorpholine
NMP	Nitroxyde-mediated polymerization
NMR	Nuclear magnetic resonance
ODA	Octadecyl acrylate
PCL	Polycaprolactone
PCPI	2-Pyridine-carbaldehyde *n*-propylimine
PEGA	Poly(ethylene glycol) acrylate
PEGMA	Poly(ethylene glycol) methyl ether methacrylate
PEOMeMA	Poly(ethylene oxide) methyl ether methacrylate
PEOPhA	Poly(ethylene oxide) phenylether acrylate
PEPDTA	1-Phenylethyl phenyldithioacetate
PS	Polystyrene
RAFT	Reversible addition–fragmentation chain transfer
RDRP	Reversible–deactivation radical polymerization
S	Styrene
$scCO_2$	Supercritical carbon dioxide
SEC	Size exclusion chromatography
SG1	*N-tert*-Butyl-*N*-(1-diethylphosphono-2,2-dimethylpropyl)nitroxide
TEMPO	2,2,6,6-Tetramethylpiperidinyl-1-oxy
TFEMA	2,2,2-Trifluoroethyl methacrylate
TFT	α,α,α-Trifluorotoluene
VAc	Vinyl acetate
VBPDE	Vinylbenzylphosphonic acid diethylester
VC2	Vinylidene chloride

REFERENCES

1. F. E. Bailey; D. W. C. Von; M. Matzner; R. H. Young; L. M. Robeson. US Pat. US4065520A.
2. Pakula, T.; Matyjaszewski, K. *Macromol. Theory Simul.* **1996**, *5* (5), 987–1006.
3. Matyjaszewski, K.; Ziegler, M. J.; Arehart, S. V.; Greszta, D.; Pakula, T. *J. Phys. Org. Chem.* **2000**, *13* (12), 775–786.
4. Davis, K. A. *Statistical, gradient, and segmented copolymers by controlled/living radical polymerizations*; Advances in polymer science; Springer: Berlin; New York, **2002**.
5. Beginn, U. *Polymer* **2006**, *47* (19), 6880–6894.
6. Zaremski, M. Y.; Kalugin, D. I.; Golubev, V. B. *Polym. Sci. Ser. A* **2009**, *51* (1), 103–122.
7. Beginn, U. *Colloid Polym. Sci.* **2008**, *286* (13), 1465–1474.
8. Mok, M. M.; Masser, K. A.; Runt, J.; Torkelson, J. M. *Macromolecules* **2010**, *43* (13), 5740–5748.

9. Mok, M. M.; Pujari, S.; Burghardt, W. R.; Dettmer, C. M.; Nguyen, S. T.; Ellison, C. J.; Torkelson, J. M. *Macromolecules* **2008**, *41* (15), 5818–5829.
10. Jouenne, S.; González-León, J. A.; Ruzette, A.-V.; Lodefier, P.; Tencé-Girault, S.; Leibler, L. *Macromolecules* **2007**, *40* (7), 2432–2442.
11. Matyjaszewski, K. *Science* **2011**, *333* (6046), 1104–1105.
12. Lefebvre, M. D.; Dettmer, C. M.; McSwain, R. L.; Xu, C.; Davila, J. R.; Composto, R. J.; Nguyen, S. T.; Shull, K. R. *Macromolecules* **2005**, *38* (25), 10494–10502.
13. Lefebvre, M. D.; Olvera de la Cruz, M.; Shull, K. R. *Macromolecules* **2004**, *37* (3), 1118–1123.
14. Wang, R.; Li, W.; Luo, Y.; Li, B.-G.; Shi, A.-C.; Zhu, S. *Macromolecules* **2009**, *42* (6), 2275–2285.
15. Mok, M. M.; Ellison, C. J.; Torkelson, J. M. *Macromolecules* **2011**, *44* (15), 6220–6226.
16. Wong, C. L. H.; Kim, J.; Torkelson, J. M. *J. Polym. Sci. Part B Polym. Phys.* **2007**, *45* (20), 2842–2849.
17. Kim, J.; Mok, M. M.; Sandoval, R. W.; Woo, D. J.; Torkelson, J. M. *Macromolecules* **2006**, *39* (18), 6152–6160.
18. Zhou, Y.-N.; Li, J.-J.; Luo, Z.-H. *J. Polym. Sci. Part Polym. Chem.* **2012**, *50* (15), 3052–3066.
19. Kim, J.; Zhou, H.; Nguyen, S. T.; Torkelson, J. M. *Polymer* **2006**, *47* (16), 5799–5809.
20. Tao, Y.; Kim, J.; Torkelson, J. M. *Polymer* **2006**, *47* (19), 6773–6781.
21. Sha, Y.; Hui, C. Y.; Kramer, E. J.; Hahn, S. F.; Berglund, C. A. *Macromolecules* **1996**, *29* (13), 4728–4736.
22. Mok, M. M.; Kim, J.; Torkelson, J. M. *J. Polym. Sci. Part B Polym. Phys.* **2008**, *46* (1), 48–58.
23. Lee, S. B.; Russell, A. J.; Matyjaszewski, K. *Biomacromolecules* **2003**, *4* (5), 1386–1393.
24. Shull, K. R. *Macromolecules* **2002**, *35* (22), 8631–8639.
25. Wang, Y.; Mattice, W. L.; Napper, D. H. *Langmuir* **1993**, *9* (1), 66–70.
26. Wong, C. L.; Kim, J.; Roth, C. B.; Torkelson, J. M. *Macromolecules* **2007**, *40* (16), 5631–5633.
27. Tito, N. B.; Milner, S. T.; Lipson, J. E. G. *Macromolecules* **2010**, *43* (24), 10612–10620.
28. Zhao, Y.; Luo, Y.-W.; Li, B.-G.; Zhu, S. *Langmuir* **2011**, *27* (18), 11306–11315.
29. Hadjichristidis, N.; Pispas, S.; Floudas, G. In *Block Copolymers: Synthetic Strategies, Physical Properties, and Applications*; John Wiley & Sons, Inc., **2002**; pp 255–267.
30. Hamley, I. W. *Block Copolymers in Solution: Fundamentals and Applications*; John Wiley & Sons, **2005**.
31. Lodge, T. P. *Macromol. Chem. Phys.* **2003**, *204* (2), 265–273.
32. Discher, D. E.; Eisenberg, A. *Science* **2002**, *297* (5583), 967–973.
33. Lefay, C.; Charleux, B.; Save, M.; Chassenieux, C.; Guerret, O.; Magnet, S. *Polymer* **2006**, *47* (6), 1935–1945.
34. Kuldová, J.; Košovan, P.; Limpouchová, Z.; Procházka, K.; Borisov, O. V. *Collect. Czechoslov. Chem. Commun.* **2010**, *75* (4), 493–505.
35. Borisova, O.; Billon, L.; Zaremski, M.; Grassl, B.; Bakaeva, Z.; Lapp, A.; Stepanek, P.; Borisov, O. *Soft Matter* **2011**, *7* (22), 10824.
36. Borisova, O.; Billon, L.; Zaremski, M.; Grassl, B.; Bakaeva, Z.; Lapp, A.; Stepanek, P.; Borisov, O. *Soft Matter* **2012**, *8* (29), 7649.

37. Borisova, O. V.; Zaremski, M. Y.; Borisov, O. V.; Billon, L. *Polym. Sci. Ser. B* **2013**, *55* (11-12), 573–576.
38. Lejeune, E.; Drechsler, M.; Jestin, J.; Müller, A. H. E.; Chassenieux, C.; Colombani, O. *Macromolecules* **2010**, *43* (6), 2667–2671.
39. Lejeune, E.; Chassenieux, C.; Colombani, O. In *Trends in Colloid and Interface Science XXIV*; Starov, V., Procházka, K., Eds.; Progress in Colloid and Polymer Science; Springer Berlin Heidelberg, **2011**; pp 7–16.
40. Colombani, O.; Lejeune, E.; Charbonneau, C.; Chassenieux, C.; Nicolai, T. *J. Phys. Chem. B* **2012**, *116* (25), 7560–7565.
41. Charbonneau, C.; Chassenieux, C.; Colombani, O.; Nicolai, T. *Macromolecules* **2011**, *44* (11), 4487–4495.
42. Charbonneau, C.; Chassenieux, C.; Colombani, O.; Nicolai, T. *Macromolecules* **2012**, *45* (2), 1025–1030.
43. Charbonneau, C.; Chassenieux, C.; Colombani, O.; Nicolai, T. *Phys. Rev. E* **2013**, *87* (6).
44. Charbonneau, C.; De Souza Lima, M. M.; Chassenieux, C.; Colombani, O.; Nicolai, T. *Phys. Chem. Chem. Phys.* **2013**, *15* (11), 3955.
45. Charbonneau, C.; Nicolai, T.; Chassenieux, C.; Colombani, O.; Miriam de Souza Lima, M. *React. Funct. Polym.* **2013**, *73* (7), 965–968.
46. Shedge, A.; Colombani, O.; Nicolai, T.; Chassenieux, C. *Macromolecules* **2014**, *47* (7), 2439–2444.
47. Kryszewski, M. *Polym. Adv. Technol.* **1998**, *9* (4), 244–259.
48. Koike, Y. *Polymer* **1991**, *32* (10), 1737–1745.
49. Matyjaszewski, K.; Shipp, D. A.; McMurtry, G. P.; Gaynor, S. G.; Pakula, T. *J. Polym. Sci. Part Polym. Chem.* **2000**, *38* (11), 2023–2031.
50. Jenkins, A. D.; Jones, R. G.; Moad, G. *Pure Appl. Chem.* **2010**, *82* (2), 483–491.
51. Szwarc, M. *Carbanions, living polymers, and electron transfer processes*; Interscience Publishers, **1968**.
52. Nakhmanovich, B. I.; Arest-Yakubovich, A. A.; Prudskova, T. N.; Litvinenko, G. I. *Macromol. Rapid Commun.* **1996**, *17* (1), 31–36.
53. Tenorio López, J. A.; Benvenuta Tapia, J. J.; Cuevas Díaz, M. del C. *Macromol. React. Eng.* **2009**, *3* (8), 473–485.
54. Quinebèche, S.; Navarro, C.; Gnanou, Y.; Fontanille, M. *Polymer* **2009**, *50* (6), 1351–1357.
55. Hashimoto, T.; Tsukahara, Y.; Tachi, K.; Kawai, H. *Macromolecules* **1983**, *16* (4), 648–657.
56. Okabe, S.; Seno, K.; Kanaoka, S.; Aoshima, S.; Shibayama, M. *Macromolecules* **2006**, *39* (4), 1592–1597.
57. Okabe, S.; Seno, K.; Kanaoka, S.; Aoshima, S.; Shibayama, M. *Polymer* **2006**, *47* (21), 7572–7579.
58. Hoogenboom, R.; Fijten, M. W.; Wijnans, S.; van den Berg, A. M.; Thijs, H. M.; Schubert, U. S. *J. Comb. Chem.* **2006**, *8* (2), 145–148.
59. Park, J.-S.; Kataoka, K. *Macromolecules* **2006**, *39* (19), 6622–6630.
60. Park, J.-S.; Kataoka, K. *Macromolecules* **2007**, *40* (10), 3599–3609.
61. Gress, A.; Völkel, A.; Schlaad, H. *Macromolecules* **2007**, *40* (22), 7928–7933.

62. Seno, K.-I.; Tsujimoto, I.; Kikuchi, T.; Kanaoka, S.; Aoshima, S. *J. Polym. Sci. Part Polym. Chem.* **2008**, *46* (18), 6151–6164.
63. Seno, K.-I.; Tsujimoto, I.; Kanaoka, S.; Aoshima, S. *J. Polym. Sci. Part Polym. Chem.* **2008**, *46* (19), 6444–6454.
64. Milonaki, Y.; Kaditi, E.; Pispas, S.; Demetzos, C. *J. Polym. Sci. Part Polym. Chem.* **2012**, *50* (6), 1226–1237.
65. Dettmer, C. M.; Gray, M. K.; Torkelson, J. M.; Nguyen, S. T. *Macromolecules* **2004**, *37* (15), 5504–5512.
66. Cypryk, M.; Delczyk, B.; Juhari, A.; Koynov, K. *J. Polym. Sci. Part Polym. Chem.* **2009**, *47* (4), 1204–1216.
67. Moad, G.; Solomon, D. H. *The chemistry of radical polymerization*; Elsevier, **2006**.
68. Harrisson, S.; Ercole, F.; Muir, B. W. *Polym Chem* **2010**, *1* (3), 326–332.
69. Odian, G. G. *Principles of polymerization*, 4th ed.; Wiley-Interscience: Hoboken, NJ, **2004**.
70. Skeist, I. *J. Am. Chem. Soc.* **1946**, *68* (9), 1781–1784.
71. IUPAC Gold Book http://goldbook.iupac.org/ (accessed Jul 29, 2015).
72. Matyjaszewski, K.; Xia, J. *Chem. Rev.* **2001**, *101* (9), 2921–2990.
73. Kamigaito, M.; Ando, T.; Sawamoto, M. *Chem. Rev.* **2001**, *101* (12), 3689–3746.
74. Hawker, C. J.; Bosman, A. W.; Harth, E. *Chem. Rev.* **2001**, *101* (12), 3661–3688.
75. Madruga, E. L. *Prog. Polym. Sci.* **2002**, *27* (9), 1879–1924.
76. Farcet, C.; Charleux, B.; Pirri, R. *Macromol. Symp.* **2002**, *182* (1), 249–260.
77. Couvreur, L.; Charleux, B.; Guerret, O.; Magnet, S. *Macromol. Chem. Phys.* **2003**, *204* (17), 2055–2063.
78. Gray, M. K.; Zhou, H.; Nguyen, S. T.; Torkelson, J. M. *Macromolecules* **2004**, *37* (15), 5586–5595.
79. Gray, M. K.; Zhou, H.; Nguyen, S. T.; Torkelson, J. M. *Polymer* **2004**, *45* (14), 4777–4786.
80. Ziegler, M. J.; Matyjaszewski, K. *Macromolecules* **2001**, *34* (3), 415–424.
81. Haddleton, D. M.; Crossman, M. C.; Hunt, K. H.; Topping, C.; Waterson, C.; Suddaby, K. G. *Macromolecules* **1997**, *30* (14), 3992–3998.
82. Uegaki, H.; Kotani, Y.; Kamigaito, M.; Sawamoto, M. *Macromolecules* **1998**, *31* (20), 6756–6761.
83. Arehart, S. V.; Matyjaszewski, K. *Macromolecules* **1999**, *32* (7), 2221–2231.
84. Arehart, S. V.; Matyjaszewski, K. *Polym. Prepr. Am. Chem. Soc. Div. Polym. Chem.* **1999**, 2, 458.
85. Chen, G.-Q.; Wu, Z.-Q.; Wu, J.-R.; Li, Z.-C.; Li, F.-M. *Macromolecules* **2000**, *33* (2), 232–234.
86. Qin, S.; Saget, J.; Pyun, J.; Jia, S.; Kowalewski, T. *Macromolecules* **2003**, *36* (24), 8969–8977.
87. Gu, B.; Sen, A. *Macromolecules* **2002**, *35* (23), 8913–8916.
88. de la Fuente, J. L.; Fernández-García, M.; Fernández-Sanz, M.; Madruga, E. L. *Macromolecules* **2001**, *34* (17), 5833–5837.
89. Brar, A. S.; Kaur, S. *J. Polym. Sci. Part Polym. Chem.* **2005**, *43* (5), 1100–1118.
90. Brar, A. S.; Kaur, S. *Polym. J.* **2005**, *37* (4), 316–323.

91. Vidts, K. R. M.; Dervaux, B.; Prez, F. E. Du. *Polymer* **2006**, *47* (17), 6028–6037.
92. Neugebauer, D.; Zhang, Y.; Pakula, T. *J. Polym. Sci. Part Polym. Chem.* **2006**, *44* (4), 1347–1356.
93. París, R.; de la Fuente, J. L. *J. Polym. Sci. Part Polym. Chem.* **2006**, *44* (18), 5304–5315.
94. He, J.; Wang, Y.; Lin, Q.; Chen, L.; Zhou, X. *J. Macromol. Sci. Part A* **2009**, *46* (4), 405–411.
95. Kim, B.-S.; Lee, H.-K.; Jeong, S.; Lee, J.-O.; Paik, H. *Macromol. Res.* **2011**, *19* (12), 1257–1263.
96. Van Steenberge, P. H. M.; D'hooge, D. R.; Wang, Y.; Zhong, M.; Reyniers, M.-F.; Konkolewicz, D.; Matyjaszewski, K.; Marin, G. B. *Macromolecules* **2012**, *45* (21), 8519–8531.
97. Matyjaszewski, K.; Tsarevsky, N. V. *Nat. Chem.* **2009**, *1* (4), 276–288.
98. Cuervo–Rodríguez, R.; Fernández–Monreal, M. C.; Férnandez–García, M.; Madruga, E. L. *J. Polym. Sci. Part Polym. Chem.* **1997**, *35* (16), 3483–3493.
99. Aerdts, A. M.; German, A. L.; van der Velden, G. P. M. *Magn. Reson. Chem.* **1994**, *32* (S1), S80–S88.
100. Kelen, T.; TÜd[Otilde]s, F. *J. Macromol. Sci. Part Chem.* **1975**, *9* (1), 1–27.
101. Tüdos, F.; Kelen, T.; Földes-berezsnich, T.; Turcsányi, B. *J. Macromol. Sci. Part Chem.* **1976**, *10* (8), 1513–1540.
102. Couvreur, L.; Lefay, C.; Belleney, J.; Charleux, B.; Guerret, O.; Magnet, S. *Macromolecules* **2003**, *36* (22), 8260–8267.
103. Charleux, B.; Nicolas, J.; Guerret, O. *Macromolecules* **2005**, *38* (13), 5485–5492.
104. Nicolas, J.; Mueller, L.; Dire, C.; Matyjaszewski, K.; Charleux, B. *Macromolecules* **2009**, *42* (13), 4470–4478.
105. Chambard, G.; Klumperman, B.; German, A. L. *Polymer* **1999**, *40* (16), 4459–4463.
106. Mignard, E.; Leblanc, T.; Bertin, D.; Guerret, O.; Reed, W. F. *Macromolecules* **2004**, *37* (3), 966–975.
107. Karaky, K.; Péré, E.; Pouchan, C.; Garay, H.; Khoukh, A.; François, J.; Desbrières, J.; Billon, L. *New J. Chem.* **2006**, *30* (5), 698.
108. Burguière, C.; Dourges, M.-A.; Charleux, B.; Vairon, J.-P. *Macromolecules* **1999**, *32* (12), 3883–3890.
109. Graeme Moad; Albert G. Anderson; Frances Ercole; Charles H. J. Johnson; Julia Krstina; Catherine L. Moad; Ezio Rizzardo; Thomas H. Spurling; San H. Thang. In *Controlled Radical Polymerization*; *ACS Symposium Series; American Chemical Society*, **1998**; Vol. *685*, pp 332–360.
110. Ananchenko, G. S.; Fischer, H. *J. Polym. Sci. Part Polym. Chem.* **2001**, *39* (20), 3604–3621.
111. Souaille, M.; Fischer, H. *Macromolecules* **2001**, *34* (9), 2830–2838.
112. Guillaneuf, Y.; Gigmes, D.; Marque, S. R. A.; Tordo, P.; Bertin, D. *Macromol. Chem. Phys.* **2006**, *207* (14), 1278–1288.
113. Mchale, R.; Aldabbagh, F.; Zetterlund, P. B. *J. Polym. Sci. Part Polym. Chem.* **2007**, *45* (11), 2194–2203.
114. Dire, C.; Belleney, J.; Nicolas, J.; Bertin, D.; Magnet, S.; Charleux, B. *J. Polym. Sci. Part Polym. Chem.* **2008**, *46* (18), 6333–6345.

115. Nicolas, J.; Dire, C.; Mueller, L.; Belleney, J.; Charleux, B.; Marque, S. R. A.; Bertin, D.; Magnet, S.; Couvreur, L. *Macromolecules* **2006**, *39* (24), 8274–8282.
116. Phan, T. N. T.; Maiez-Tribut, S.; Pascault, J.-P.; Bonnet, A.; Gerard, P.; Guerret, O.; Bertin, D. *Macromolecules* **2007**, *40* (13), 4516–4523.
117. Kalugin, D. I.; Zaremski, M. Y.; Golubev, V. B.; Garina, E. S. *Polym. Sci. Ser. B* **2011**, *53* (5-6), 307–312.
118. Goto, A.; Fukuda, T. *Macromolecules* **1999**, *32* (3), 618–623.
119. Moad, G.; Rizzardo, E.; Thang, S. H. *Aust. J. Chem.* **2006**, *59* (10), 669–692.
120. Moad, G.; Rizzardo, E.; Thang, S. H. *Aust. J. Chem.* **2009**, *62* (11), 1402–1472.
121. Graeme Moad; Y.K. Chong; Roger Mulder; Ezio Rizzardo; San H. Thang. In *Controlled/Living Radical Polymerization: Progress in RAFT, DT, NMP & OMRP*; ACS Symposium Series; American Chemical Society, **2009**; Vol. *1024*, pp 3–18.
122. Boyer, C.; Stenzel, M. H.; Davis, T. P. *J. Polym. Sci. Part Polym. Chem.* **2011**, *49* (3), 551–595.
123. Chiefari, J.; Chong, Y. K. (Bill); Ercole, F.; Krstina, J.; Jeffery, J.; Le, T. P. T.; Mayadunne, R. T. A.; Meijs, G. F.; Moad, C. L.; Moad, G.; Rizzardo, E.; Thang, S. H. *Macromolecules* **1998**, *31* (16), 5559–5562.
124. Rizzardo, E.; Chiefari, J.; Chong, B. Y.; Ercole, F.; Krstina, J.; Jeffery, J.; Le, T.; Mayadunne, R. T.; Meijs, G. F.; Moad, C. L.; others. In *Macromolecular Symposia*; Wiley Online Library, **1999**; Vol. *143*, pp 291–307.
125. Rixens, B.; Severac, R.; Boutevin, B.; Lacroix-Desmazes, P. *J. Polym. Sci. Part Polym. Chem.* **2006**, *44* (1), 13–24.
126. Ribaut, T.; Lacroix-Desmazes, P.; Fournel, B.; Sarrade, S. *J. Polym. Sci. Part Polym. Chem.* **2009**, *47* (20), 5448–5460.
127. Collins, S.; Yoda, K.; Anazawa, N.; Birkinshaw, C. *Polym. Degrad. Stab.* **1999**, *66* (1), 87–94.
128. Van Herk, A. M. *Macromol. Theory Simul.* **2000**, *9* (8), 433–441.
129. Zubov, V. P.; Valuev, L. I.; Kabanov, V. A.; Kargin, V. A. *J. Polym. Sci. [A1]* **1971**, *9* (4), 833–854.
130. Percec, V.; Popov, A. V.; Ramirez-Castillo, E.; Monteiro, M.; Barboiu, B.; Weichold, O.; Asandei, A. D.; Mitchell, C. M. *J. Am. Chem. Soc.* **2002**, *124* (18), 4940–4941.
131. Gody, G.; Boullanger, P.; Ladavière, C.; Charreyre, M.-T.; Delair, T. *Macromol. Rapid Commun.* **2008**, *29* (6), 511–519.
132. Narain, R.; Housni, A.; Gody, G.; Boullanger, P.; Charreyre, M.-T.; Delair, T. *Langmuir* **2007**, *23* (26), 12835–12841.
133. Sun, X.; Luo, Y.; Wang, R.; Li, B.-G.; Liu, B.; Zhu, S. *Macromolecules* **2007**, *40* (4), 849–859.
134. Kotani, Y.; Kamigaito, M.; Sawamoto, M. *Macromolecules* **1998**, *31* (17), 5582–5587.
135. Karaky, K.; Péré, E.; Pouchan, C.; Desbrières, J.; Derail, C.; Billon, L. *Soft Matter* **2006**, *2* (9), 770.
136. Arehart, S. V.; Greszta, D.; Matyjaszewski, K. *Polym. Prepr. Am. Chem. Soc. Div. Polym. Chem.* **1997**, *38*, 705.
137. Greszta, D.; Matyjaszewski, K. *Polym. Prepr. Am. Chem. Soc. Div. Polym. Chem.* **1997**, *38*, 709.

138. Min, K.; Li, M.; Matyjaszewski, K. *J. Polym. Sci. Part Polym. Chem.* **2005**, *43* (16), 3616–3622.

139. Min, K.; Kwon Oh, J.; Matyjaszewski, K. *J. Polym. Sci. Part Polym. Chem.* **2007**, *45* (8), 1413–1423.

140. Haddleton, D. M.; Crossman, M. C.; Hunt, K. H.; Topping, C.; Waterson, C.; Suddaby, K. G. *Macromolecules* **1997**, *30* (14), 3992–3998.

141. Kaszás, G.; Földes-Berezsnich, T.; Tüdös, F. *Eur. Polym. J.* **1984**, *20* (4), 395–398.

142. Fu, Y.; Cunningham, M. F.; Hutchinson, R. A. *Macromol. Symp.* **2007**, *259* (1), 151–163.

143. Wang, R.; Luo, Y.; Li, B.-G.; Zhu, S. *AIChE J.* **2007**, *53* (1), 174–186.

144. Zhao, Y.; Luo, Y.-W.; Ye, C.; Li, B.-G.; Zhu, S. *J. Polym. Sci. Part Polym. Chem.* **2009**, *47* (1), 69–79.

145. Zhou, Y.-N.; Luo, Z.-H. *Polym. Chem.* **2013**, *4* (1), 76–84.

146. D'hooge, D.; Van Steenberge, P.; Reyniers, M.-F.; Marin, G. *Polymers* **2014**, *6* (4), 1074–1095.

147. Kim, J.; Gray, M. K.; Zhou, H.; Nguyen, S. T.; Torkelson, J. M. *Macromolecules* **2005**, *38* (4), 1037–1040.

148. Kim, J.; Torkelson, J. M. *Polym. Prepr. Am. Chem. Soc. Div. Polym. Chem.* **2007**, *48*, 221.

149. Woo, D.; Kim, J.; Suh, M.-H.; Zhou, H.; Nguyen, S. T.; Lee, S.-H.; Torkelson, J. M. *Polymer* **2006**, *47* (10), 3287–3291.

150. Brandrup, J.; Immergut, E. H.; Grulke, E. A.; Abe, A.; Bloch, D. R. In *Polymer Handbook*; Wiley: New-York, **1999**; pp II – 263.

151. Buzin, A. I.; Pyda, M.; Costanzo, P.; Matyjaszewski, K.; Wunderlich, B. *Polymer* **2002**, *43* (20), 5563–5569.

152. Karaky, K.; Billon, L.; Pouchan, C.; Desbrières, J. *Macromolecules* **2007**, *40* (3), 458–464.

153. Karaky, K.; Clisson, G.; Reiter, G.; Billon, L. *Macromol. Chem. Phys.* **2008**, *209* (7), 715–722.

154. Cherifi, N.; Issoulie, A.; Khoukh, A.; Benaboura, A.; Save, M.; Derail, C.; Billon, L. *Polym. Chem.* **2011**, *2* (8), 1769.

155. Wang, L.; Broadbelt, L. J. *Macromolecules* **2009**, *42* (21), 8118–8128.

156. Wang, L.; Broadbelt, L. J. *Macromolecules* **2009**, *42* (20), 7961–7968.

157. Inoue, Y.; Watanabe, J.; Takai, M.; Yusa, S.; Ishihara, K. *J. Polym. Sci. Part Polym. Chem.* **2005**, *43* (23), 6073–6083.

158. Wang, H.; Zhou, H.; Chen, Y.; Zhang, C. *Colloid Polym. Sci.***2014**, *292* (11), 2803–2809.

159. Chen, Y.; Zhang, Y.; Wang, Y.; Sun, C.; Zhang, C. *J. Appl. Polym. Sci.* **2013**, *127* (3), 1485–1492.

160. Escalé, P.; Ting, S. R. S.; Khoukh, A.; Rubatat, L.; Save, M.; Stenzel, M. H.; Billon, L. *Macromolecules* **2011**, *44* (15), 5911–5919.

161. Sun, X.; Luo, Y.; Wang, R.; Li, B.-G.; Zhu, S. *AIChE J.* **2008**, *54* (4), 1073–1087.

162. Ye, Y.; Schork, F. J. *Ind. Eng. Chem. Res.* **2009**, *48* (24), 10827–10839.

163. Liu, Y.; Huglin, M. B.; Mao, R.; Holmes, P. A. *Polymer* **1996**, *37* (22), 5069–5075.

164. Steinhauer, W.; Hoogenboom, R.; Keul, H.; Moeller, M. *Macromolecules* **2013**, *46* (4), 1447–1460.

165. Benoit, D.; Hawker, C. J.; Huang, E. E.; Lin, Z.; Russell, T. P. *Macromolecules* **2000**, *33* (5), 1505–1507.

166. Luo, Y.; Liu, X. *J. Polym. Sci. Part Polym. Chem.* **2004**, *42* (24), 6248–6258.
167. Kaneyoshi, H.; Matyjaszewski, K. *Macromolecules* **2005**, *38* (20), 8163–8169.
168. Hurtgen, M.; Debuigne, A.; Fustin, C.-A.; Jérôme, C.; Detrembleur, C. *Macromolecules* **2011**, *44* (12), 4623–4631.
169. Fineman, M.; Ross, S. D. *J. Polym. Sci.* **1950**, *5* (2), 259–262.
170. Toppet, S.; Slinckx, M.; Smets, G. *J. Polym. Sci. Polym. Chem. Ed.* **1975**, *13* (8), 1879–1887.
171. Semchikov, Y. D. *Makromol. Chem. Macromol. Symp.* **1996**, *111*, 317.
172. Semchikov, Y. D.; Izvolenskii, V. V.; Smirnova, L. A.; Kopylova, N. A.; Sveshnikova, T. G. *Vysok. Soedin. Ser. A* **1993**, *35*, 495.
173. Semchikov, Y. D.; Smirnova, L. A.; Bulgakova, S. A.; Knyazeva, T. E.; Sherstyanykh, V. I.; Slavnitskaya, N. N. *Dokl. Akad. Nauk. SSSR* **1988**, *298*, 411.
174. Semchikov, Y. D.; Smirnova, L. A.; Knyazeva, T. E.; Bulgakova, S. A.; Voskoboinik, G. A.; Sherstyanykh, V. I. *Vysok. Soedin. Ser. A* **1984**, *26*, 704.
175. Semchikov, Y. D.; Knyazeva, T. E.; Smirnova, L. A.; Bazhenova, N. N.; Slavnitskaya, N. N. *Vysok. Soedin. Ser. B* **1981**, *23*, 483.
176. Harwood, H. J. *Makromol Chem. Macromol. Symp.* **1987**, *10-11*, 331.
177. Plochocka, K. *J Macromol. Sci. Chem.* **1981**, *20*, 67.
178. Borisov, O. V.; Zhulina, E. B.; Leermakers, F. A. M.; Müller, A. H. E. In *Self Organized Nanostructures of Amphiphilic Block Copolymers I*; Müller, A. H. E., Borisov, O., Eds.; Advances in Polymer Science; Springer Berlin Heidelberg, **2011**; pp 57–129.
179. Zhulina, E. B.; Borisov, O. V. *Macromolecules* **2005**, *38* (15), 6726–6741.
180. Semenov, A. N.; Joanny, J.-F.; Khokhlov, A. R. *Macromolecules* **1995**, *28* (4), 1066–1075.

4

ELECTROSTATICALLY ASSEMBLED COMPLEX MACROMOLECULAR ARCHITECTURES BASED ON STAR-LIKE POLYIONIC SPECIES

Dmitry V. Pergushov and Felix A. Plamper

Department of Chemistry, M.V. Lomonosov Moscow State University, Russia; Institute of Physical Chemistry II, RWTH Aachen University, Germany

4.1 INTRODUCTION

One of the real challenges in current polymer science is the directed preparation of novel macromolecular architectures with certain desired functionalities and characteristics. Among different approaches attempted so far, macromolecular self- and co-assembly processes in multicomponent polymer systems appear to be the most promising. Over the last decades, scientists have recognized the potential suitability of ionic polymers as simple building blocks for the construction of complex macromolecular architectures founded on a bottom-up strategy. In particular, it has been observed that the mutual interaction of oppositely charged polyelectrolytes in aqueous mixtures can generate interpolyelectrolyte complexes (IPECs) [1], which are also often referred in the literature as complex coacervates [2] or polyion (or polyanion-polycation) complexes [3, 4]. Now such macromolecular co-assemblies stabilized via a cooperative system of interpolymer salt bonds are considered to

Macromolecular Self-Assembly, First Edition. Edited by Laurent Billon and Oleg Borisov.

present a particular class of (multi)functional complex macromolecular architectures that is of interest not only from a fundamental point of view but also ready to be explored for use in ecology, medicine, biotechnology, and so forth. For example, IPECs have already shown promise as available and efficient binders/flocculants of dispersed systems [5], for targeted gene delivery [6, 7], as enzyme carriers [7], for preparation and further stabilization of metal/metal oxide nanoparticles [8, 9]. Furthermore, IPECs formed at the interfaces (polyelectrolyte multilayers on surfaces or colloids) via the layer-by-layer technique have proven to be effective in smart coating applications and in microcapsules [10, 11].

The significant progress in the controlled synthesis of macromolecules achieved in the last few years, and which included the synthesis of well-defined ionic polymers (or their precursors), suggests the possibility to involve new types of polyionic species in electrostatically driven co-assembly process (also referred to as interpolyelectrolyte complexation [1]), such as ionic (block) co-/terpolymers and even their self-assembled structures, ionic polymers of nonlinear topology (star-shaped polyelectrolytes, cylindrical or spherical polyelectrolyte brushes, ionic dendrimers), crosslinked ionic (co)polymers, such as (micro/nano)gels, thereby considerably enhancing the complexity and increasing the functionality of the resultant macromolecular co-assemblies.

The aim of this chapter is to give a general overview of the IPECs that are based on polyionic species with star-like topology and to demonstrate the important effects of such topology on the properties of the obtained macromolecular co-assemblies. Star-like topology can be realized through (1) the synthetic routes of “grafting-from” or “grafting-to” techniques [12] leading to polyelectrolyte stars (either homoarm or heteroarm [miktoarm]; see Figure 4.1), or (2) self-assembly (micellization) of linear precursors (ionic amphiphilic diblock copolymers) generating star-like micelles with polyelectrolyte coronae (Figure 4.1C) [13]. Also this chapter shows the different morphological types of IPECs based on a star-like polyionic species, going from simple core-corona structures (Section 4.2) to “onion-like” (core-shell-corona) structures (Section 4.3), and further to the relatively rare, for IPECs, vesicular structures (Section 4.4).

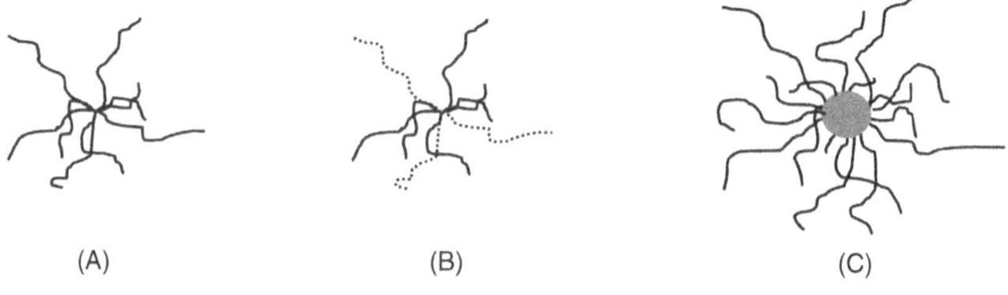

Figure 4.1 Schematic representation of a homoarm star (A), a heteroarm (miktoarm) star (B), and a star-like micelle (C).

4.2 CORE-CORONA CO-ASSEMBLIES OF HOMOPOLYELECTROLYTE STARS COMPLEXED WITH LINEAR POLYIONS

Homopolyelectrolyte stars (Figure 4.1A) can host linear polyions bearing the opposite charge, thereby forming water-soluble IPECs, provided that their charged groups are present in a certain excess compared to the charged groups of the linear polymeric component [14–17]. Figure 4.2 shows a typical turbidimetric titration curve of an aqueous solution of a star-shaped homopolyelectrolyte with an aqueous solution of an oppositely charged linear homopolyelectrolyte. A window of Z-values $Z < Z_M$ ($Z = c_{GPE}/c_{HPE}$ is the charge-to-charge stoichiometry of mixtures of oppositely charged polymeric components, with c_{GPE} and c_{HPE} being the molar concentrations of ionic groups of the linear guest polyelectrolyte (GPE) and the star-shaped host polyelectrolyte (HPE), respectively) corresponds to a homogeneous system, strongly suggesting the formation of water-soluble IPECs. Direct confirmation that water-soluble IPECs are formed comes from analysis of such homogeneous mixtures ($Z < Z_M$) by analytical ultracentrifugation [15].

Based on results of dynamic/static light scattering measurements, the structural picture of the formed macromolecular co-assemblies of star-shaped HPE with rather large number of arms hosting linear GPE has been proposed [14–17]. Indeed, a dominant fraction of the complex particles formed in homogeneous mixtures of oppositely charged polymeric components shows a hydrodynamic radius in the

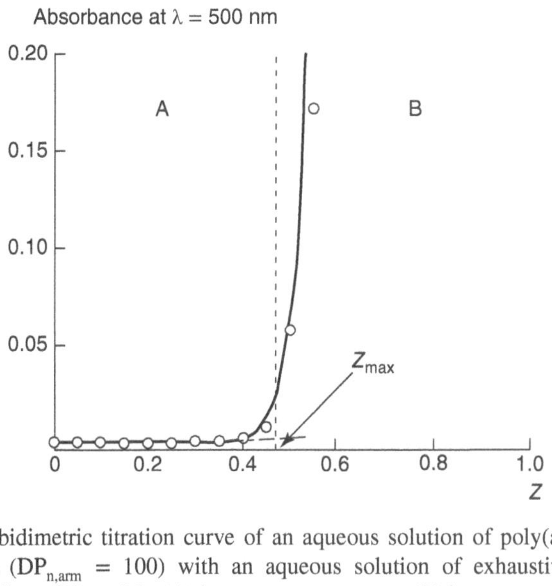

Figure 4.2 Turbidimetric titration curve of an aqueous solution of poly(acrylic acid) star having 21 arms ($DP_{n,arm}$ = 100) with an aqueous solution of exhaustively quaternized poly(2-vinylpyridine) (DP_n = 36): (A) homogeneous system; (B) heterogeneous system. Conditions: 0.1 mol/L NaCl, pH 7. Reprinted from [15] with permission from Springer.

nanometer range being very close to that of single (uncomplexed) star-shaped macromolecules, and a weight-average molecular weight indicating only one star-shaped macromolecule incorporated into each of the complex species [15–17]. These findings suggest that chains of the linear GPE penetrate into the oppositely charged star-shaped HPE and are likely to settle in the vicinity of its branching point, thereby forming an inhomogeneous core-corona structure (Figure 4.3). Hence a sufficiently hydrophobic core is assembled from electrostatically coupled monomer units of the oppositely charged polymeric components in apparently about equivalent amounts, namely in the stoichiometric 1:1 ratio. A hydrophilic (ionic) corona is built up from the essentially free fragments of the star-shaped HPE, which are not complexed with the linear GPE that only partly compensates the charge of the homopolyelectrolyte star. Thus the corona grants solubility to the whole complex species and protects that species from aggregation.

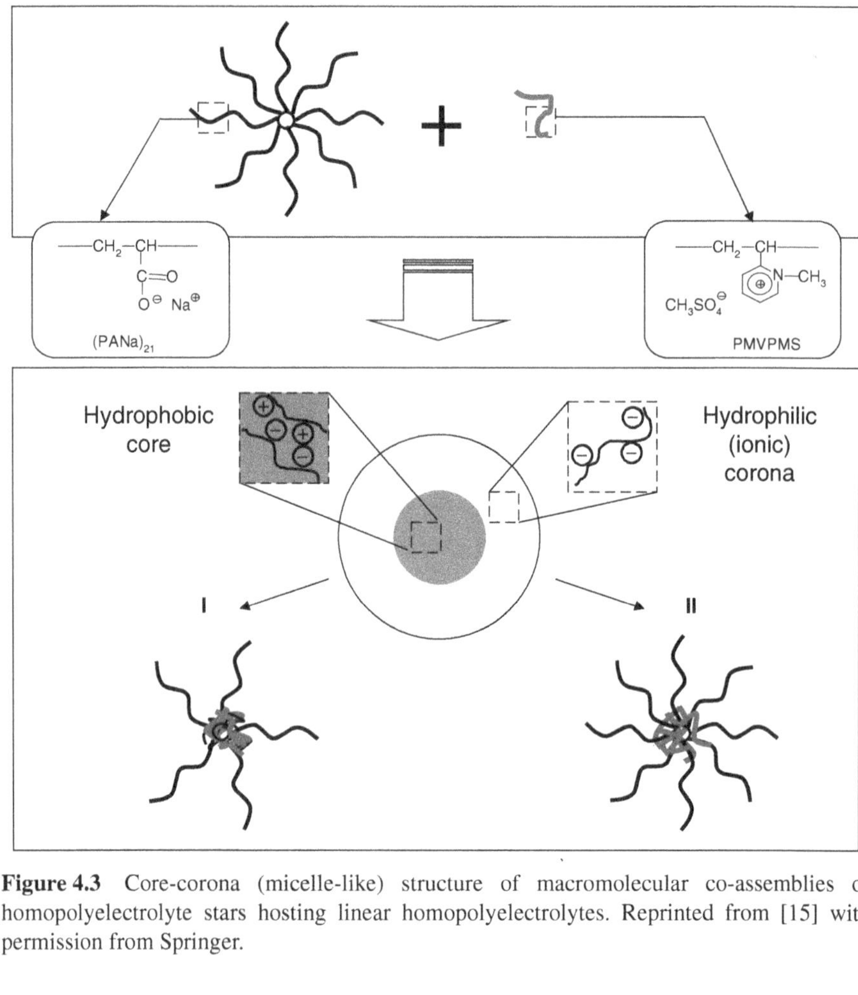

Figure 4.3 Core-corona (micelle-like) structure of macromolecular co-assemblies of homopolyelectrolyte stars hosting linear homopolyelectrolytes. Reprinted from [15] with permission from Springer.

Remarkably, the results from molecular dynamics simulations provide the evident support for the presence of a compartmentalized (micelle-like) structure of the complex species but also reveal some other important features of their internal structural organization [18]. A snapshot of a typical conformation of a single complex species clearly shows its inhomogeneous structure of a core-corona type (Figure 4.4) that results from a pronounced non-uniform distribution of chains of the linear GPE within the volume occupied by the oppositely star-shaped HPE. Additionally, clearly seen in the figure is the distinct fractionation of the arms of the homopolyelectrolyte star between the two populations. Note that in one of the populations, the arms are completely embedded in a complex core of the star, whereas in the other population, the arms remain virtually free (uncomplexed) and so build up the corona.

Use of bis-hydrophilic (also often called as double hydrophilic) diblock copolymers, which comprise an ionic block and a hydrophilic non-ionic one, in

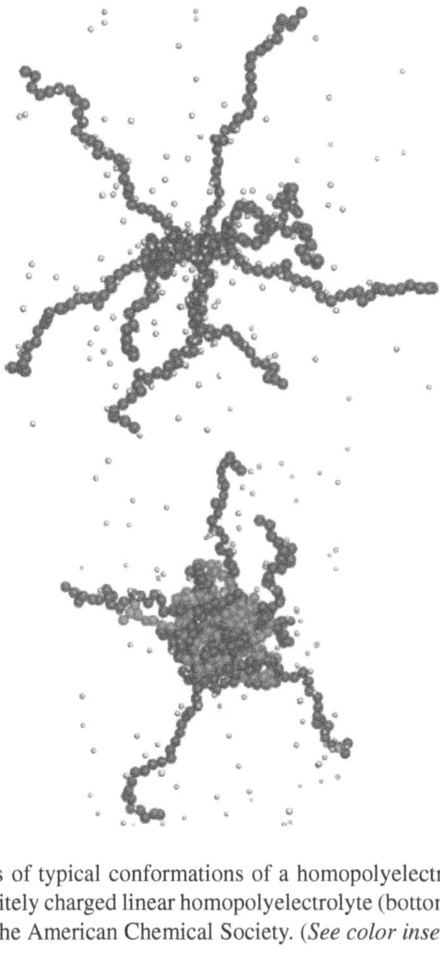

Figure 4.4 Snapshots of typical conformations of a homopolyelectrolyte star (top) and its complex with an oppositely charged linear homopolyelectrolyte (bottom). Reprinted from [18] with permission from the American Chemical Society. (*See color insert for color representation of this figure*).

interpolyelectrolyte complexation could have prevented the macroscopic phase separation (observed at $Z > Z_M$ for the described systems comprising a homopolyelectrolyte star complexed with the oppositely charged linear homopolyelectrolyte). With the presence of the non-ionic block of the bis-hydrophilic diblock copolymer, the preparation would have boosted the water-soluble macromolecular co-assemblies, even at a 1:1 charge-to-charge ratio of the polymeric components in the aqueous mixture [2, 19–21]. Besides the enhanced solubility of the formed complex species in aqueous media, the desired properties and additional functionalities, such as biocompatibility and thermosensitivity, could then be imparted through the non-ionic block of the bis-hydrophilic diblock copolymer, as would be of definite interest for medical and biotechnological applications.

The macromolecular co-assemblies formed in such systems are supposed to have a compartmentalized (micelle-like) structure with a rather hydrophobic core. The core is assembled from electrostatically coupled monomer units of the oppositely charged polymeric components in apparently about equivalent amounts, namely in the stoichiometric 1:1 ratio. At $Z = 1$, the complex species have a hydrophilic corona formed by the non-ionic blocks of the bis-hydrophilic diblock copolymer.

Indeed, the results of analytical ultracentrifugation, dynamic/static light scattering, and cryogenic transmission electron microscopy indicate that such water-soluble core-corona macromolecular co-assemblies result from interpolyelectrolyte complexation of a homopolyelectrolyte star with the oppositely charged bis-hydrophilic diblock copolymer at $Z = 1$ [22]. In contrast with the above-considered systems, that is, complex species based on a homopolyelectrolyte star hosting chains of the oppositely charged linear homopolyelectrolyte, in most cases (but not always) they are of distinctly aggregated nature as follows from the results of dynamic/static light scattering and analytical ultracentrifugation measurements, incorporating several star-shaped macromolecules whose charges are fully compensated by the ionic blocks of the bis-hydrophilic diblock copolymer (Figure 4.5).

4.3 CORE-SHELL-CORONA CO-ASSEMBLIES OF STAR-LIKE MICELLES OF IONIC AMPHIPHILIC DIBLOCK COPOLYMERS COMPLEXED WITH LINEAR POLYIONS

The self-assemblies of a core-corona type (micelles) generated in aqueous solutions of ionic amphiphilic diblock copolymers very much resemble star-shaped polyelectrolytes with a large number of arms (the latter is given by the aggregation number of micelles) provided that the hydrophobic core is considerably smaller than the ionic corona [13]. Such star-like micelles (Figure 4.1C) with polyelectrolyte coronae are naturally expected to host polyions bearing the opposite charge, thereby generating IPECs.

Figure 4.6 demonstrates that aqueous mixtures of the oppositely charged star-like micelles acting as HPE and the linear homopolyelectrolyte acting as GPE remain homogeneous at $Z < Z_M$ while they become heterogeneous at $Z > Z_M$. Analysis of the homogeneous mixtures ($Z < Z_M$) by analytical ultracentrifugation confirms the

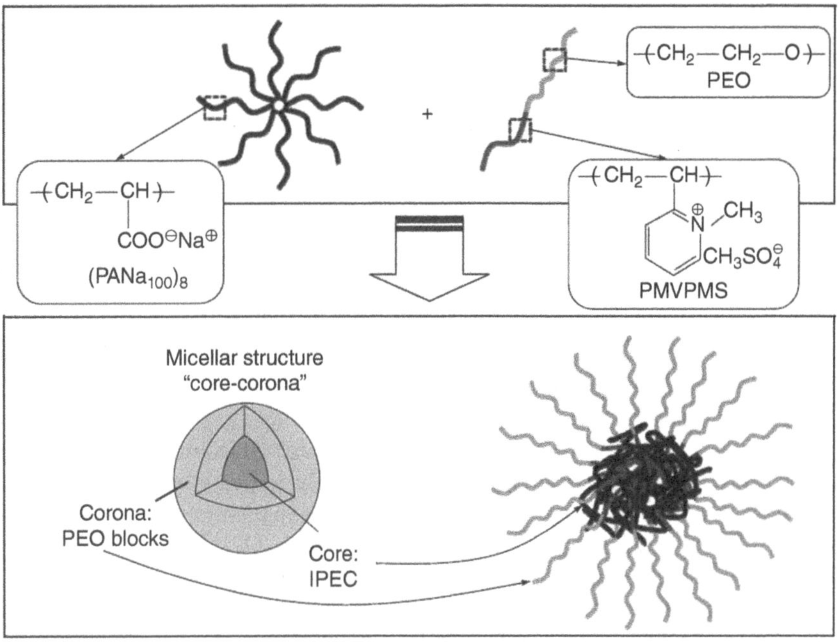

Figure 4.5 Core-corona (micelle-like) structure of macromolecular co-assemblies of a homopolyelectrolyte star complexed with an oppositely charged bis-hydrophilic diblock copolymer. Reprinted from [22] with permission from Springer.

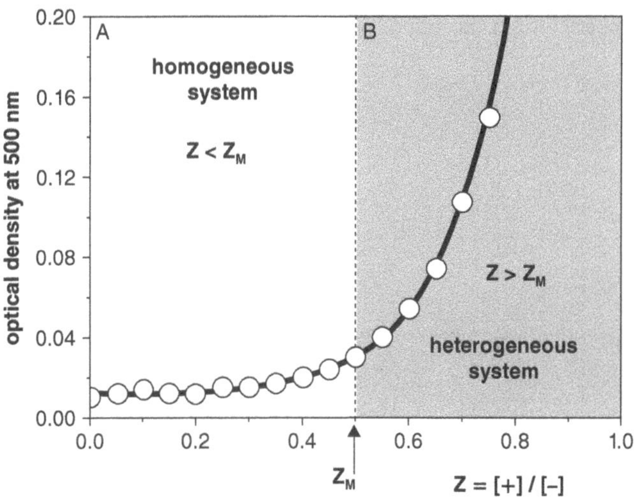

Figure 4.6 Turbidimetric titration curve of an aqueous solution of star-like polyisobutylene-*block*-poly(methacrylic acid) micelles ($DP_{n,PIB}$ = 20, $DP_{n,PMAA}$ = 100) with an aqueous solution of exhaustively quaternized poly(4-vinylpyridine) (DP_n = 285): (A) homogeneous system; (B) heterogeneous system. Conditions: 0.1 mol/L NaCl, pH 9. Reprinted from [23] with permission from the American Chemical Society.

formation of water-soluble IPECs that are considerably enriched by ionic groups of the micelle-forming polymeric component [23]. Thus this trend is very similar to that discussed in the previous section for the formation of macromolecular co-assemblies between the oppositely charged star-shaped HPE and linear GPE. Further examination of water-soluble IPECs based on star-like micelles of ionic amphiphilic diblock copolymers by various techniques — including fluorescence spectroscopy (with the use of pyrene as a polarity probe), small-angle neutron and X-ray scattering, dynamic light scattering, and cryogenic transmission electron microscopy — suggests that such macromolecular co-assemblies have a peculiar core-shell-corona (or "onion-like"/multilayer) structure (Figure 4.7) [16, 23–28]. Specifically, each of the water-soluble complex species contains a hydrophobic core formed by nonpolar blocks of the ionic amphiphilic diblock copolymer (A in Figure 4.7). This core is surrounded by a compact complex inner shell assembled from oppositely charged fragments of the polymeric components (ionic blocks of the ionic amphiphilic diblock copolymer and linear homopolyelectrolyte) in apparently about 1:1 charge-to-charge ratio (B in Figure 4.7). The ionic corona is built up from excess fragments of the ionic blocks of the ionic amphiphilic diblock copolymer (Figure 4.7C), which are not complexed with the linear GPE (the linear GPE only partly compensates the charge of the polyelectrolyte corona of the star-like micelle).

Detailed analyses of the scattering curves obtained for such systems by small-angle neutron scattering (Figure 4.8) [24, 27] strongly suggest that the aggregation number of the original ionic amphiphilic diblock copolymer micelles hardly changes when they host the oppositely charged linear GPE, though for some systems the original self-assemblies can exhibit "dynamic" properties; that is, they can change their aggregation numbers upon variations in the conditions of the surrounding solution [29]. Further, interpolyelectrolyte complexation of such "dynamic" micelles with the oppositely charged linear GPE does not make them

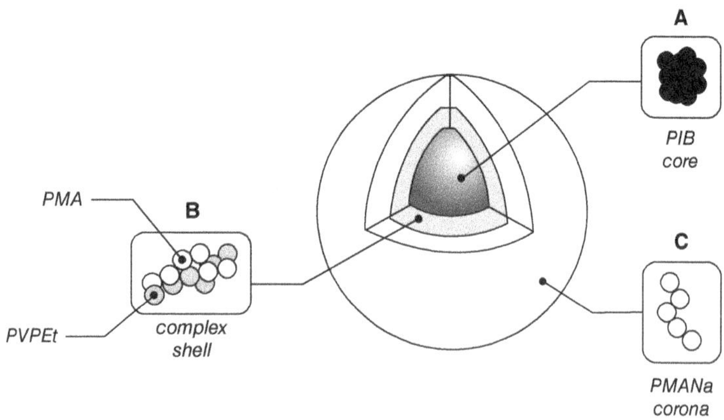

Figure 4.7 Core-shell-corona ("onion-like") structure of macromolecular co-assemblies of a star-like micelle of ionic amphiphilic diblock copolymers hosting an oppositely charged linear homopolyelectrolyte. Reprinted from [24] with permission from Elsevier.

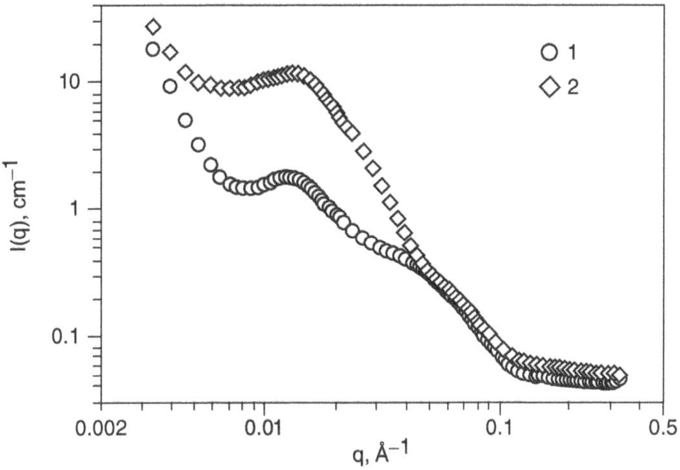

Figure 4.8 Small-angle neutron scattering intensity as a function of the scattering vector for aqueous solutions of star-like polyisobutylene-*block*-poly(methacrylic acid) micelles ($DP_{n,PIB}$ = 20, $DP_{n,PMAA}$ = 100) (*1*) and their water-soluble IPEC with exhaustively quaternized poly(4-vinylpyridine) (DP_n = 285) at Z = 0.4 (2). Conditions: 0.1 mol/L NaCl, pH 9. Reprinted from [24] with permission from Elsevier.

"frozen" as their aggregation numbers remain still sensitive to variations in the pH of the surrounding solution [27].

As in the case of macromolecular co-assemblies of homopolyelectrolyte stars complexed with linear polyions (see the previous section), the hydrodynamic sizes of water-soluble IPECs based on star-like micelles are quite close to those of the corresponding original self-assemblies [24–27]. This observation again strongly suggests intra-coronal disproportionation of the polyelectrolyte blocks of the micelle-forming polymer component [16]. A certain fraction of such blocks is completely embedded in the complex shell, whereas the remaining polyelectrolyte blocks only contribute a small number of their monomer units in the shell domain and form ionic corona, which provides solubility of the resultant core-shell-corona macromolecular co-assemblies in aqueous media.

4.4 VESICULAR CO-ASSEMBLIES OF BIS-HYDROPHILIC MIKTOARM STARS COMPLEXED WITH LINEAR POLYIONS

To prepare still more advanced macromolecular co-assemblies, bis-hydrophilic miktoarm stars can be involved in interpolyelectrolyte complexation instead of star-shaped homopolyelectrolytes. Such stars often have a non-centrosymmetric architecture, for instance, if their polyelectrolyte and non-ionic hydrophilic arms are segregated to different domains rather than homogeneously mixed with each other. Naturally, such miktoarm stars can host oppositely charged linear polyions.

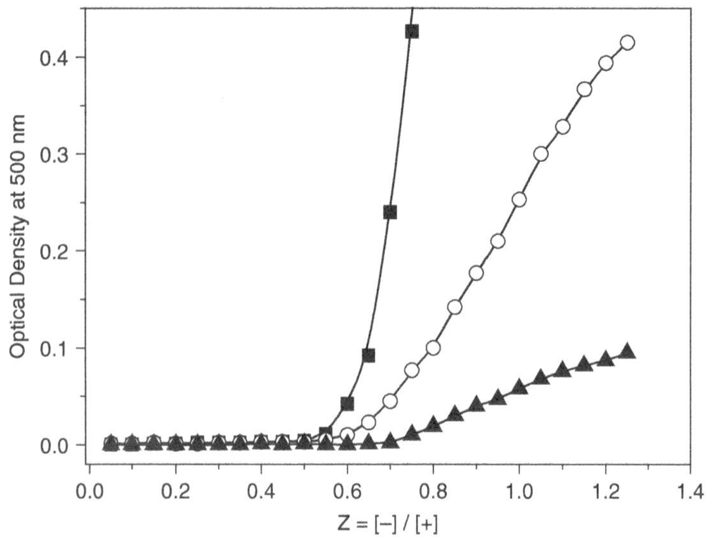

Figure 4.9 Turbidimetric titration curves of aqueous solutions of miktoarm stars poly(ethylene oxide)-quaternized poly(2-dimethylaminoethyl methacrylate) (1 PEO arm, $DP_{n,PEO}$ = 113, 2.5 PDMAEMAQ arms, $DP_{n,PDMAEMAQ}$ = 110) (○), poly(ethylene oxide)-quaternized poly(2-dimethylaminoethyl methacrylate) (1 PEO arm, $DP_{n,PEO}$ = 113, 2.5 PDMAEMAQ arms, $DP_{n,PDMAEMAQ}$ = 40) (▲), and for a comparison an aqueous solution of homopolyelectrolyte star quaternized poly(2-dimethylaminoethyl methacrylate) (3.1 PDMAEMAQ arms, $DP_{n,PDMAEMAQ}$ = 100) (■) with an aqueous solution of poly(sodium styrenesulfonate) (DP_n = 20). Conditions: 0.3 mol/L NaCl. Reprinted from [30] with permission from Elsevier.

Typically, water-soluble macromolecular co-assemblies are formed in a rather broad window of lower Z-values, that is, at $Z < Z_M$, whereas stable colloidal systems are obtained at Z closer to 1 (Figure 4.9, triangles) [30]. Remarkably, the complex species observed at $Z < Z_M$ comprise only one bis-hydrophilic miktoarm star (HPE) per particle as found by combining analytical ultracentrifugation and dynamic light scattering and apparently can keep the non-centrosymmetric feature of the original star-shaped polymeric component, thereby generating water-soluble macromolecular co-assemblies of Janus-like type [30]. Latter are rather rare for IPECs and mostly reported for micelle-like macromolecular co-assemblies formed by two oppositely charged bis-hydrophilic diblock copolymers with different non-ionic hydrophilic blocks undergoing phase separation in a corona, which results into double-faced micellar IPECs [31–33]. The formation of Janus-like structures from bis-hydrophilic miktoarm stars complexed with oppositely charged linear homopolyelectrolytes is corroborated by ^{1}H 2D NOESY nuclear magnetic resonance [30].

At Z = 1, electrostatically driven co-assembly between oppositely charged bis-hydrophilic miktoarm star and linear homopolyelectrolyte can lead to a rare vesicular structure [34] despite the intuitive expectation of core-corona structures.

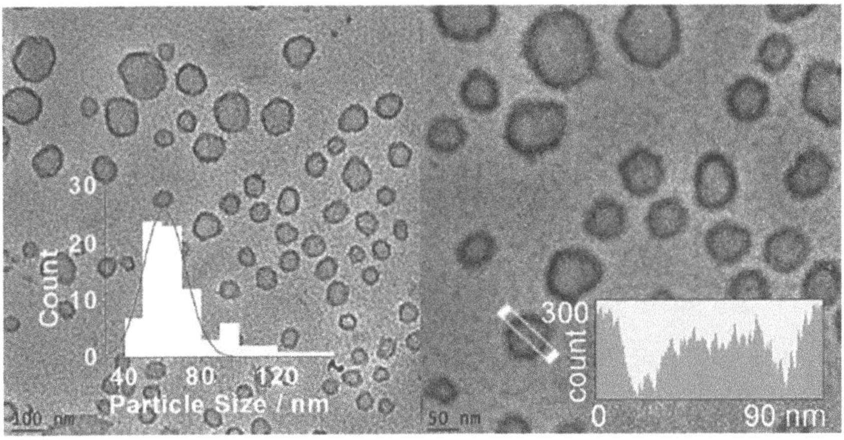

Figure 4.10 Cryogenic transmission electron microscopy images of vesicular IPECs formed by poly(ethylene oxide)-quaternized poly(2-dimethylaminoethyl methacrylate) (1 PEO arm, $DP_{n,PEO}$ = 113, 2.5 PDMAEMAQ arms, $DP_{n,PDMAEMAQ}$ = 40) and poly(sodium styrenesulfonate) (DP_n = 20) at Z = 1. Conditions: 0.3 mol/L NaCl. Insets show particle diameter histogram and a gray-value histogram representing the transmitted electrons. Reprinted from [34] with permission from Wiley.

Such vesicular macromolecular co-assemblies (polymersomes) with the overall size in the nanometer range are clearly visualized by cryogenic transmission electron microscopy (Figure 4.10). The polymersomes comprise an insoluble IPEC domain (dark rings in Figure 4.10) containing essentially equivalent amounts of ionic groups of the oppositely charged polymeric components, which is sandwiched between two poly(ethylene oxide) brushes (not seen in Figure 4.10 because of their low contrast), which grant solubility (colloidal stability) to the whole macromolecular co-assembly in aqueous media. Schematic representation of vesicular IPECs based on bis-hydrophilic miktoarm stars complexed with oppositely charged chains of a linear homopolyelectrolyte is shown in Figure 4.11.

It is assumed that such complex polymersomes result from a hierarchical assembly process. First, the bis-hydrophilic miktoarm stars get complexed with chains of the oppositely charged linear homopolyelectrolyte (electrostatically driven co-assembly), and then the formed primary complex species, each containing one macromolecule of the star-shaped polymeric component, self-assemble further, producing vesicular IPECs [30, 34].

Remarkably, such unilamellar IPEC vesicles were found before only for certain specific oppositely charged polypeptide-derived diblock copolymers undergoing electrostatically driven co-assembly in aqueous media [35–39]. In those publications, additional hydrogen bonding could be suggested, which helps stratifying the IPEC layers, thereby leading the self-assembly toward polymersomes.

At the same time, theoretical considerations (generalized theoretical mean-field approach [40]) on electrostatically driven co-assembly of bis-hydrophilic miktoarm

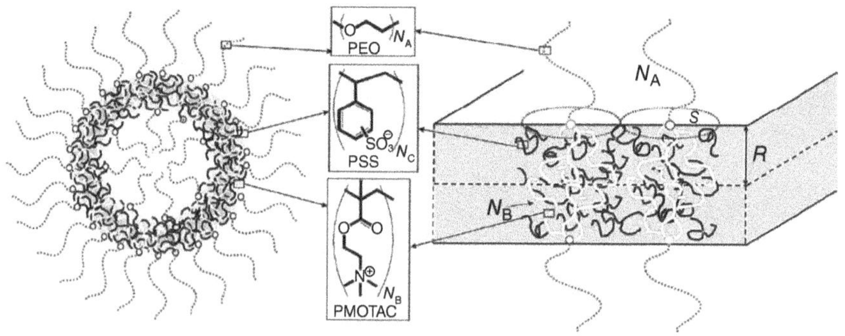

Figure 4.11 Schematic representation of vesicular IPECs based on miktoarm star poly(ethylene oxide)-quaternized poly(2-dimethylaminoethyl methacrylate) complexed with poly(sodium styrenesulfonate) at Z = 1. Reprinted from [34] with permission from Wiley.

stars with oppositely charged linear homopolyelectrolytes provide evidence that the star-shaped architecture of one of the polymeric components forces the morphology of the formed IPECs into a vesicular one [34]. Remarkably, the lamellar (vesicular) morphology is thermodynamically stable if

$$N_B > \frac{N_A^{8/5} \cdot \phi^{3/2} \cdot \upsilon_A^{9/10} \cdot a^{9/5}}{k^2 \cdot (V_B + V_C)^{3/2}} \left(\frac{k_B \cdot T}{\gamma \cdot a^2}\right)^{3/5} \left(\frac{2 \cdot 3^{2/5}}{\pi^2}\right)^{1/2}$$

where N_B and k are the length and the number of the core-forming (polyelectrolyte) arms B, respectively, N_A is the length of the corona-forming (non-ionic hydrophilic) arms A, ϕ is the polymer volume fraction in the IPEC, υ_A is the excluded volume parameter of the monomer unit of the corona-forming (non-ionic hydrophilic) arms A, a is the length of the monomer unit (taken the same for all monomer units), V_B and V_C are volumes of the monomer units of the core-forming (polyelectrolyte) arms B and the linear homopolyelectrolyte C, respectively, γ is surface tension at the core-corona interface, T is the absolute temperature, and k_B is Boltzmann constant, herein the core means a water-insoluble IPEC domain. This condition indicates that an increase in the number k leads to a widening of the region of thermodynamic stability of vesicles. In particular, it implies that if diblock copolymers comprising a non-ionic hydrophilic block of length N_A and one IPEC-forming polyelectrolyte block of length $k{\cdot}N_B$ assemble (at equilibrium) into spherical micelles, the miktoarm stars comprising the same non-ionic hydrophilic arm of length N_A and k IPEC-forming polyelectrolyte arms each of length N_B may form vesicles, provided that k is sufficiently large to satisfy inequality in the equation above.

A simple physical explanation of this finding is the following [34]. The transition from spherical to cylindrical and further to lamellar morphology of the self-assembled aggregates is driven by a concomitant decrease in stretching and a corresponding gain in the conformational entropy of the core-forming blocks. In the specific case considered above, the polyelectrolyte arms' charge is fully compensated by chains of the oppositely charged linear homopolyelectrolyte, thereby rendering them insoluble. At given volume ($\approx k{\cdot}N_B$) of the IPEC domain, the conformational entropy penalty for stretching of k arms, each of length N_B, is larger than that for stretching of one arm of length $k{\cdot}N_B$. As a result, the equilibrium is shifted toward the planar-like (vesicular) morphology.

Remarkably, use of the miktoarm stars can allow a better fine-tuning of the thickness/permeability of the vesicle wall (e.g., by reducing the length of the single arms to degrees of polymerization, where the corresponding linear diblock copolymers would already form nonvesicular co-assemblies). In sum, the architecture of miktoarm stars provides a facile transition from monostar co-assemblies to aggregates with a very high number of the miktoarm stars (vesicular co-assemblies) in a rather narrow window of Z-values.

4.5 CONCLUSIONS

Star-like polyionic species (homopolyelectrolyte stars, star-like micelles with polyelectrolyte coronae, bis-hydrophilic miktoarm stars containing polyelectrolyte arms and non-ionic hydrophilic arm(s)) can host chains of polyelectrolytes bearing the opposite charge, forming macromolecular co-assemblies of various morphologies, such as core-corona and core-shell-corona ones. Under the certain conditions, even vesicular structures can be obtained as a result of matched co- and self-assembly processes. All such various compartmentalized complex macromolecular architectures comprising different domains may be highly in demand for such applications as nanocontainers and nanoreactors in drug delivery, diagnostics, and catalysis.

ACKNOWLEDGMENT

This work was partially supported by the Russian Foundation for Basic Research (project number 15-03-06974).

REFERENCES

1. V. A. Kabanov, *Russ. Chem. Rev. 74*, 3–20 (**2005**).
2. J. van der Gucht, E. Spruijt, M. Lemmers, and M. A. Cohen Stuart, *J. Colloid Interface Sci. 361*, 407–422 (**2011**).
3. A. S. Michaels and R. G. Miekka, *J. Phys. Chem. 65*, 1765–1773 (**1961**).
4. A. Nakajima and H. Sato, *Bull. Inst. Chem. Res., Kyoto Univ. 47*, 177–183 (**1969**).
5. V. A. Kabanov, A. B. Zezin, V. A. Kasaikin, A. A. Yaroslavov, and D. A. Topchiev, *Russ. Chem. Rev., 60*, 288–291 (**1991**).
6. A. V. Kabanov and V. A. Kabanov, *Adv. Drug Deliv. Rev. 30*, 49–60 (**1998**).
7. Y. Lee and K. Kataoka, *Soft Matter 5*, 3810–3817 (**2009**).
8. A. B. Zezin, V. B. Rogacheva, V. I. Feldman, P. Afanasiev, and A. A. Zezin, *Adv. Colloid Interface Sci. 158*, 84–93 (**2010**).

9. D. V. Pergushov, A. A. Zezin, A. B. Zezin, and A. H. E. Müller, *Adv. Polym. Sci. 255*, 173–225 (**2014**).
10. G. Decher, *Science 277*, 1232–1237 (**1997**).
11. F. Caruso, R. A. Caruso, and H. Möhwald, *Science 282*, 1111–1114 (**1998**).
12. Y. Xu, F. Plamper, M. Ballauff, and A. H. E. Müller, *Adv. Polym. Sci. 228*, 1–38 (**2010**).
13. S. Förster, V. Abetz, and A. H. E. Müller, *Adv. Polym. Sci. 166*, 173–210 (**2004**).
14. D. V. Pergushov, I. A. Babin, F. A. Plamper, A. B. Zezin, and A. H. E. Müller, *Langmuir 24*, 6414–6419 (**2008**).
15. D. V. Pergushov, I. A. Babin, F. A. Plamper, H. Schmalz, A. H. E. Müller, and A.B. Zezin *Doklady Phys. Chem. 425*, 57–61 (**2009**).
16. D. V. Pergushov, O. V. Borisov, A. B. Zezin, and A. H. E. Müller, *Adv. Polym. Sci. 241*, 131–161 (**2011**).
17. D. V. Pergushov, I. A. Babin, A. B. Zezin, and A. H. E. Müller, *Polym. Int. 62*, 13–21 (**2013**).
18. S. V. Larin, A. A. Darinskii, E. B. Zhulina, and O. V. Borisov, *Langmuir 25*, 1915–1918 (**2009**).
19. A. Harada and K. Kataoka, *Macromolecules 28*, 5294–5299 (**1995**).
20. A. V. Kabanov, T. K. Bronich, V. A. Kabanov, K. Yu, and A. Eisenberg, *Macromolecules 29*, 6797–6802 (**1996**).
21. M. A. Cohen Stuart, N. A. M. Besseling, and R. G. Fokkink, *Langmuir 14*, 6846–6849 (**1998**).
22. I. A. Babin, D. V. Pergushov, A. Wolf, F. A. Plamper, H. Schmalz, A. H. E. Müller, and A.B. Zezin *Doklady Phys. Chem. 441*, 219–223 (**2011**).
23. D. V. Pergushov, E. V. Remizova, J. Feldthusen, A. B. Zezin, A. H. E. Müller, and V. A. Kabanov, *J. Phys. Chem. (B) 107*, 8093–8096 (**2003**).
24. D. V. Pergushov, E. V. Remizova, M. Gradzielski, P. Lindner, J. Feldthusen, A. B. Zezin, A. H. E. Müller, and V. A. Kabanov, *Polymer 45*, 367–378 (**2004**).
25. E. A. Lysenko, P. S. Chelushkin, T. K. Bronich, A. Eisenberg, V. A. Kabanov, and A.V. Kabanov *J. Phys. Chem. (B) 108*, 12352–12359 (**2004**).
26. J. E. Laaser, Y. Jiang, S. R. Petersen, T. M. Reineke, and T. P. Lodge *J. Phys. Chem. (B) 119*, 15919–15928 (**2015**).
27. M. Burkhardt, M. Ruppel, S. Tea, M. Drechsler, R. Schweins, D. V. Pergushov, M. Gradzielski, A. B. Zezin, and A. H. E. Müller, *Langmuir 24*, 1769–1777 (**2008**).
28. D. V. Pergushov, F H. Schacher, and A. H. E. Müller, *Chem. Soc. Rev. 41*, 6888–6901 (**2012**).
29. M. Burkhardt, N. Martinez-Castro, S. Tea, M. Drechsler, I. Babin, I. Grishagin, R. Schweins, D. V. Pergushov, M. Gradzielski, A. B. Zezin, and A. H. E. Müller, *Langmuir 23*, 12864–12874 (**2007**).
30. A. P. H. Gelissen, D V. Pergushov, and F. A. Plamper, *Polymer 54*, 6877–6881 (**2013**).
31. I. K. Voets, A. de Kaizer, P. de Waard, P. M. Frederik, P. H. H. Bomans, H. Schmalz, A. Walther, S. M. King, F. A. M. Leermakers, and M. A. Cohen Stuart, *Angew. Chem. Int. Ed. 45*, 6673–6676 (**2006**).
32. I. K. Voets, R. Fokkink, A. de Kaizer, R. P. May, P. de Waard, and M. A. Cohen Stuart, *Langmuir 24*, 12221–12227 (**2008**).

33. I. K. Voets, R. Fokkink, T. Hellweg, S. M. King, P. de Waard, A. de Kaizer, and M. A. Cohen Stuart, *Soft Matter 5*, 999–1005 (**2009**).

34. F. A. Plamper, A. P. Gelissen, J. Timper, A. Wolf, A. B. Zezin, W. Richtering, H. Tenhu, U. Simon, J. Mayer, O. V. Borisov, and D. V. Pergushov, *Macromol. Rapid Commun. 34*, 855–860 (**2013**).

35. A. Koide, A. Kishimura, K. Osada, W.-D. Jang, Y. Yamasaki, and K. Kataoka, *J. Am. Chem. Soc. 128*, 5988–5989 (**2006**).

36. A. Kishimura, A. Koide, K. Osada, Y. Yamasaki, and K. Kataoka, *Angew. Chem. Int. Ed. 46*, 6085–6088 (**2007**).

37. W.-F. Dong, A. Kishimura, Y. Anraku, S. Chuanoi, and K. Kataoka, *J. Am. Chem. Soc. 131*, 3804–3804 (**2009**).

38. Y. Anraku, A. Kishimura, M. Oba, Y. Yamasaki, and K. Kataoka, *J. Am. Chem. Soc. 132*, 1631–1636 (**2010**).

39. K. Osada, H. Cabral, Y. Mochida, S. Lee, K. Nagata, T. Matsuura, M. Yamamoto, Y. Anraku, A. Kishimura, N. Nishiyama, and K. Kataoka, *J. Am. Chem. Soc. 134*, 13172–13175 (**2012**).

40. O. V. Borisov, E. B. Zhulina, F. A. M. Leermakers, and A. H. E. Müller, *Adv. Polym. Sci. 241*, 57–129 (**2011**).

5

SOLUTION PROPERTIES OF ASSOCIATING POLYMERS

Olga Philippova
Physics Department, Moscow State University, Russia

5.1 INTRODUCTION

Water-soluble associating polymers represent hydrophilic polymers containing some amount of hydrophobic groups attracting each other with the formation of reversible junctions (hydrophobic domains). When the polymers are uncharged, they can have only up to 1 mol% of hydrophobic groups without losing their solubility in water. For charged polymers (polyelectrolytes), this amount can be much higher. The solution properties of uncharged associating polymers were recently reviewed by Aseyev *et al.* [1]. In this chapter, I focus mainly on associating polyelectrolytes (APs). In aqueous medium, they dissociate leaving charged groups on polymer chains and small counterions in solution. At the aggregation of such macromolecules triggered by hydrophobic sites, similarly charged chains approach each other, which induces an electrostatic repulsion between them and a loss of translational entropy of the counterions entrapped within the aggregates in order to partially compensate their charge [2]. Thus the properties of APs in solution are governed by a balance between the attraction of associating groups and the repulsion of polyelectrolyte chains related to the electrostatic interaction of their like charges and the restriction in the motion of the counterions.

An AP was first synthesized by Strauss and Jackson in 1951 by quaternization of poly(2-vinylpyridine) with *n*-dodecyl bromide [3]. But extensive studies of the aggregation behavior of such polyelectrolytes started only around 20 years ago. It

Macromolecular Self-Assembly, First Edition. Edited by Laurent Billon and Oleg Borisov.

was demonstrated that associating groups provide many peculiar properties to the solutions of APs. In dilute regime, they induce the formation of stable self-assembled nano-objects (unimolecular globules, clusters, etc.) of various shapes that are promising, in particular, for drug delivery [4–6]. In a semidilute regime, they provide very high viscosity and even induce the formation of physical gels. Because of this property, the APs are widely used as thickeners for various aqueous formulations, among these paints, cosmetics, food, and fluids for oil recovery [7, 8].

In this chapter, I focus on the experimental studies of the solution properties of APs. First, I describe the structure of typical APs. Then I consider the behavior of APs in dilute solutions. Because of the potential biomedical applications of AP systems, particular attention will be paid to APs based on nontoxic, biodegradable, and biocompatible polymers, such as chitosan. In the final part of the chapter, the behavior of APs in semidilute solutions will be described.

5.2 STRUCTURES OF ASSOCIATING POLYELECTROLYTES

APs can have a variety of structures. Some are depicted in Figure 5.1. In the simplest case, the associating groups are located at the ends of polyelectrolyte chains (telechelic polymers) [9, 10]. However, they can be situated as well in the side chains or just in the backbone. In the latter cases, a different distribution of hydrophobes along the backbone is possible: random, microblock, or sequence designed (Figure 5.1). Sometimes two types of associating hydrophobic groups are present in the same macromolecule of an AP.

Over the last decade there began a tendency to move away from synthetic APs with a simple random or microblocky structure to more complicated systems. One such system represents the APs on the basis of the polymers being of natural origin, among which are the ionic polysaccharides. These APs, which are rather complex, often contain several types of associating groups. An example is hydrophobically modified (HM) chitosan, which has both *N*-acetyl-D-glucosamine units remaining upon incomplete deacetylation of chitin and *n*-alkyl side chains specially introduced to increase its hydrophobicity (Figure 5.2) [11, 12]. Two kinds of hydrophobic domains are consequently formed in aqueous solutions of this polyelectrolyte: (1) hydrophobic domains inherent to the chitosan itself (i.e., domains composed of *N*-acetyl-D-glucosamine units) and (2) hydrophobic domains typical for different associating polymers with hydrophobic side chains [11]. Another tendency of the last ten years is the use of synthetic APs with a sophisticated distribution [12–18] of associating groups targeted to get a desired self-assembled structure. These APs, which have a specific sequence design of associating groups, can be prepared either by chemical modification of the polyelectrolyte or by using modern techniques of controlled polymerization.

5.3 ASSOCIATING POLYELECTROLYTES IN DILUTE SOLUTIONS

Although the polymer concentration in a dilute regime is lower than the overlap concentration for the parent nonassociating polyelectrolyte, the aggregation of

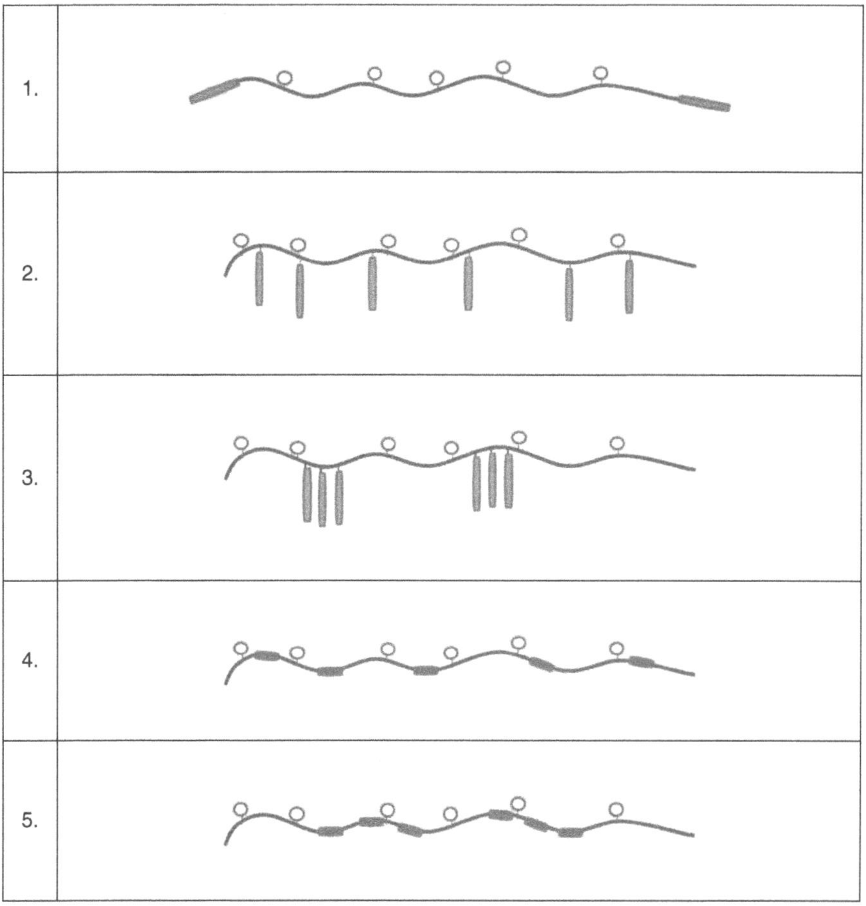

Figure 5.1 Schematic representation of typical associating polyelectrolytes. Circles denote charged groups, and rectangles the associating groups: (1) Telechelic polyelectrolytes with associating groups at the ends of the backbone; (2, 3) grafted polyelectrolytes with associating groups in the side chains; (4, 5) linear polyelectrolytes with associating groups in the main chain.

hydrophobic moieties can proceed both within the same polymer chain (intramolecularly) and/or between different chains (intermolecularly). The predominance of either intra- or intermolecular aggregation depends primarily on the chemical structure of AP, namely on the content of associating and charged units, the size and the location of associating groups, their distribution along the chain, and the nature of the spacer (if any) separating the associating group from the backbone.

When the amount and the size of hydrophobic groups in one polymer chain are sufficient for the formation of at least one stable hydrophobic domain, intramolecular aggregation can occur. For example, it has been roughly estimated [19] that in the

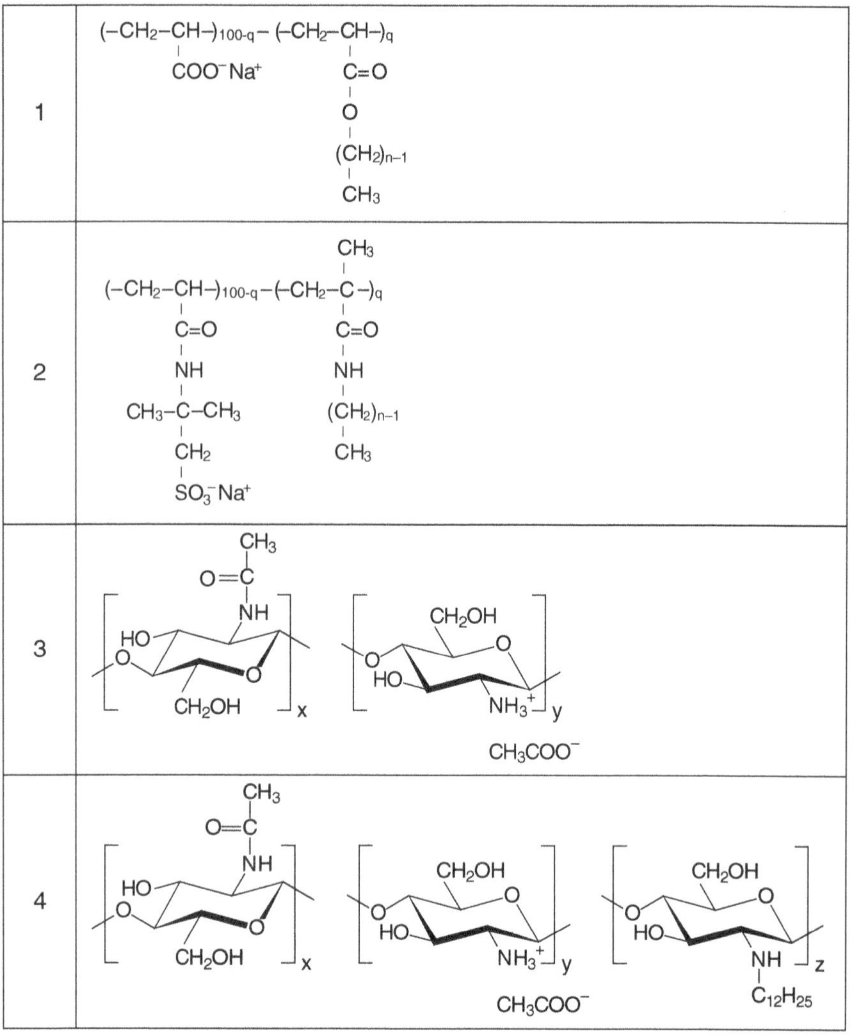

Figure 5.2 Chemical structure of some associating polyelectrolytes: (1) HM poly(sodium acrylate), (2) HM poly(sodium 2-(acrylamide)-2-methylpropanesulfonate), (3) chitosan, and (4) HM chitosan.

case of the association of *n*-dodecyl groups (C12) a hydrophobic domain would contain around 40 groups (this number was obtained by dividing the volume of a sphere with the radius equal to the contour length of the C12 chain by the volume of one C12 group). Therefore, to be susceptible to purely intramolecular aggregation, the chains would have to contain a few tens of the C12 groups. Besides the content, the distribution of hydrophobic units along the backbone [20, 21] and the nature of the connecting group between hydrophobes and polyelectrolyte main chain [22–24] can affect the character of the association processes. For instance, short amide spacers facilitate intramolecular aggregation [22] because they form hydrogen bonds that stabilize compact unimolecular globules. Conversely, long flexible hydrophilic spacers whose backbone less affects the associating groups favor intermolecular contacts [23, 24].

5.3.1 Intramolecular Association

As is mentioned above, intramolecular aggregation requires the presence of a significant amount of hydrophobic units in one macromolecule. For example, HM poly(sodium 2-(acrylamide)-2-methylpropanesulfonate) (PAMPS), which has a random distribution of C12 hydrophobic side groups (Figure 5.2), aggregates intramolecularly if the content of C12 groups lies in the range 20 to 50 mol% [25, 26]. Beyond these limits, the aggregation is intermolecular [25, 26].

The intramolecular aggregation causes a decrease of viscosity as compared with that of the solution of the parent polyelectrolyte without hydrophobes [27]. This is due to the shrinking of polymer coils with the formation of the globules. The globules can take different forms: spherical, necklace, cylindrical, and so on.

5.3.1.1 Spherical Globules In many systems spherical intramolecular aggregates are observed [15, 22, 24–26, 28–31]. The spheres can have different internal structures, and key among these being core-shell and polycore structures. Core-shell spherical globules consist of a compact middle hydrophobic domain surrounded by more or less expanded hydrophilic corona of loops composed of charged segments [30, 32, 33]. In particular, such globules are observed when the number and size of hydrophobic groups are large enough to form a single spherical hydrophobic domain and when there are rather long and flexible hydrophilic segments capable of folding back and forming loops in the shell. The number of hydrophobic groups in the core of such globules can be much smaller than in the micelles of the low-molecular-weight surfactants with the same type of associating groups, since the loops when crowding each other can limit the growth of the globules [32, 33]. The resulting flowerlike globules can be observed, for example, in HM chitosan with 2 mol% of the C12 side groups (ca. 25 per one macromolecule) [30]. It was estimated [34] that the formation of flowerlike aggregates is hardly possible when the polymer contains 4 mol% of hydrophobic units randomly distributed along the chitosan chain. In this case, the mean distance between neighboring hydrophobic groups along the chain was 12 nm. However, taking into account that the persistent length l_p of chitosan is equal to 5–22 nm [35], it is difficult to imagine that a significant number of loops can be formed from such short chain fragments.

In other systems, intramolecular aggregates were observed to have polycore structures [22, 25, 26, 28, 32]. According to Morishima [25], polycore globules are formed in two steps. In the first step, a large amount of small hydrophobic domains develops along the AP chain through the association of few neighboring hydrophobic groups. In small domains, a significant number of these hydrophobes are partially exposed to water. In the second step, the chains fold up into a dense spherical globule, which further reduces the contact of hydrophobes with water. When the amount of hydrophobic groups is not too high (e.g., 20–30 mol% of C12), the aggregation stops at the first stage, and loosely folded polycore aggregates are formed. For a dense spherical globule to form, a high content of hydrophobes is required (40–50 mol%) [25]. Such polymer aggregates (called "unimolecular micelles" [26, 28]) were observed, for example, for various derivatives of PAMPS containing random distributions of the hydrophobic side groups [22, 26, 28].

The behavior of associating polymers in solution, particularly in water, is also highly dependent on the monomer sequences. Two decades ago the concept of protein-like polymers was put forth [13, 14]. Such polymers mimic globular proteins and have a special sequence of hydrophobic H and polar P units (primary structure) that allows them to form stable water-soluble globules. A special monomer sequence can be created by computer modeling by making first a globule from the hydrophobic monomer H and then changing the monomer units on the surface of the globule to polar ones P. The resulting polymer, which "remembers" its native conformation, is able to form stable globules with dense hydrophobic core stabilized by the long dangling loops of hydrophilic units P. The polymer "inherits" some of the properties of the "parent globule" because of the presence of some long-range correlations in the distribution of H and P units in its primary structure.

This concept, as originally developed by computer modeling, was in fact recently proved experimentally using an AP based on a copolypeptoid [15]. In this study, first a protein-like polymer sequence was generated computationally for a particular pair of hydrophobic H and hydrophilic P monomers, which were later used for the synthesis. Then the copolymer with the predicted sequence was prepared by a stepwise submonomer synthesis [36] on a commercial robotic synthesizer using *N*-methylglycine (or sarcosine) as the hydrophobic monomer H (80%) and *N*-(2-carboxyethyl)glycine (20%) as the polar monomer P. For comparison, there was chosen a repeating sequence of the same monomer composition (80% H and 20% P) with a minimized length of any continuous region of H or P monomers. The self-assembly of the APs was studied at pH 8, which is where the hydrophilic units P have a negative charge. It was shown that APs with a specially designed protein-like sequence fold into a compact globule with a good hydrophobic core, whereas the AP with a repeating sequence remains in a loosely collapsed state with a lower density. The authors claim that by this approach, it may be possible to get water-soluble polymer globules with a defined composition of core- and shell-residing sequences and with a controlled size and density, and thus to create a new class of drug delivery vehicles [37].

Polyelectrolyte chains with specially designed sequences [17] of monomer units were observed to produce very sophisticated spherical globules with encapsulated catalytic chambers [16]. Such self-assembly of the AP would resemble the behavior of enzymes in nature.

5.3.1.2 Necklace Globules When the number of hydrophobic groups in one macromolecule is very high (e.g., for very long-chain polymers), several stable hydrophobic cores can form within a single macromolecule. In this case, often a necklace structure of the aggregate is observed [38–42]. In this structure, several spherical micelles (beads) are connected by hydrophilic strings. The strings lessen the electrostatic repulsion between the charged beads, but this proceeds at the expense of the attraction energy, since in the strings the hydrophobic groups do not associate [43]. The size of each globular bead is controlled by a balance between the excess interfacial free energy of a bead and the electrostatic repulsion of charges on its surface.

Necklace globules were observed in dilute aqueous solutions of partially sulfonated polystyrene by small-angle neutron scattering (SANS), indicating contributions to the form factor of both spherical objects and extended chain parts [41]. The spherical objects become dominant at decreasing levels of sulfonation, as the electrostatic repulsion weakens and the necklaces transition to spherical globules.

Necklace globules with different numbers of beads per chain were imaged by atomic force microscopy (AFM) for partially protonated poly(2-vinyl pyridine) and poly(methacryloyloxyethyl dimethylbenzylammonium chloride) [38, 39]. These globules appear as an intermediate state at the conformational transition from coils to compact spherical globules induced by the increase of the ionic strength of the solution. It was observed that coils, necklaces, and compact spherical globules can coexist, which indicates that the phase transition is of the first-order type [39]. The length of the necklace globules and the number of beads in them decrease, whereas the diameter of beads increases at increasing ionic strength [39], which is consistent with theory and computer modeling data [44–47]. These changes result mainly from the screening of electrostatic repulsion, which causes similarly charged beads to come closer together, due to the decreasing length of the strings, and to grow in size.

5.3.1.3 Other Globules Most widespread are the spherical and necklace globules, but several other forms of intramolecular aggregates of APs can be obtained. For instance, when the lengths of the hydrophilic sequences in the AP chain are fairly short, cylindrical globules form. Such globules could also be expected for stiff-chain polyelectrolytes [48–50]. But, in fact, such polyelectrolytes often prefer to form intermolecular aggregates [51–55] instead of single-molecule globules because the bending of stiff macromolecules causes some essential energy losses [49].

The formation of globules showing a specific ordered organization was demonstrated recently for sequence-controlled synthetic APs. These polyelectrolytes can fold into chiral globules [56] or form α-helices or β-strands [57].

5.3.2 Intermolecular Association

When the content of hydrophobic groups in one macromolecule is not sufficient to form even a single hydrophobic domain of optimum size, the association proceeds mainly intermolecularly even in a dilute regime. In this case, clusters consisting of several macromolecules are formed. The formation of clusters is only possible if the attraction between hydrophobic groups is so strong that it can overcome both the energy of Coulombic repulsion of chains and the entropy losses of counterions that are localized within the clusters where they partially compensate the charge [2]. The higher the polyelectrolyte concentration, the less critical are the entropy losses of the counterions. The intermolecular aggregation starts at some critical aggregation concentration (cac), when the entropy losses of counterions are small enough to allow the hydrophobic association to dominate. The values of the cac can be easily estimated, for example, by the fluorescence probe method using pyrene as the probe [58], by conductivity, by NMR [29, 59], or by electron spin-echo resonance (ESR) [60] techniques. It should be noted that the cac allows the intermolecular aggregation to be discriminated from purely intramolecular aggregation, since in the last case hydrophobic domains are present at any concentration of the polymer. As to the values of cac, they decrease with increasing content [24, 61, 62] and size of hydrophobic groups [62–64] (i.e., at the enhancement of hydrophobic attraction), and with the addition of salt, but they are almost insensitive to the whole length of macromolecules [59].

5.3.2.1 Clusters A convenient way to study the intermolecular clusters in dilute solutions is by light scattering [19, 30, 34, 65]. Above cac, the time relaxation function becomes bimodal with a fast relaxation corresponding to the diffusion of single macromolecules and a slow relaxation corresponding to the diffusion of clusters, thus indicating the coexistence of clusters and nonaggregated chains [19, 30, 34, 65]. The fraction of nonaggregated chains decreases with increasing polymer concentration, content of hydrophobic groups, and at the addition of salt [19, 34].

5.3.2.1.1 Size and Aggregation Number of Clusters The competition of hydrophobic attraction leading to the growth of clusters and electrostatic repulsion coupled with losses in the translational entropy of counterions limiting their growth leads to the formation of clusters of optimum size [2, 66, 67]. Interestingly, such clusters are monodispersed in size and their dimensions can be widely varied by changing the balance between attractive and repulsive interactions. By applying a combination of static and dynamic light scattering, the aggregation numbers of clusters N_c (i.e., the number of macromolecules in a single cluster) can be estimated [19, 30, 34]. It was found that N_c increases when the content of the associating groups is increased [19, 34]. For example, N_c increases from 10 to 20 when passing from chitosan to its hydrophobic derivative with 4 mol% of the hydrophobic C12 groups [19].

The effect of the main chain length of macromolecules on their aggregation in an aqueous solution was studied by light scattering using samples of chitosan and HM chitosan [34]. It was observed that the radii of clusters are independent of the chain length of their constituent macromolecules. This implies that, within a cluster, each polymer chain has more than two binding sites with other chains, so the memory of the lengths of the initial macromolecules disappears. Although the radii of clusters remain invariable, the number of macromolecules N_c in one cluster decreases with increasing length of the individual chains. This phenomenon can be understood as follows: The energy of association of one aggregate is determined by the number of attracting groups. Because the content of such groups in short polymers is smaller, a larger number of these macromolecules is necessary to provide the required gain in the energy of association [66, 67]. This suggestion is confirmed by the fact that the number of associating groups in one cluster is constant independently of the length of macromolecules involved in the aggregation [34]. Quite different behavior is observed in systems with a small contribution of electrostatic interactions (at low content of the charged units or/and at high salt concentration). At these conditions, longer polyelectrolyte chains demonstrate a stronger tendency [68, 69] toward aggregation, which has been attributed to a lower loss of entropy. Thus, if the screening of charges in the system is incomplete, the length of macromolecules has no effect on the size of clusters or on the content of associating groups in them. However, if electrostatic factors are suppressed, an increase in the length of individual chains facilitates aggregation.

5.3.2.1.2 Shape and Inner Organization of Clusters The clusters have been visualized by different methods: transmission electron microscopy (TEM) [34, 70–75], scanning electron microscopy (SEM) [76], and AFM [77]. The most reliable images were those obtained by SEM in a humid environment [76], which allows the aggregates to be observed in an undisturbed state directly in the solvent medium. In most studies the observed clusters have a spherical form [34, 70–76]. But also in some cases, "cigar-like" [77] or long cylindrical clusters [51–55] have been detected.

The details of the inner structure of clusters have been derived from data obtained by light scattering, SANS, and fluorescence spectroscopy. Several models of the structural organization of clusters have been proposed: flowerlike micelles, nanogels, among others.

HM polyelectrolytes with very low content of hydrophobes (e.g., telechelics [53, 54] or polymers with a low molar percentage of hydrophobic side chains [30]) tend to form intermolecular flowerlike micelles. These may be either single micelles or several micelles linked via long tie chains into a very loose structure [78, 79]. For example, in HM chitosan, such interconnected flowerlike micelles were observed at as low a content of hydrophobic units as 2 mol%. For polyelectrolytes with a larger number of associating groups (and therefore shorter hydrophilic segments), a nanogel model was proposed [70, 80]. According to this model, the clusters consist of a network of polymer chains linked together by hydrophobic domains assembling hydrophobic groups of different macromolecules. Unlike the interconnected flowerlike micelles, such nanogels contain a negligibly small number of loops and a significant number of chains connecting the hydrophobic domains.

A modified model of nanogels, which can be considered as core-shell nanogels, has been proposed [19, 34] to describe the structure of clusters of HM chitosan containing 2–4 mol% of *n*-alkyl side groups. It was shown [19, 34] that, for these clusters, the values of R_g/R_h are 0.46–0.62, which is typical [81] for microgels with a surface layer much looser than the core and, hence, with a much higher hydrodynamic radius R_h than the gyration radius R_g. The appearance of such a strongly swollen surface layer in the clusters of HM chitosan was explained as being caused by (1) the surface layer of the nanogels being loosely crosslinked, because of the lack of associating groups from the side of solution [2, 67], and (2) a significant number of free chain ends being present on the surface of the nanogels. Indeed, the macromolecules of HM chitosan that contain a small number of hydrophobic groups have sufficiently long terminal chain fragments free of hydrophobes. The majority of these free ends may be expelled to the surface of clusters because they do not have associating groups; moreover, they have a large number of charged units that may be repelled from the like-charged cluster. According to this model, there is, in addition to the core composed of hydrophobically crosslinked chains, a strongly swollen shell covered by dangling chains. This hydrophilic shell is believed to stabilize the aggregates in an aqueous medium.

Cylindrical clusters were observed for stiff polyelectrolytes, such as for alkyl-substituted poly(*p*-phenylene)sulfonates [51–55].

5.3.2.1.3 Hydrophobic Domains inside the Clusters Within the cluster, individual polymer chains are linked via hydrophobic domains. In the case of intermolecular aggregation, these domains start to form when the polymer concentration reaches the cac value. Often only a part of hydrophobic groups is involved in the aggregation [58, 59, 82–84]. The fraction of aggregated hydrophobic groups can be directly determined by NMR-spectroscopy by integration of separate signals of free and aggregated groups [82–84]. The fraction of aggregated hydrophobes becomes higher at increasing content and size of hydrophobic groups and salt concentration as all these factors favor hydrophobic association [59].

The aggregation numbers of hydrophobic domains in solutions of APs can be determined by several techniques, including fluorescence spectroscopy, pulse field gradient spin-echo NMR, and sometimes ESR. It has been shown that the aggregation number of the domains N_d changes only a little with the content of hydrophobic groups. For example, for HM poly (sodium acrylate) (Figure 5.2), it remains equal to ca. 45, when the content of C12 hydrophobic substituents increases from 5 to 15 mol% [59]. The aggregation numbers of hydrophobic domains in uncharged HM polyacrylamide with the same associating C12 groups (N_d = 60–80) [58] are somewhat higher than those in highly charged HM poly(sodium acrylate). This difference may be due to the electrostatic repulsion between charged acrylate groups in the corona of hydrophobic domains, which reduces the size of the domains. It should be noted that the aggregation numbers of hydrophobic domains in APs can be smaller than in the corresponding low molecular weight surfactants which was attributed [59] to steric hindrances imposed by polymer backbone. Therefore the size of domains formed by a given type of hydrophobic group seems to be controlled by electrostatic repulsion and steric hindrances imposed by polymer chains connected to the domain.

Inside the clusters, the hydrophobic domains are not identical. NMR data unambiguously show [83] that there are two types of hydrophobic domains that differ in the exchange rate with free hydrophobes: strong domains with slow exchange rate and weak domains with faster exchange rate. The latter are called "small preaggregates" [83, 84]. To my opinion, the small domains can be located mainly near the surface of the cluster. This may be because of (1) enhanced electrostatic repulsion in this area due to higher local concentration of charged segments tending to be in contact with water and (2) a lack of neighboring chains providing hydrophobes. Yet, in the interior of the cluster, more energetically favorable ("strong") hydrophobic domains with an optimum content of hydrophobic groups are being formed. For a complete realization of hydrophobic attraction in the system, the aggregates must be as large as possible with the smaller fraction of "weak" hydrophobic domains in the surface layer. So, the formation of stronger hydrophobic domains, which exist mainly in the interior part of the cluster, may be the molecular basis for the growth of clusters induced by the enhancement of a hydrophobic aggregation.

5.3.2.2 ***Macroscopic Gel*** To gain in attraction energy, the clusters can coalesce, forming a dense macroscopic gel. As in the dilute regime, the amount of polymer in the solution is insufficient to form a gel in the whole volume of the system, so the smaller gel will form and it will coexist with a large volume of the external solution. Such a coalescence of clusters can occur, for example, due to an increasing concentration of associating groups. Along with this "normal" gelation (the clusters form a dense gel with the increase in the number of associating groups), there has also been observed so-called anomalous gelation [67], when the clusters coalesce into a loose gel at the decrease in the number of associating groups. The anomalous gelation phenomenon is of a polyelectrolyte nature. When the number of associating groups decreases, the hydrophilic segments linking these groups become longer, resulting in a higher swelling of the cluster. At some degree of swelling, the total volume of the clusters becomes so large (and the volume of external solution so small), that for the counterions it becomes preferable not to be located in a single cluster (and for the part of them—in a rather small volume of the external solution) but to travel in the whole large volume of the highly swollen macrogel that was formed by the coalescence of the clusters. Thus, in the dilute regime, the phase separation that occurs with the formation of the macroscopic gel can take place at both low and high contents of hydrophobic groups, whereas in the intermediate cases, clusters are favored.

5.4 ASSOCIATING POLYELECTROLYTES IN SEMIDILUTE SOLUTIONS

In semidilute solutions, the aggregation of hydrophobic groups belonging to different macromolecules leads to the formation of an intermolecular network. The association of chains in a semidilute regime is more straightforward [1] than in the dilute regime because of the lower entropy losses of the counterions: in the semidilute regime, the counterions can displace throughout the whole volume of the system and not be confined in the restricted area of clusters [2].

In semidilute solutions, APs show much higher viscosity, as compared to the parent polymers without associating groups (Figure 5.3). This is due to the intermolecular crosslinking of polymer chains by hydrophobic domains. Each hydrophobic domain usually contains a few dozen associating groups belonging to different macromolecules [58], so that it acts as multifunctional crosslink in the network. When the concentration of APs increases, new hydrophobic domains are formed, but the aggregation number of each domain N_d (i.e., the number of

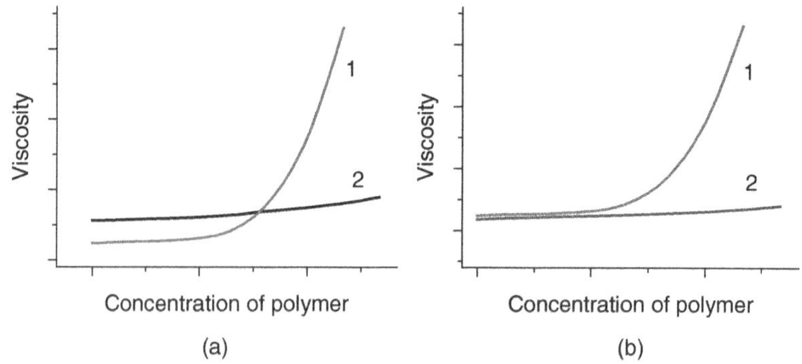

Figure 5.3 Typical dependences of viscosity on polymer concentration for associating polyelectrolyte (1) and corresponding polymer without associating groups (2) in the presence (a) and absence (b) of intramolecular aggregation in a dilute solution.

hydrophobic groups in it) remains unchanged. This is similar to the behavior of low molecular weight surfactants just above the critical micelle concentration: the increasing surfactant concentration induces the increase of the number of micelles, but their size and aggregation number do not change. Similar to the micelles of low molecular weight surfactants, the hydrophobic domains are "living" objects as they are being incessantly reformed: there is always an exchange of hydrophobic groups between different domains, and there are also some free nonaggregated groups present in the system [83]. At the same time, the dynamics of exchange of the hydrophobic tails between hydrophobic domains in APs is much slower [83] than in the low-molecular-weight surfactants because they are linked to a polymeric chain and therefore keep this chain with them at the displacement.

The rheological properties of solutions of associating polymers including APs were recently reviewed by C. Chassenieux et al. [85]. In this section, I consider mainly the responsiveness of these properties to various parameters, which is due to the reversible character of hydrophobic domains formed by weak noncovalent interactions. The responsiveness opens some rich possibilities for fine-tuning the rheological properties of the solutions. Due to the reversible character of the junctions, the semidilute solutions of APs exhibit many peculiar phenomena. The first set of phenomena concerns shear-responsive behavior, which, depending on the conditions, can lead either to thinning or thickening of the solutions.

A pronounced shear-thinning behavior is very common for most, if not all, APs. This behavior is due to the disruption of the labile crosslinks (hydrophobic domains) under the shear [86]. At the highest shear rates (above ca. 800–1500 s^{-1}) the viscosity tends toward that of the corresponding parent polymer without hydrophobes [8, 86, 87]. An important point is that for APs the shear-thinning is completely reversible, and at rest, the solution recovers its initial high viscosity. This property is highly exploited commercially, in particular, in various fluids for oil recovery (low viscosity at pumping and gelation at rest) and for painting (low viscosity at the application of paint on a surface with further jellification preventing dithering). It should be noted that high viscosity at rather low polymer concentrations can be attained not only for

APs but also for very-high-molecular-weight polymers that do not contain associating groups. But in this instance, the shear-thinning is not fully reversible because of the possibility of chain destruction under the shear. This is why APs are more promising as viscosity modifiers in comparison with very-high-molecular-weight polymers.

In contrast to shear-thinning, an opposite effect—shear thickening—is rather rarely observed in APs [8, 27, 86, 88, 89]. Shear thickening is attributed to shear-induced switching from intra- to intermolecular interactions of associating groups [90]. At a certain shear rate, the shear forces extend so considerably the polymer coils that they disrupt the intramolecular hydrophobic associations. The released hydrophobic groups associate intermolecularly, thus intensifying the viscosity [86]. A determining factor for a good thickening ability is the distribution of hydrophobic groups [8]. Shear thickening disappears with increasing concentration of polymer, which favors the formation of intermolecular hydrophobic domains in the initial solutions (before the application of shear); therefore, no significant switching from intra- to intermolecular interactions can be achieved under the shear.

In some APs, a small shear thickening is observed just before a wide shear-thinning region on the flow curves (Figure 5.4a) [8, 86]. But the most interesting is a giant shear-thickening effect, at which the viscosity of the solution of APs abruptly increases by few orders of magnitude at some critical shear rate (Figure 5.4b). Such a huge effect was observed in the aqueous solutions of a HM poly(*N*,*N*′-dimethylacrylamide-co-acrylic acid) [27, 88, 89]. This polymer has a high molecular weight and exhibits pronounced intramolecular aggregation in dilute solutions. It was shown that for the abrupt giant shear thickening to occur, the solution of the associating polymer would need to be on the threshold of phase separation and the concentration of the polymer be close to cac [89]. In fact, the giant shear-induced thickening is a gelation process occurring at a critical shear rate [89]. It was observed that reducing the stress after gelation leads to the stiffening of the gels. This behavior was explained [89] by the formation of additional intermolecular associations, when the bridging chains in a gel retract upon the reduction of stress.

Keeping in mind that the aggregation of APs is governed by the interplay of hydrophobic attraction and electrostatic repulsion, it appears that we can control the rheological properties by increasing or decreasing the attractive or repulsive forces in accord with content, size, and distribution of hydrophobic groups; content of the charged groups; or addition of low molecular weight salt or some organic co-solvents preventing aggregation. It was demonstrated that the higher the content of the hydrophobic groups and the longer their length (i.e., the higher the attraction energy), the more pronounced is the increase of viscosity (Figure 5.3). At a constant hydrophobe level, a microblocky distribution of hydrophobic groups along the chain enhances the viscosity. The longer the hydrophobic microbocks along the chain, the more pronounced is the increase of viscosity [8, 86, 87]. However, at too strong an attraction of hydrophobic groups (e.g., due to their high content), phase separation occurs.

As for the charged units, the electrostatic repulsion that they induce has two opposite effects on viscosity [8]. On the one hand, the viscosity may decrease because of the lower degree of association. On the other hand, the viscosity may increase because of the coil expansion. Also charged units may enable a larger number of associating groups to be introduced in the AP without losing its water solubility. Therefore, for each type of AP, an optimum composition (content, size, and distribution of hydrophobic groups; content of charged groups) capable of providing higher viscosity can be found.

Since APs contain charged units, their viscosity is highly sensitive to the addition of low-molecular-weight salt, which can screen electrostatic repulsion and strengthen the hydrophobic interactions. The screening of the electrostatic repulsion encourages similarly charged macromolecules to associate, and thus viscosity increases. At the same time, the screening of the repulsion causes the polyelectrolyte coils to shrink, and thus viscosity decreases. Therefore, salt can have different impacts on viscosity, depending on which of these effects dominates at given conditions (molecular structure of polymer, its concentration, temperature, etc.).

The rheological properties of APs can also be fine-tuned by adding some organic co-solvents that weaken the hydrophobic attraction between associating groups. This approach was exploited recently [91–95] to create a system that blocks selectively the water flow in the oil well. For this purpose, the authors used an organic co-solvent (e.g., ethanol) that is perfectly soluble in water but insoluble in hydrocarbons (oil). Addition of this co-solvent to a semidilute aqueous solution of an HM partially hydrolyzed polyacrylamide had a pronounced effect on viscosity, reducing it as a result of a disruption to the hydrophobic domains and a shrinking of the polymer coils (ethanol is a poor solvent for polyacrylamide). When this fluid with low viscosity was brought in contact with water, ethanol, which is perfectly soluble in water, diffused out of the polymer solution. Once the substance preventing gelation was removed, instantaneously a strong gel formed on the interface between water and polymer solution. This gel served as a plug to prevent the water flow. However, when in contact with oil, the polymer solution maintained a low viscosity because ethanol, being insoluble in oil, remains in the polymer solution and continues

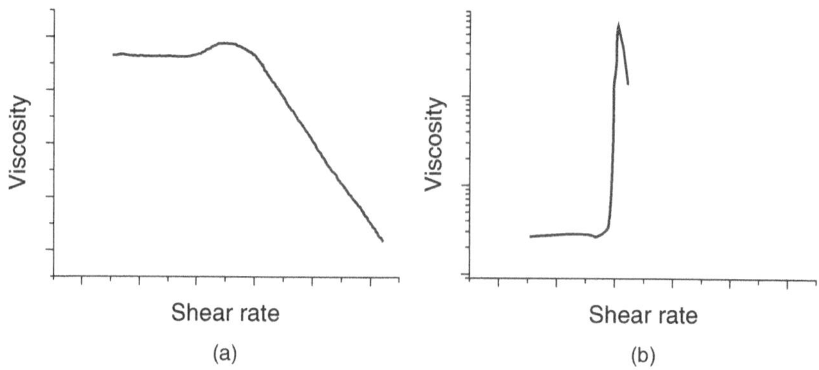

Figure 5.4 Schematic representation of "ordinary" (a) and "giant" (b) shear thickening preceding shear thinning in solutions of associating polyelectrolytes.

to prevent gelation. Thus, by exploiting the reversible character of hydrophobic junctions, a smart system was created [91–95]. This system is able to recognize the place of water inflow in the well and to block it without hindering the flow of oil.

Clearly, the reversibility of hydrophobic associations can be exploited in tailoring the properties of AP solutions, and this is quite promising for various applications of these systems.

5.5 CONCLUSIONS

Competitive interactions of APs appear to be easily tunable by different parameters. This makes APs very attractive for the development of responsive nano-objects in dilute solutions and "smart" rheology modifiers acting in a semidulute regime. Among these systems, in the last decade particular attention was paid to the creation of various self-assembled nanostructures in dilute solutions. Taking into account that the folding and the self-assembly of APs depends essentially on the sequence of monomer units, one of the most promising approaches is to design the nanostructures with the required properties in two steps: (1) determine the specific monomer sequence that can ensuring the formation of the desired structure (e.g., by computer modeling) and (2) prepare the targeted polymer by controlled polymerization techniques. In this way one can obtain quite sophisticated nano-objects with the desired internal structure capable of being applied, for example, for specific molecular recognition or catalysis.

REFERENCES

1. V. Aseyev, H. Tenhu, F. M. Winnik. *Adv. Polym. Sci. 196*, 1–85 (**2006**).
2. I. I. Potemkin, K. B. Zeldovich, A. R. Khokhlov, *Polymer Sci., Ser. C 42*, 154–171 (**2000**).
3. U. P. Strauss, E. G. Jackson, *J. Polym. Sci. 6*, 649–659 (**1951**).
4. K. S. Soppimatha, T. M. Aminabhavia, A. R. Kulkarnia, W. E. Rudzinski, *J. Controlled Release 70*, 1–20 (**2001**).
5. N. Bhattarai, J. Gunn, M. Zhang, *Adv. Drug Delivery Rev. 62*, 83–99 (**2010**).
6. R. T. Chacko, J. Ventura, J. Zhuang, S. Thayumanavan, *Adv. Drug Delivery Rev. 64*, 836–851 (**2012**).
7. M. A. Winnik, A. Yekta, *Curr. Opin. Colloid Interface Sci. 2*, 424–436 (**1997**).
8. F. Candau, J. Selb, *Adv. Colloid Interface Sci. 79*, 149–172 (**1999**).
9. C. Tsitsilianis, I. Iliopoulos, *Macromolecules 35*, 3662–3667 (**2002**).
10. R. Zhang, T. Shi, L. An, Q. Huang, *Macromolecules 45*, 555–562 (**2012**).
11. O. E. Philippova, E. V. Volkov, N. L. Sitnikova, A. R. Khokhlov, J. Desbrières, M. Rinaudo, *Biomacromolecules 2*, 483–490 (**2001**).
12. M. Rinaudo, *Progress Polym. Sci. 31*, 603–632 (**2006**).
13. A. R. Khokhlov, P. G. Khalatur, *Physica A 249*, 253–261 (**1998**).
14. A. R. Khokhlov, P. G. Khalatur, *Phys. Rev. Lett. 82*, 3456–3459 (**1999**).

15. H. K. Murnen, A. R. Khokhlov, P. G. Khalatur, R. A. Segalman, R. N. Zuckermann, *Macromolecules 45*, 5229–5236 (**2012**).
16. N. Giuseppone, J. F. Lutz, *Nature 473*, 40–41 (**2011**).
17. J. F. Lutz, M. Ouchi, D. R. Liu, M. Sawamoto, *Science 341*, 1238149 (**2013**).
18. J. F. Lutz, *ACS Macro Letters 3*, 1020–1023 (**2014**).
19. E. V. Korchagina, O. E. Philippova, *Langmuir 28*, 7880–7888 (**2012**).
20. Y. Chang, C. L. McCormick, *Macromolecules 26*, 6121–6126 (**1993**).
21. Y. Hu, R. S. Armentrout, C. L. McCormick, *Macromolecules 30*, 3538–3546 (**1997**).
22. Y. Morishima, S. Nomura, T. Ikeda, M. Seki, M. Kamachi, *Macromolecules 28*, 2874–2881 (**1995**).
23. F. S. Hwang, T. E. Hogen-Esch, *Macromolecules 28*, 3328–3335 (**1995**).
24. T. Noda, A. Hashidzume, Y. Morishima, *Macromolecules 34*, 1308–1317 (**2001**).
25. H. Yamamoto, Y. Morishima, *Macromolecules 32*, 7469–7475 (**1999**).
26. H. Yamamoto, I. Tomatsu, A. Hashidzume, Y. Morishima, *Macromolecules 33*, 7852–7861 (**2000**).
27. A. Lele, A. Shedge, M. Badiger, P. Wadgaonkar, C. Chassenieux, *Macromolecules 43*, 10055–10063 (**2010**).
28. S. Yusa, A. Sakakibara, T. Yamamoto, Y. Morishima, *Macromolecules 35*, 10182–10188 (**2002**).
29. H. Morimoto, A. Hashidzume, Y. Morishima, *Polymer 44*, 943–952 (**2003**).
30. C. Esquenet, E. Buhler, *Macromolecules 34*, 5287–5294 (**2001**).
31. O. Trotsenko, Y. Roiter, S. Minko, *Langmuir 28*, 6037–6044 (**2012**).
32. O. V. Borisov, A. Halperin, *Macromolecules 29*, 2612–2617 (**1996**).
33. O. V. Borisov, A. Halperin, *Curr. Opin. Colloid Interface Sci. 3*, 415–421 (**1998**).
34. E. V. Korchagina, O. E. Philippova, *Biomacromolecules 11*, 3457–3466 (**2010**).
35. O. E. Philippova, E. V. Korchagina, *Polymer Sci., Ser. A 54*, 552–572 (**2012**).
36. A. M. Rosales, H. K. Murnen, R. N. Zuckermann, R. A. Segalman, *Macromolecules 43*, 5627–5636 (**2010**).
37. M. E. Davis, Z. Chen, D. M. Shin, *Nat. Rev. Drug Discovery 7*, 771–782 (**2008**).
38. S. Minko, A. Kiriy, G. Gorodyska, M. Stamm, *J. Am. Chem. Soc. 124*, 3218–3219 (**2002**).
39. A. Kiriy, G. Gorodyska, S. Minko, J. Werner, P. Štěpánek, M. Stamm, *J. Am. Chem. Soc. 124*, 13454–13462 (**2002**).
40. L. J. Kirwan, G. Papastravou, M. Borkovec, *Nanoletters 4*, 149–152 (**2004**).
41. M. N. Spiteri, C. E. Williams, F. Boué, *Macromolecules 40*, 6679–6691 (**2007**).
42. W. Essafi, W. Raissi, A. Abdelli, F. Boué. *J. Phys. Chem. B 118*, 12271–12281 (**2014**).
43. A. V. Dobrynin, M. Rubinstein, *Macromolecules 33*, 8097–8105 (**2000**).
44. A. V. Dobrynin, M. Rubinstein, S. P. Obukhov, *Macromolecules 29*, 2974–2979 (**1996**).
45. J. Jeon, A. V. Dobrynin, *Macromolecules 40*, 7695–7706 (**2007**).
46. A. V. Dobrynin, Solutions of Charged Polymers. In: K. Matyjaszewski and M. Möller (eds.) *Polymer Science: A Comprehensive Reference*, Amsterdam: Elsevier, *1*, 81–132 (**2012**).
47. A. V. Lyulin, B. Dünweg, O. V. Borisov, A. A. Darinskii, *Macromolecules 32*, 3264–3278 (**1999**).

48. V. V. Vasilevskaya, V. A. Markov, P. G. Khalatur, A. R. Khokhlov, *J. Chem. Phys. 124*, 144914 (**2006**).
49. V. V. Vasilevskaya, V. A. Markov, G. ten Brinke, A. R. Khokhlov, *Macromolecules 41*, 7722–7728 (**2008**).
50. A. A. Glagoleva, V. V. Vasilevskaya, K. Yoshikawa, A. R. Khokhlov, *J. Chem. Phys. 139*, 244901 (**2013**).
51. O. E. Philippova, R. Rulkens, B. I. Kovtunenko, S. S. Abramchuk, A. R. Khokhlov, G. Wegner, *Macromolecules 31*, 1166–1179 (**1998**).
52. R. Rulkens, G. Wegner, T. Thurn-Albrecht, *Langmuir 15*, 4022–4025 (**1999**).
53. M. Bockstaller, W. Kohler, G. Wegner, G. Fytas, *Macromolecules 34*, 6353–6358 (**2001**).
54. A. Kroeger, J. Belack, A. Larsen, G. Fytas, G. Wegner, *Macromolecules 39*, 7098–7106 (**2006**).
55. A. Kroeger, A. Deimedo, J. Belack, I. Liberwirth, G. Fytas, G. Wegner, *Macromolecules 40*, 105–115 (**2007**).
56. T. Mes, R. Van der Weegen, A. R. A. Palmans, E. W. Meijer, *Angew. Chem. Int. Ed. 50*, 5085–5089 (**2011**).
57. J. Huang, A. Heise, *Chem. Soc. Rev. 42*, 7373–7390 (**2013**).
58. Yu. A. Shashkina, Yu. D. Zaroslov, V. A. Smirnov, O. E. Philippova, A. R. Khokhlov, T. A. Pryakhina, N. A. Churochkina, *Polymer 44*, 2289–2293 (**2003**).
59. F. Petit-Agnely, I. Iliopoulos, R. Zana, *Langmuir 16*, 9921–9927 (**2000**).
60. C. Senan, J. Meadows, P. T. Shone, P. A. Williams, *Langmuir 10*, 2471–2479 (**1994**).
61. S. Ercelen, X. Zhang, G. Duportail, C. Grandfils, J. Desbrières, S. Karaeva, V. Tikhonov, Y. Mely, V. Babak, *Colloids Surf. B 51*, 140–148 (**2006**).
62. G.B. Jiang, D. Quan, K. Liao, H. Wang, *Carbohydr. Polym. 66*, 514–520 (**2006**).
63. M. Rinaudo, R. Auzely, C. Vallin, I. Millagaliev, *Biomacromolecules 6*, 2396–2407 (**2005**).
64. O. Ortona, G. D'Errico, G. Mangiapia, D. Ciccarelli, *Carbohydr. Polym. 74*, 16–22 (**2008**).
65. O.E. Philippova, E. V. Korchagina, E. V. Volkov, V. A. Smirnov, A. R. Khokhlov, M. Rinaudo, *Carbohydr. Polym. 87*, 687–694 (**2012**).
66. I. I. Potemkin, V. V. Vasilevskaya, A. R. Khokhlov, *J. Chem. Phys. 111*, 2809–2817 (**1999**).
67. I. I. Potemkin, S. A. Andreenko, A. R. Khokhlov, *J. Chem. Phys. 115*, 4862–4872 (**2001**).
68. G. Lamarque, J. M. Lucas, C. Viton, A. Domard, *Biomacromolecules 6*, 131–142 (**2005**).
69. N. Kubota, Y. Eguchi, *Polym. J. (Tokyo) 29*, 123–127 (**1997**).
70. T. Noda, Y. Morishima, *Macromolecules 32*, 4631–4640 (**1999**).
71. M. Bodnar, J. F. Hartmann, J. Borbely, *Biomacromolecules 7*, 3030–3036 (**2006**).
72. C. Zhang, Y. Ding, Q. Ping, L. Yu, *J. Agric. Food Chem. 54*, 8409–8416 (**2006**).
73. G.-B. Jiang, D. Quan, K. Liao, H. Wang, *Carbohydrate Polym. 66*, 514–520 (**2006**).
74. Y.-S. Wang, L.-R. Liu, Q. Jiang, Q.-Q. Zhang, *Eur. Polym. J. 43*, 43–51 (**2007**).
75. Z. Zhao, M. He, L. Yin, J. Bao, L. Shi, B. Wang, C. Tang, C. Yin, *Biomacromolecules 10*, 565–572 (**2009**).
76. S. Popa Nita, P. Alcouffe, C. Rochas, L. David, A. Domard, *Biomacromolecules 11*, 6–12 (**2010**).
77. M. A. Pigaleva, I. V. Portnov, A. A. Rudov, I. V. Blagodatskikh, T. E. Grigoriev, M. O. Gallyamov, I. I. Potemkin, *Macromolecules 47*, 5749–5758 (**2014**).

78. S. Yusa, M. Kamachi, Y. Morishima, *Langmuir 14*, 6059–6067 (**1998**).
79. D. Baigl, M. Sferrazza, C.E. Williams, *Europhys. Lett. 62*, 110–116 (**2003**).
80. Y. S. Wang, L. R. Liu, Q. Jiang, Q.-Q. Zhang, *Eur. Polym. J. 43*, 43–51 (**2007**).
81. W. Burchard, *Adv. Polym. Sci. 143*, 113–194 (**1999**).
82. F. Petit, I. Iliopoulos, R. Audebert, *Polymer 39*, 751–753 (**1998**).
83. F. Petit-Agnely, I. Iliopoulos, *J. Phys. Chem. B 103*, 4803–4808 (**1999**).
84. I. Furó, I. Iliopoulos, P. Stilbs, *J. Phys. Chem. B 104*, 485–494 (**2000**).
85. C. Chassenieux, T. Nicolai, L. Benyahia, *Curr. Opin. Colloid Interface Sci. 16*, 18–26 (**2011**).
86. E. Volpert, J. Selb, F. Candau, *Macromolecules 29*, 1452–1463 (**1996**).
87. A. Hill, F. Candau, J. Selb, *Macromolecules 26*, 4521–4532 (**1993**).
88. A. Cadix, C. Chassenieux, F. Lafuma, F. Lequeux, *Macromolecules 38*, 527–536 (**2005**).
89. J. Wang, L. Benyahiya, C. Chassenieux, J.-F. Tassin, T. Nicolai, *Polymer 51*, 1964–1971 (**2010**).
90. T. A. Witten, M. H. Cohen, *Macromolecules 18*, 1915–1918 (**1985**)
91. Y. D. Zaroslov, O. E. Filippova, I. V. Blagodatskikh, A. R. Khokhlov, Patent GB 2393962 A (13.07.2004).
92. Y. D. Zaroslov, O. E. Filippova, I. V. Blagodatskikh, A. R. Khokhlov, Patent US 7151078 (19.12.2006).
93. Y. D. Zaroslov, O. E. Filippova, I. V. Blagodatskikh, A. R. Khokhlov, Patent US 7287588 (30.10.2007).
94. Y. D. Zaroslov, O. E. Filippova, I. V. Blagodatskikh, A. R. Khokhlov, Patent RU 2276675 C2 (20.05.2006).
95. O. E. Philippova, A. R. Khokhlov, *Petroleum Chemistry 50*, 266–270 (**2010**).

6

MACROMOLECULAR DECORATION OF NANOPARTICLES FOR GUIDING SELF-ASSEMBLY IN 2D AND 3D

CHRISTIAN KUTTNER, MUNISH CHANANA, MATTHIAS KARG, AND ANDREAS FERY

Leibniz-Institut für Polymerforschung Dresden e.V., Institute of Physical Chemistry and Polymer Physics, Germany; Technische Universität Dresden, Physical Chemistry of Polymeric Materials and Cluster of Excellence Centre for Advancing Electronics Dresden (cfaed), Germany; University of Bayreuth, Physical Chemistry I and Physical Chemistry II, Germany; Heinrich Heine University Düsseldorf, Physical Chemistry I, Germany; ETH Zürich, Institute of Building Materials, Switzerland

6.1 INTRODUCTION

Modern functional materials aim at combining different compounds in order to exploit *synergistic effects* or even trigger the *emergence of novel properties*, which neither of the compounds in pure form possesses. Typically, this requires hierarchical organization of nano-/microscale building blocks along the lines of rational design concepts, rather than simple blending of compounds. In fact, in many ways, artificial functional materials approach biological systems, though still far from their complexity [1–3]. Typical examples of such *materials by design* are optical or mechanical metamaterials [4–7], but also modern photovoltaic [8, 9] and optoelectronic devices [10], and nanocomposites [11–15].

In this context, macromolecules play a key role in different perspectives: First, they cover a broad range of length scales between a few nanometers up to several hundred nanometers. As illustrated in Figure 6.1, the critical distances for several important

Macromolecular Self-Assembly, First Edition. Edited by Laurent Billon and Oleg Borisov.

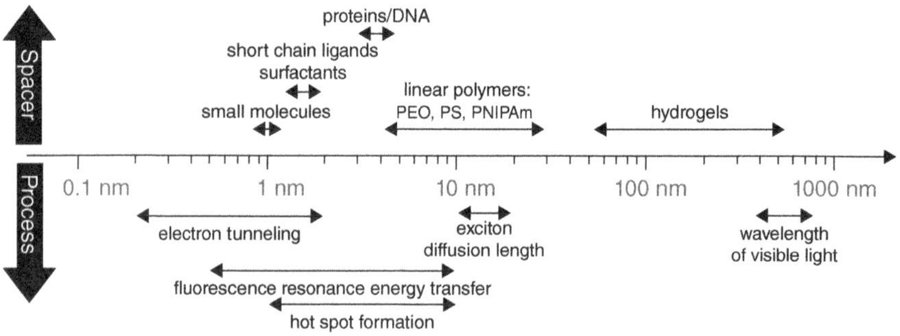

Figure 6.1 Schematic overview showing the correlation of different spacer sizes and physical processes covering the length scales from 0.1 to 1000 nm.

optical and electronic processes fall within this regime. Second, they offer a rich variety of ways for controlling mesostructure formation, for example, as microphase separation for block-copolymer systems [16] or stimuli-responsive assemblies (fibers) for structural protein systems [17–21]. Finally, polymers can be sensitive to a broad range of stimuli such as solvent quality, temperature, pH, salt-concentration, and redox potential [22–25].

In this chapter, we focus on decorating nanoparticles with synthetic as well as biomacromolecules for controlling interparticle distances and relative orientations in assemblies. We introduce synthetic approaches toward core/shell architectures and show how macromolecular ligands allow controlling order in 2D and 3D assemblies. We also illustrate how symmetries can be broken in order to advance from simple hexagonally close-packed assemblies to systems of higher complexity. We put special emphasis on core/gel-shell nanoparticles as a versatile class allowing for large interparticle separations and on protein-modified nanoparticles that are particularly well suited for achieving complex symmetries in particle assemblies. Finally, we point out several examples for functionality, not with the aim of completely covering a specific field of applications but rather for the sake of illustrating the underlying concepts for *functional material design.*

6.2 GUIDING ASSEMBLY BY DECORATION WITH ARTIFICIAL MACROMOLECULES

The state-of-the-art synthesis routes for macromolecular building blocks provide us with excellent structural and compositional control. Parameters relevant for guiding assembly such as molecular weight, wettability/solvent compatibility, steric properties, and conformation can be controlled to a large degree. In this section, we will introduce an overview over the different strategies for coupling macromolecules to nanoparticles in a well-defined fashion, covering both crosslinked and non-crosslinked polymeric coatings. In the following, we will illustrate, how these coatings can be employed for controlling assembly in two and three dimensions, starting from hexagonal structures up to linear assemblies.

6.2.1 Decoration of Nanoparticles

Nearly all types of nanoparticles feature an organic low-molecular-weight ligand shell, which is inevitable for their synthesis and ensures colloidal stability. The most prominent stabilizers comprise citrate, fatty acids, and other surfactant molecules. Usually these stabilizing molecules are present in huge excess in the as-prepared nanoparticle dispersions. For assembly purposes, this excess concentration has to be reduced by purification procedures such as centrifugation/redispersion, precipitation/redispersion, dialysis, and filtration. Due to the low molecular weight of these stabilizers, only short interparticle distances are achievable in assemblies of the respective nanoparticles. Therefore, for better interparticle distance control, the original stabilizing agents have to be replaced by macromolecules. By controlling the polymer size and architecture, interparticle distances in 2D and 3D assemblies can be finely tuned (Figure 6.1).

The main challenge in decorating nanoparticles with polymers is to anchor the polymer onto the nanoparticle surface. Depending on the surface chemistry of the nanoparticles, different anchoring groups can be employed to attach the polymer robustly onto the nanoparticle surface. In order to remove the initial low-molecular-weight ligand from the surface and bind polymers strongly to the particle surface, these anchoring groups are required to exhibit a larger binding affinity toward the particle surface (large binding constant K) [26]. In general, for metal nanoparticles such as gold or silver nanoparticles, sulfurous groups such as thiols, disulfides, and amine groups are commonly used as anchor groups [27]. For metal oxide nanoparticles, carboxylates, phosphonates, and catechol groups can be employed for attaching polymers to the particle surface [28]. In particular, catechol groups show a high binding affinity toward iron oxide surfaces, even in aqueous dispersion [29–31]. For silica particles, silane chemistry offers a vast library of molecules such as mono-, di-, and trialkoxy or halogenated silanes, thus enabling many different synthetic procedures to anchor organic macromolecules very robustly onto the silica surfaces [32–34]. For organic particles and surfaces, organic coupling reactions can be employed to covalently couple macromolecules on them. Some prominent examples are amide bond coupling (carboxylate + amine), thiol-ene coupling reaction (thiol + unsaturated species) [35, 36], urea-reaction (iso(thio)cyanate + amine), acetal and hemiacetal reactions (ketone or aldehyde + alcohol), imine reaction (ketone or aldehyde + amine), and other click-chemistry approaches (e.g., azide + alkyne) [37–41].

Also multiple anchoring groups, as present in diblock copolymers, yield a robust and permanent attachment of the desired polymer to the particle surface. For more detailed information, the reader is referred to the reviews cited here [42, 43]. Once the right anchoring group has been identified, the polymer chemistry offers many synthetic strategies to functionalize nanoparticles with polymers [41]. With the aid of various living polymerization techniques developed thus far, a plethora of polymers can be anchored on various types of nanoparticles.

Figure 6.2 gives a schematic representation of available pathways to decorate nanoparticles with macromolecules. Depending on the strategy, different architectures and thicknesses of the polymer layer can be realized. Thus complex polymer coronas have been achieved, such as brush-like architectures with end-tethered chains or gel-like structures with a crosslinked polymer matrix surrounding the nanoparticle cores. The major strategies for the polymer decoration of particles are direct polymer ligand exchange, surface-initiated polymer grafting, and gel-encapsulation upon crosslinking, which will be discussed below in detail.

The first strategy involves the anchoring of pre-synthesized polymer chains (1), which bear specific anchoring/coordination groups for the nanoparticle (see Figure 6.2). When the anchoring group is presented on a terminal position of a polymer chain, the polymer can be tethered onto nanoparticles with one end, yielding brush-type polymer architectures [42]. Hence, this *grafting-onto* technique starts with the design and synthesis of polymeric ligands, with only one terminal

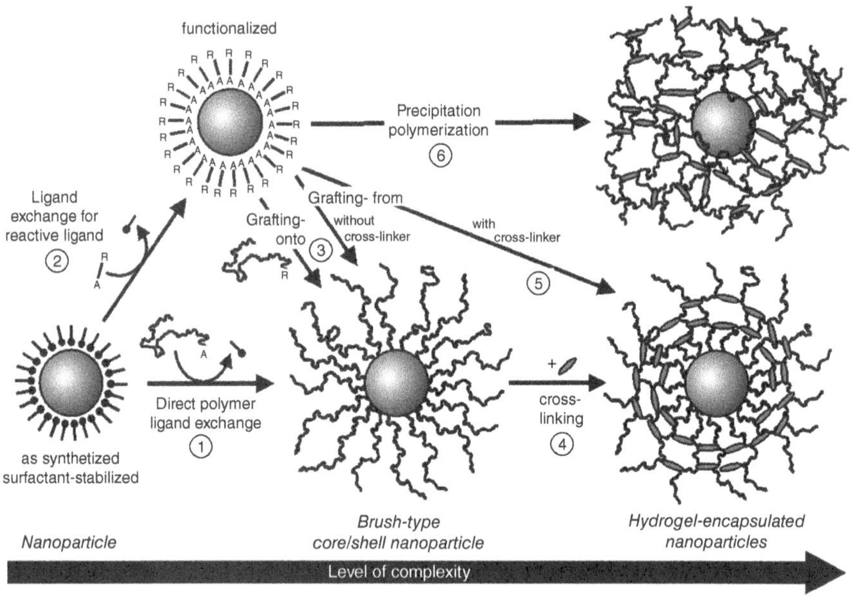

Figure 6.2 Decoration of nanoparticles with polymers with increasing level of complexity (*left to right*): As-synthetized nanoparticles are decorated with linear polymer chains by either (1) direct polymer ligand exchange or by (2) functionalization with reactive ligands and subsequent (3) polymer grafting (from/onto) to form brush-type core/shell nanoparticles. Hydrogel-encapsulation can be achieved by (4) crosslinking of brush-type core/shell nanoparticles, (5) grafting-from polymerization with crosslinker, or directly via (6) precipitation polymerization of functionalized nanoparticles. (A for anchoring and R for reactive functional group).

group serving as the anchoring group. Controlled living polymerization techniques, in particular, the atom-transfer radical polymerization (ATRP) [44], ring-opening metathesis polymerization (ROMP) [45, 46], nitroxide-mediated living radical polymerization (NMP) [47], and reversible addition-fragmentation chain transfer (RAFT) polymerization [48, 49] have proved to be very useful, since the polymerization initiator molecule can be selected such that it bears both functionalities, that is, the polymerization initiating moiety as well as the surface anchoring moiety on the same molecule.

The resulting end-functionalized polymers can be tethered to the particle surface, replacing the original stabilizing agent from the surface in a ligand exchange process. The exchange of the ligand can be driven by concentration effects when the *new* polymer ligand is presented in huge excess and/or making use of the higher binding affinity of the anchor groups of the *new* ligand. This approach yields polymer-decorated nanoparticles. Depending on the exchange protocol including the type of macromolecular ligand and the particle purification, a complete exchange can be achieved [50]. The direct polymer ligand exchange method is rather simple and allows for the precise control of the layer thickness simply by adjusting the molecular weight of the polymer during the polymer synthesis step [30, 51–54].

Brush-like polymer decoration of nanoparticles can also be achieved by growing polymers directly from the particle's surface [55]. As depicted in Figure 6.2, this *grafting-from* approach starts with the attachment of reactive molecules onto the nanoparticle surface (2). This reactive group can then be used in the subsequent polymerization of monomers (3). Such a surface-initiated polymerization yields polymer brushes of higher grafting densities with polymer chains extending outward from the nanoparticle surface. The most common polymerization techniques applied for this are again living radical polymerization techniques such as ATRP, RAFT, ROMP, and NMP, owed to their versatility toward monomers and uniform polymer growth. However, the control over the polymer chain lengths is rather difficult in comparison to *grafting-onto* method, due to the increased rate of chain growth termination caused by the high steric crowding on the particle surface. The main advantage of this technique is the facile purification of the final polymer-decorated nanoparticles via centrifugation steps [55].

The achievable interparticle distances from such polymer-encapsulated nanoparticles with brush-like polymer shells range from a few nanometers to a few tens of nanometers. Larger shell thicknesses, and hence particle separations, are accessible by gel-like polymer shells. Such gel encapsulations enable particle-to-particle separations up to hundreds of nanometers. The most prominent example for such gel networks are hydrogels, which consist of hydrophilic, crosslinked polymer networks swollen with water. The degree of swelling of a hydrogel can be enormous (>95%), and thus leads to the unique viscoelastic properties of hydrogels. In contrast to the rather small polymer brushes, the introduction of a hydrogel shell significantly increases the effective particle volume. This can be of particular interest for 3D assembly, since larger volume fractions can be achieved as compared to as-prepared nanoparticles without a thick hydrogel coating. Figure 6.2 illustrates the schematically different synthetic strategies used to obtain gel-like polymer architectures around the particle cores. In general, *grafting-from* and *grafting-onto* approaches typically lead to core/shell particles with nanoparticle cores and brush-type polymer shells (3). If such brush-type shells are crosslinked, polymer network shells can be obtained (4). In a *grafting-from* approach, crosslinked shells can be grown directly onto the particles, simply by the addition of a bi- or multifunctional crosslinker as a comonomer in the polymerization (5). Besides these two approaches, another common approach to generate core/shell particles with hydrogel shells is based on precipitation polymerization (Figure 6.2). In this case the polymer shell is built up by precipitation of growing oligomer and polymer chains onto the nanoparticle surface (6). Reactive groups on the nanoparticle surface can chemically help to anchor the growing hydrogel to the nanoparticle. These examples only highlight the different concepts applicable to encapsulate nanoparticles by hydrogels. Additional routes may be suited for this task, for example, as in the mini-emulsion polymerization and microfluidic approaches.

The encapsulation of nanoparticles into hydrogel shells has been shown for many nanoparticle compositions, sizes, and even shapes. Already in 1994 Makino *et al.* coated poly(styrene-*co*-*N*-isopropylacrylamide) (PS-*co*-PNIPAm) latex particles with temperature-responsive shells of crosslinked poly(*N*-isopropylacrylamide) (PNIPAm) [56]. These authors synthesized the copolymer cores in a first step using a soap-free emulsion polymerization of styrene with 10 wt-% *N*-isopropylacrylamide (NIPAm) as water-soluble comonomer. The resulting latex spheres were then coated by crosslinked hydrogel shells using a seeded polymerization procedure with NIPAm and the crosslinker *N*,*N*′-methylenebisacrylamide. The final core/shell particles had single latex cores and thermo-responsive hydrogel shells homogeneously encapsulating the cores. In a very similar fashion Okubo *et al.* have synthesized core/shell particles with polystyrene (PS) and PS-*co*-PNIPAm cores, and copolymer shells composed of dimethylaminoethyl methacrylate and ethylene glycol dimethacrylate, as well as NIPAm and *N*,*N*′-methylenebisacrylamide [57]. In this work, the authors have used seeded emulsion polymerization to encapsulate the pre-synthesized latex cores. This two-step protocol resulted in core/shell particles with single nanoparticle cores and homogeneous polymer shells. Dingenouts *et al.* studied the volume phase transition behavior and the form factor of such core/shell particles using small-angle X-ray scattering (SAXS) [58]. In this particular case polystyrene core particles with hydrogel shells composed of chemically crosslinked PNIPAm were analyzed. The scattering data revealed liquid-like local concentration fluctuations of the hydrogel shell. In addition, the authors found that the linking of the hydrogel shell to the latex cores reduces the swelling capacity of the shell. At the same time, a full chain collapse at high temperatures is also prevented.

Extending the concept of seeded polymerization, Zha *et al.* have used silica nanoparticles as seed particles to yield inorganic/organic core/shell particles [59]. To guarantee a covalent attachment of the polymerizing hydrogel network onto the silica cores, these cores were functionalized with 3-(trimethoxysilyl)propyl methacrylate. This functionalization led to the formation of terminal double bonds on the nanoparticle surface. These double bonds could then be used to bind the hydrogel shell obtained by copolymerizing NIPAm with the crosslinker

N,*N*′-methylenebisacrylamide in the presence of the seed particles. The resulting core/shell particles showed the expected volume phase transition behavior of NIPAm-based hydrogels as determined by dynamic light scattering (DLS). Furthermore, the authors could show that dissolution of the silica cores allows for the preparation of hollow hydrogel shells. Following the protocol by Zha *et al.*, silica cores of a broad range of sizes were incorporated into crosslinked hydrogel shells by Karg *et al.* [60–62]. These authors studied the inorganic/organic core/shell particles by different scattering techniques to investigate the swelling properties and network structure of the hydrogel shells as well as the form factor of the silica cores. Particles with a defined core/shell structure and core sizes up to 170 nm in diameter could be successfully synthesized. In all cases, the hydrogel shells showed the typical volume phase transition behavior of PNIPAm. In order to achieve core/shell particles with metal nanoparticle cores bearing a much higher refractive index than silica, the same authors have used silica-coated gold nanoparticles as core material [60]. Due to the silica coating on the gold particles, the same surface chemistry as for functionalization of pure silica nanoparticles could be applied. This way, gold/silica/PNIPAm core/shell/shell particles with thermo-responsive outer shells could be realized.

The direct encapsulation of gold nanoparticles of 17 nm in diameter into responsive hydrogel shells has been presented by Kim *et al.* [63]. These authors used surface-initiated ATRP to polymerize PNIPAm shells on gold nanoparticles. This was achieved by attaching a disulfide containing ATRP initiator on the gold nanoparticle surface. Using NIPAm as monomer and ethylene diacrylate as crosslinking comonomer, the authors were also able to produce core/shell particles with a crosslinked shell apart from brush-type core/shell particles with a non-crosslinked shell. Larger gold nanoparticle cores of 67 nm in diameter were encapsulated by shells of PNIPAm with different degrees of crosslinking by Contreras-Cáceres *et al.* [64]. In their protocol the authors used cetyltrimethylammonium bromide (CTAB) stabilized gold nanoparticles prepared in a seeded-growth fashion, and they polymerized a first layer of PS crosslinked with divinylbenzene around these cores. This first polymer layer improved the colloidal stability of the gold particles and made their surface compatible for the precipitation polymerization of NIPAm with the crosslinker *N*,*N*′-methylenebisacrylamide. The latter precipitation polymerization led to core/shell particles with thermo-responsive hydrogel shells. Furthermore, the authors demonstrated that the porous hydrogel shell allows chemical overgrowth of the gold cores. Applying different amounts of the stabilizer CTAB in the seeded-growth of the gold cores larger cores of around 90 nm in diameter with either a spherical or a star-like shape could be prepared. The same authors have also shown that the concentration of the monomer NIPAm during the polymerization can be used to realize different shell thicknesses and different crosslinking densities [65]. These parameters could be used to tune the response behavior of the core/shell particles.

A more general approach for the encapsulation of metal nanoparticles was introduced by Fernández-López *et al.* [66] These authors used a layer-by-layer strategy to coat CTAB-stabilized gold nanoparticles by an allyl-containing poly(acrylic acid).

The incorporated allyl groups could then be employed for the attachment of a hydrogel shell formed by precipitation polymerization. In this manner, the authors were able to encapsulate spherical, decahedral, and star-shaped nanoparticles into PNIPAm hydrogel shells.

Recently, Karg *et al.* have shown that gold nanoparticles of 14 nm in diameter can be homogeneously encapsulated by PNIPAm-based hydrogels without the need of an intermediate polymer coating [67, 68]. These authors have directly functionalized as-prepared gold nanoparticles with 3-butenylamine and then performed a precipitation polymerization of NIPAm with the crosslinker *N,N′*-methylenebisacrylamide in presence of the functionalized gold nanoparticles. Tuning the monomer feed concentration in the polymerization step and varying the crosslinker amount and addition procedure, core/shell particles with various shell thicknesses, shell morphologies, and swelling capacities were achieved.

Most examples in the literature so far employed hydrogel shells composed of NIPAm with the crosslinker *N,N′*-methylenebisacrylamide prepared via precipitation polymerization. Hence the resulting core/shell particles show the typical volume phase transition behavior of PNIPAm in water. This is an interesting aspect for assembly of such core/shell particles because the shell thickness becomes controllable by temperature. Also a response to pH and ionic strength can be easily programmed in the polymer shell by introducing chargeable functionalities such as carboxyl or amine groups. These groups can be introduced, for example, by copolymerization with respective comonomers. Thus, depending on the chemical composition, the physical properties of the polymer shell can be a function of external parameters such as temperature, pH, and ionic strength among others such as electric field and light. The responsiveness of such core/shell particles has led to an enormous interest in the last decades due to their potential in applications such as sensing, actuation, smart membranes, and as drug delivery systems. Since the main interest in this book chapter lies on the self-assembly of polymer-encapsulated nanoparticles, the reader is referred to these excellent review articles for further information on responsive core/shell materials and their potential applications [69–74].

6.2.2 Distance Control in 2D and 3D

As depicted in Figure 6.1, for several optical and electronic phenomena, interparticle distances in particle assemblies have to match certain well-defined length scales. Therefore, it is crucial to control the interparticle distances or pitch of nanoparticle lattices on surfaces or in bulk matrices. Molecular and macromolecular ligands allow for covering interparticle spacings between 1 nanometer and several 100 nanometers (see Figure 6.3A). Here the main parameter in order to control the distance in 2D and 3D assemblies is the thickness of the macromolecular corona.

Directly from synthesis, small-molecule stabilizers such as citrate yield interparticle distances of only 1 nm. Figure 6.3B shows a colloidal monolayer of 14 nm-sized gold nanoparticles crystallized into hexagonally close-packed lattices [75]. The spacing corresponds well with the size of two overlapping citrate molecules in-between

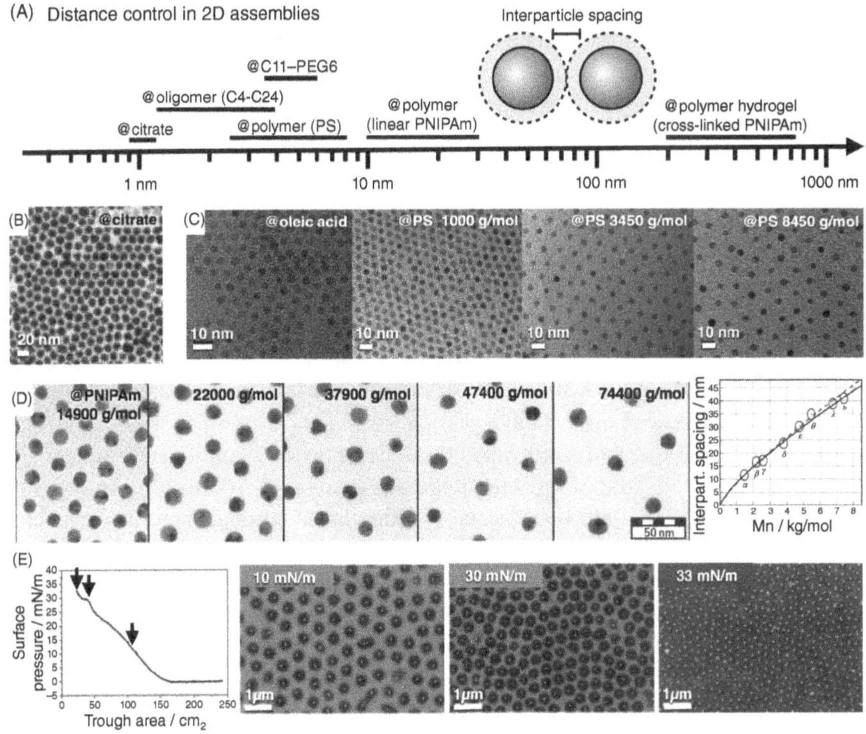

Figure 6.3 Distance control in 2D assemblies: (A) Length scale of interparticle spacings with corresponding examples of macromolecular ligands: citrate (from synthesis), short chain oligomers/surfactants, surfactant–polymer hybrids, linear polymers of different molecular weights, and hydrogel shells. (B) As-synthesized citrate-stabilized gold nanoparticles. Adapted with permission from [75]. Copyright 1993 ACS. (C) Fe_3O_4 nanoparticles with oleic acid and modified with PS ligands. Adapted with permission from [26]. Copyright 2014 ACS. (D) Gold nanoparticles with shells of linear PNIPAm ligands. Adapted with permission from [76]. Copyright 2013 ACS. (E) Ordered arrays of gold nanoparticles with crosslinked PNIPAm shells assembled at the air/water interface at different (transfer) surface pressures. Surface pressure-area isotherm and corresponding monolayer assemblies after transfer to solid substrates. Adapted with permission from [77]. Copyright 2012 ACS .

two particles separated by steric repulsion, if all three carboxyl groups are coordinated toward the surface [75].

To increase the spacing, the original ligand shell can be substituted for oligomeric ligands, which are surfactants with chain lengths of up to twenty carbons. Consequently interparticle separations with oligomeric ligands are inherently limited by their short persistence length and their inherent tendency to intercalate when situated in close contact. Especially in the longer alkyl chains with coordination groups of moderate binding affinity, some interpenetration of contacting ligands may occur, leading to a decrease in the interparticle spacing. Thus surfactants can provide spacings up to about 4 nm [78]. The interparticle separation increases linearly with the number of CH_2 groups, though the size of the head group and of the methyl-terminating group should be taken into account [75]. The surfactant-oligomer hybrids, composed of a short oligomeric anchoring chain (HS-$C_{11}H_{22}$-) and a small number of repeating units of poly(ethylene glycol) (-PEG_6-COOH), can only provide for about 4–6 nm of spacing [79]. Larger separations inherently require a transition from oligomers to polymers.

Figure 6.3C shows the application of PS ligands to widen the interparticle gaps in a colloidal 2D assembly of 5 nm-sized Fe_3O_4 nanoparticles [26]. The first TEM image shows the iron oxide particles with their original oleic acid corona, as received from synthesis. The next images show assemblies of the same core particles but with PS ligands of different molecular weights (M_n = 1000, 3450, and 8450 g/mol) introduced by direct polymer ligand exchange using poly(ethyleneimine) anchors. Here the shortest PS spacer with merely 1000 g/mol, which corresponds to about 10 repeating units, shows an even smaller particle spacing (7.2 nm) than the original stabilizing corona of oleic acid. Only when the ligands become significantly longer, 30 and 80 repeating units, the 2D spacing increases to about 8–10 nm [26]. In order to yield well-defined and controllable distances, the ligand corona has to be dense and homogeneous [26, 80].

Another example of molecular-weight-dependent changes in spacing is shown in Figure 6.3D. Ebeling and Vana prepared 14 nm-sized gold nanoparticles with linear PNIPAm chains prepared by RAFT polymerization [76]. The resulting self-assembled lattices of gold cores show interparticle spacings from 11 to 41 nm, depending on the molecular weight of the ligand. The authors indicated that the spacing does not increase strictly linear with increasing molecular weight of the PNIPAm ligand (see Figure 6.3D, plot to the right). The observed deviation from the model is due to imperfections of the (partially coiled) polymer brush and variations in the grafting density. Besides the end-tethered ligands, the authors presented also ligand motifs resulting in more complex (e.g., crosslinked) binding scenarios [76].

One might expect that the step toward even wider spacings can be simply done by further increasing the size of the polymeric ligands. However, at high molecular weights the binding strength of only one anchoring group may not suffice to keep the polymer chain tethered to the particle surface [81]. Pletsch *et al.* reported on Ag particles modified by direct polymer ligand exchange of the original poly(vinylpyrrolidone) (PVP) shell for thiol-terminated PS chains with six different molecular weights between 21,200 to 217,200 g/mol [82]. Especially for higher molecular weights, a tendency toward aggregation was found even though the NMR results and elementary analysis indicated a quantitative ligand exchange. This colloidal instability may result from the irreversible desorption of the polymer ligand from the particle surface, owed to the imbalance of ligand size and anchoring strength (Au-thiol binding) above a threshold molecular weight.

In order to achieve homogeneous particle arrays of larger spacing, this colloidal instability needs to be avoided. As already discussed in the previous section, one measure to counteract such instabilities is to crosslink the ligand shell forming a stable core/shell particle. Karg *et al.* have shown that monolayer of gold/PNIPAm core/shell particles can be prepared by different methods such as spin-coating [83]. Using hydrogel shells of two different thicknesses, the authors show that the surface coverage of gold nanoparticles can be tuned. The monolayers were characterized by imaging as well as by spectroscopic techniques. The optical properties of the monolayer studied by ellipsometry revealed a larger area fraction of gold particles for the monolayer prepared from core/shell particles with thinner hydrogel shells. Hence, it could be shown that the hydrogel shell allows adjustment of the interparticle separation of the nanoparticle cores.

Capitalizing on the soft properties of hydrogels Vogel *et al.* achieved a gold nanoparticle monolayer with different interparticle distances using a Langmuir–Blodget approach [77]. These authors used gold cores of 60 nm in diameter homogeneously encapsulated in a soft PNIPAm hydrogel shell by precipitation polymerization. Figure 6.3E shows ordered 2D assemblies of the Au@PNIPAm gel particles, which were assembled at the air/water interface and transferred onto solid substrates. Due to the deformability of the soft hydrogel shell, spacings of 650–300 nm, controlled by the surface pressure, could be obtained. The hexagonal ordering is preserved throughout the isotherm and even after transfer of the monolayer onto substrates allowing for precise tuning of the interparticle distances.

Recently, Clara-Rahola *et al.* have shown that the temperature sensitivity of PNIPAm can be used to tune the surface coverage of 2D arrays of PNIPAm-encapsulated gold nanoparticles [84]. In this work the particle arrays were prepared by drop-casting and drying the dilute particle dispersions on indium tin oxide (ITO)-coated glass substrates at different temperatures. The authors showed that the polymer shells can be removed by post-treatment of the particle arrays with plasma etching, hence allowing the preparation of a metal nanoparticle monolayer with tunable interparticle separation.

The effect of surface charge on the assembly of core/shell particles with PS cores was investigated by Lu *et al.* [85]. Core/shell particles with positive and negative surface charge were prepared by simply changing the type of radical initiator used in the polymerization of the hydrogel shell. Ordered structures were obtained for assemblies of charged particles on oppositely charged substrates; however, the degree of order was significantly lower for substrates with an equal sign of charge.

Apart from distance control in two dimensions, the self-assembly can be extended to 3D organizations [87]. Due to the relatively large dimensions of hydrogel shells employed for encapsulating nanoparticle cores, the effective particle volume is strongly increased. This significantly lowers the amount of material needed to achieve volume fractions, which are above the crystallization threshold. Karg *et al.* have studied the phase behavior of gold nanoparticles with PNIPAm hydrogel shell and identified a large range of concentrations where crystalline samples are found [86].

Figure 6.4A shows a self-organized colloidal crystal of such core/shell particles. The hexagonal packing and high degree of order can be depicted from the shown electron micrograph. For studying the crystal structure of 3D assembled polymer-encapsulated nanoparticles, small angle scattering experiments are ideally

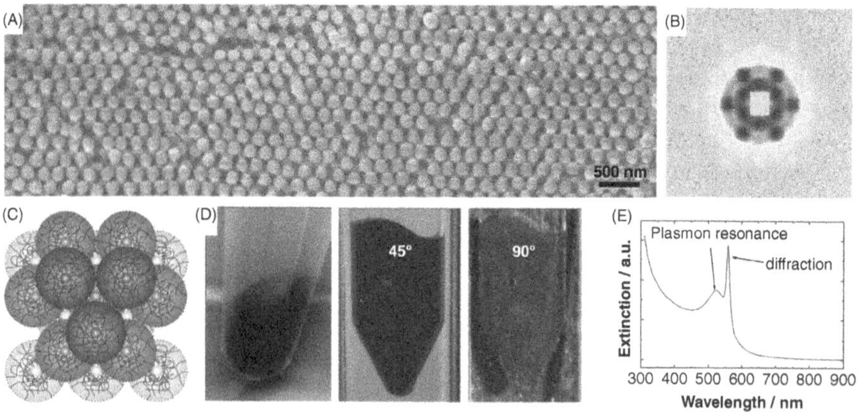

Figure 6.4 Distance control in 3D assemblies: (A) SEM image of a colloidal crystal of gold nanoparticles with shells of crosslinked PNIPAm prepared by precipitation polymerization [Karg *et al.*, unpublished data]. (B) 2D detector image from small-angle neutron scattering of a crystalline Au/PNIPAm core/shell sample measured on the D11 instrument of the Institute Laue–Langevin at a sample-to-detector distance of 28 m [Karg *et al.*, unpublished data]. (C) The pronounced Bragg peaks can be attributed to a face-centered cubic structure [86]. (D) Photographs of the crystalline samples under different angles of illumination [86]. The strong Bragg diffraction can be seen with the naked eye. (E) UV/vis extinction spectrum of a photonic crystal of gold/PNIPAm core/shell particles showing the typical localized surface plasmon resonance of spherical gold nanoparticles and a diffraction peak [86]. (C–E) Adapted with permission from [86]. Copyright 2011 Wiley. (*See color insert for color representation of this figure*).

suited. Figure 6.4B shows a 2D detector image obtained from small angle neutron scattering (SANS) investigation of a colloidal crystal of hydrogel encapsulated gold nanoparticles. The detector image shows strong Bragg peaks, indicating the high degree of order in the colloidal crystal. In this particular case, the Bragg peaks correlate well with a face-centered cubic lattice (Figure 6.4C). For sufficiently large lattice constants, achieved by large hydrogel shell thicknesses, such crystals are characterized by strong and narrow diffraction peaks, which are located in the visible wavelength range. As shown in Figure 6.4D, the crystalline sample exhibits a strong purple iridescence upon top illumination (90°), whereas at an incident angle of 45° the reflection appears deep red. The UV/vis extinction spectra reveal the typical localized surface plasmon resonance of the gold cores (519 nm) and the lattice diffraction (560 nm). Increasing the volume fraction led to a blue shift of the diffraction peak due to a decreasing lattice constant [86]. Furthermore, the authors have shown that temperature can be used to melt and anneal the crystals, due to the volume phase transition behavior of the thermo-responsive PNIPAm shell.

The examples in this section have illustrated that polymer decoration of nanoparticles enables the preparation of 2D and 3D assemblies where the polymer shell thickness can be employed to control the interparticle distance. Furthermore, it was shown that the soft and deformable properties of polymers allow for the reduction of the interparticle separation when for example the volume fraction is increased. The control of the particle gap is prerequisite for the development of optically functional materials that show, for example, strong coupling to external electromagnetic fields as a consequence of collective plasmonic oscillations [87]. As a result, frequency-selective responses of high spectral quality are achievable. Such features increase the potential of colloidal crystals as well as other examples of 3D assemblies like composite and inverse opals to open up new pathways to sophisticated photonic applications [88].

6.2.3 Breaking the Symmetry

After the discussion of macromolecule-guided self-assembly to fabricate close-packed structures in 2D and 3D, the following section will focus on the controlled formation of linear assemblies. The means by which the aggregation of nanoparticles can be directed toward the formation of chains and rings is by breaking the symmetry.

Figure 6.5A introduces the concept of employing the reversibility of surface coordination for controlled aggregation. Ehlert *et al.* showed that reduction of the ligand excess in the direct polymer exchange may result in formation of short multiplet chains and percolated networks from 4 nm CdSe particles stabilized by PS ligands [26]. Following the mass action law, the substitution of ligands is determined by the ligands' binding affinities (equilibrium-binding constant K) and the available ligand concentrations at the particle surface and in solution. Therefore the stability of polymer-stabilized nanoparticles can be reduced in a controlled fashion by using ligands with lower binding strength and/or reducing the excess of the polymer ligands in the ligand exchange [26]. An alternative approach is the application of a competing ligand with higher binding affinity in order to reduce the brush density at the particle surface and thus its colloidal stability.

Kumacheva *et al.* reported on the assembly behavior of bifunctional nanoparticles site-specifically functionalized with low- or high-molecular-weight ligands [89]. Figure 6.5B shows gold nanorods with a surfactant bilayer of CTAB at the longitudinal side and PS end-caps at both ends. The end-caps consist of thiol-terminated linear PS chains of 12,000 g/mol, which were attached by partial ligand exchange based on different crystallographic facets of gold [89]. Upon aggregation in a selective solvent mixture, such amphiphilic nanoparticles form linear aggregates by end-to-end aggregation. However, bifunctionality is not mandatory to break the symmetry in nanoparticle self-assemblies, and also nanoparticles with an isotropic/uniform macromolecule form linear end-to-end assemblies.

Figure 6.5C shows the solution-based self-assembly of silver nanocubes and spheres into dimers and linear oligomers [90]. The 50 nm-sized nanoparticles were stabilized by thiol-terminated PS ligands of 5,000 g/mol attached by ligand exchange. The self-assembly upon controlled aggregation was triggered by the reduction of the solvent quality for PS. Surprisingly, predominantly linear aggregation takes place. End-on attachment was explained as due to attractive hydrophobic/poor solvency forces and repulsive electrostatic forces of charged nanoparticles in a

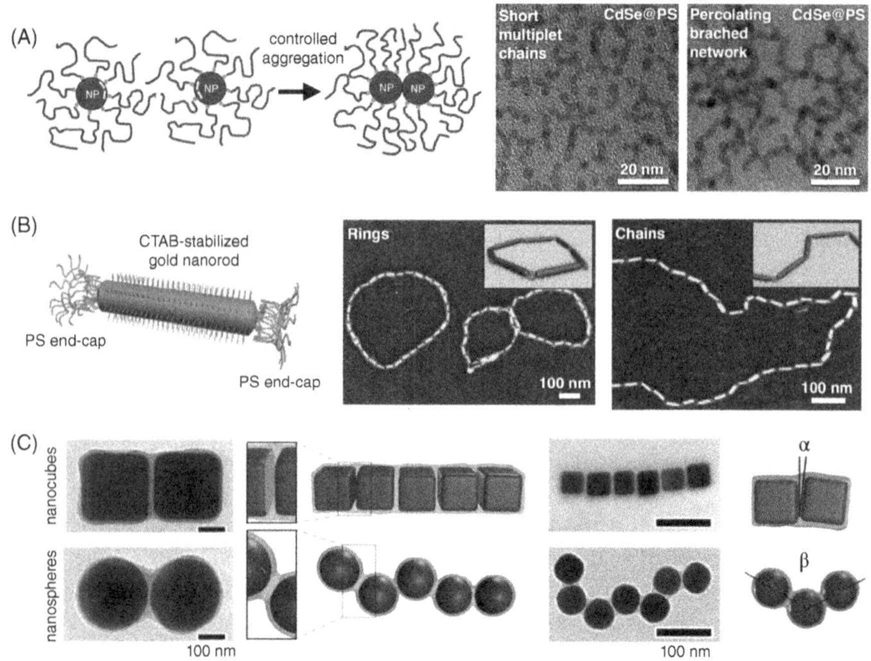

Figure 6.5 Controlled aggregation of polymer-decorated nanoparticles into linear assemblies: (A) Controlled aggregation of CdSe nanoparticles decorated with PS brushes forming short chain multiplets and percolating branched networks. Adapted with permission from [26]. Copyright 2014 ACS. (B) Self-assembled rings and chains of PS end-capped CTAB-stabilized gold nanorods in selective solvents. Adapted with permission from [89]. Copyright 2007 NPG. (C) TEM images of self-assembled dimers and linear chains of nanocubes and nanospheres with PS ligands. The schematic depictions show the contact situation and define the contacting angle α for cubes and β for sphere assemblies. Adapted with permission from [90]. Copyright 2014 ACS.

solvent with a high dielectric constant. In comparison to nanospheres, nanocube assemblies exhibited significantly higher collinearity because of the preferential face-to-face configuration that allows for larger contact areas and/or overlap volumes of two ligand shells in contact (see Figure 6.5C). The authors found that 75% of neighboring cubes show a junction angle α, the angle between the faces of the adjacent cubes, of less than 10°. Face-to-face orientation provides the maximum screening of interactions of PS with the poor solvent. By reducing the particle size to 25 nm but using the same PS ligand, the junction angle is increased significantly (only 30% with $\alpha < 10°$) and thus reduced face-to-face orientation. Regarding longer nanoparticle oligomers, this trend translates to a high collinearity for nanocube assemblies (70%, $\beta > 160°$). Consequently, nanosphere assemblies suffer from the small contact area/overlap volume and show higher tendency to bend or form kinks (45%, $\beta > 160°$) independent of the particle-to-shell size ratio [90].

6.3 GUIDING ASSEMBLY BY DECORATION WITH BIOMACROMOLECULES

While we have focused on nanoparticle modification by artificial macromolecules in the previous chapter, we will now turn our attention toward biomacromolecules, which are of particular interest for further increasing structural, chemical, and biological complexities. Biological molecules exhibit extremely well-defined molecular structures, shapes, and chemical activity as a result of the defined placement of specific functional groups in space and in time. Consequently, hierarchical 3D structures from molecular scale to macroscale and complex enzymatic cascade reactions are achieved. These features have motivated many researchers of different scientific communities to take advantage of the complexity already present in these materials and manipulate them for directing the formation of intricate nanostructures. Thereby, the physicochemical interactions in these biological macromolecules that give rise to the structural and functional diversity in nature are adopted for guiding the self-assembly of colloidal nanoparticles because of their specificity and programmable nature [91]. The two main classes of such biological macromolecules are the polynucleic acids (DNA and RNA) and polypeptides (proteins, enzymes). In the following, we will discuss the strategies to show how nanoparticles can be decorated with such biopolymers to subsequently induce self-assembly into sophisticated nanostructures in 2D and 3D.

6.3.1 DNA-Assisted Assembly

DNA-assisted assembly has opened a wide field of well-defined nanostructures due to the unique properties of DNA, in particular its specific base-pairing interactions known as the Watson–Crick base-pairing [92]. Figure 6.6A shows a schematic representation of an unfolded paired strand of DNA coordinated by the local interactions of its four building blocks (adenine, thymine, guanine, and cytosine) that are translated into the characteristic double helical structure and folding. Here the nucleic base adenine pairs exclusively with thymine, and guanine pairs with cytosine via intermolecular hydrogen bonds (see Figure 6.6B), allowing for highly specific recognition and reversible bonding [92, 93].

Seeman [93] was one of the first researchers in the DNA technology field who suggested DNA as a structural material for self-assembly using base-pairing interactions [92]. By programming structural design in the DNA and using them as DNA scaffolds or site-recognition entities, complex DNA-assisted nanoassemblies can be obtained [96–102]. The basic concept for DNA-assisted assembly of nanoparticles comprises of attaching a DNA single strand orthogonally to a nanoparticle surface and hybridizing it with a complementary strand, which is attached to another particle or surface [103]. In 1996, Mirkin *et al.* and Alivisatos *et al.* were the first researchers who used DNA-hybridization to assemble nanoparticles into nanoparticle arrays [94] and distinct dimers and trimers [95]. They applied thiol-functionalized complementary strands of DNA and attached them to different gold nanoparticles, owed to the thiophilic character of gold (see Figure 6.6C). In this respect, DNA strands with terminal

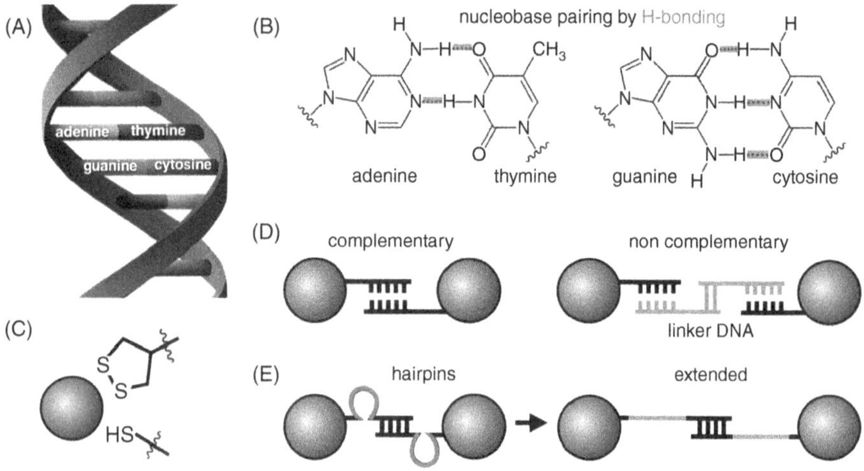

Figure 6.6 Principles of DNA ligation and hybridization as schematic representations: (A) Helical segment of double-stranded DNA with (B) Watson–Crick base-pairing. (C) Decoration of nanoparticles by ligands with terminal thiols of disulfides. (D) Hybridization strategies utilizing single-stranded DNA: direct linking of complementary ligands versus hybridization of noncomplementary ligands via a linker segment. (E) Simplified transition of hairpin segments into extended DNA ligands [94, 95].

thiol or disulfide proved to be the most applicable for ligation on gold nanoparticles. Consequently, the nanoparticles with the complementary strands could be assembled upon the hybridization of the DNA strands (see Figure 6.6D). In general, two hybridization strategies found application: (1) nanoparticle decoration with complementary strands for selective but immediate coupling (Figure 6.6C); and (2) nanoparticle decoration with noncomplementary strands, which require an additional linker strand with the respective inverse motifs as a mediator (Figure 6.6D). The latter strategy, upon application of distinct linker agents, allows for highly selective addressing of specific nanoparticles based on their ligation. Hairpin segments bring additional functionality for structural control. Hairpin loop segments arise from back folding a single strand of DNA, owing to intrastrand base-pairing of complementary base regions. The resulting loop structure is a key feature in the folding of RNA secondary structures. Upon unpairing, such hairpin loops allow for the subsequent extension of ligand-coordinated assemblies, yielding variations of interparticle separations (see Figure 6.6E) [94, 95].

Since then, a huge number of different approaches for the fabrication of nanostructures by DNA self-assembly were developed [96–102]. Simple as well as complex 2D and 3D nanostructures such as oligomeric nanoparticle chains [104], chains of nanoparticle pairs (tetramers and hexamers) [105], 3D networks, nanoparticle assemblies in triangular [106], square [106], hexagonal [107], and chiral configurations [108] have been reported. In addition, particle assemblies in core/satellite nanostructures consisting of a central nanoparticle linked to satellite nanoparticles were also achieved by DNA-hybridization approaches. Here the core and the satellite particles can be varied in their material composition such as by semiconductor quantum dots [109] or metals such as gold or silver nanoparticles [110, 111], and in shape such as isotropic and anisotropic [112, 113]. For the DNA decoration of the various nanoparticles, ligation strategies vary from single thiol and dithiol bonds for metal and semiconductor particles and classical amide bonds using amine and activated carboxylic groups [103]. The combination of the nanoparticle building blocks can be performed in dispersion or on functionalized substrates. For example, core/satellite nanostructures consisting of a gold core and gold satellites [114, 115], silver core and gold satellites [110, 116], and semiconductor quantum dot core with gold satellites were obtained in dispersion, using DNA [109].

Besides the configurational and compositional variations, also switchable and reconfigurable assemblies can be achieved using sophisticated DNA designs and combinations. Using hairpin-structured DNA, reconfigurable and switchable Au-core/Au-satellite nanoclusters are obtained, where the interparticle distance between the core and the satellite nanoparticles can be changed. The DNA-tether between the particles was designed to assemble in a compact (hairpin) state and to reconfigure to an extended state by activated strand substitution. These types of nanoassemblies can be used as molecularly driven plasmonic switches [110, 115].

Interparticle distances in DNA assisted assemblies can also be tuned by using DNA of various lengths. For example, Cheng *et al.* have employed such a DNA-based route to create freestanding nanoparticle superlattices (see Figure 6.7A) [117]. The authors have capped gold nanoparticles of 13 nm in diameter with single-stranded DNA using thiol-linkers and were able to prepare freestanding nanoparticle films trough dewetting on a microhole substrate. Using DNA strands of different lengths, the authors were also able to tune interparticle distance within the nanoparticle sheets. Although DNA is solely used as a ligand here and base-pairing was not involved in the lattice formation, this example nicely shows how DNA can be used as a versatile capping agent to achieve homogeneous, ordered nanoparticle structures with large dimensions [117].

Figure 6.7B shows the pH-responsive core/satellite assemblies for use in living cell plasmonic imaging. These nanostructures/clusters are assembled from a complementary recognition of binary gold nanoparticles. The coupling of 50 nm cores carrying a guanidine-rich motif ligand and the satellites (14 nm in diameter) with the inverse motif (see Figure 6.7B, bottom left) was archived avoiding the formation of macroscopic aggregates (core : satellite = 1:200, at pH 8) [118]. Upon assembly, a red shift of the LSPR of 14 nm relative to that of the initial Au core nanoparticle dispersion was observed. Upon changing the pH to 5.0, the LSPR band showed a sharp blue shift, indicating the release of satellite nanoparticles (disassembly). The pH-triggered assembly and disassembly was also monitored by measuring scattering spectra from single nanoclusters [118].

The DNA-assisted approach was also used to combine different shaped nanoparticle building blocks. Core/satellite nanoclusters consisting of a triangular Au nanoprism core and spherical Au satellites were reported [112]. Asymmetric gold Janus nanoclusters were synthesized by a stepwise surface-encoding protocol. The

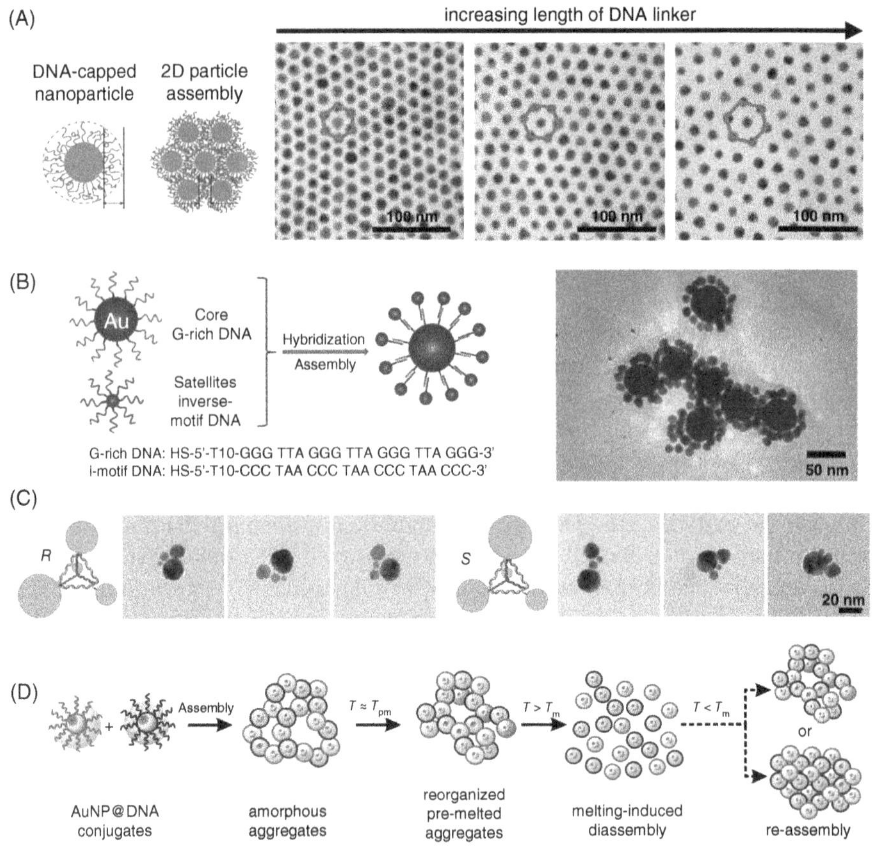

Figure 6.7 Examples of DNA-mediated assembly of nanoparticles upon DNA hybridization: (A) Distance control in freestanding superlattices of gold nanoparticles capped with DNA linker of variable length. Adapted with permission from [117]. Copyright 2009 NPG. (B) core/satellite structures of particles functionalized with complementary DNA ligands. TEM image show the dense population of satellites around the core particles. Adapted with permission from [118]. Copyright 2013 RSC. (C) Chiral pyramids of DNA-nanocrystal conjugates. TEM images show that the assembled enantiomers exhibit different chirality (left *R*, right *S*). Adapted with permission from [108]. Copyright 2009 ACS. (D) DNA-guided amorphous aggregation and subsequent reorganization into nanoparticle crystals. Adapted with permission from [119]. Copyright 2008 NPG.

predominant penta-valent core/satellite structures were obtained after different immobilization (on functionalized substrate), hybridization, and release steps using different sized DNA-encoded Au nanoparticles [111, 120].

Figure 6.7C shows chiral pyramids of DNA-nanoparticle conjugates using gold nanoparticles. Initially, when nanoparticles of equal sizes join, pyramidal assemblies are formed. The pyramidal geometry was confirmed by SAXS. The use of specific DNA ligands allows for control over the spatial position and order of these nanoparticles. Consequently, by joining tetraeders of nanoparticles that all differ in size, chiral nanostructures are feasible. The corresponding TEM images show the successful assembly of enantiomers of different chirality (left *R*, right *S*) [108].

Figure 6.7D shows the DNA-guided crystallization of binary nanoparticles into aggregates of variable long-range order. Gold nanoparticles with complementary DNA decorations were assembled by hybridization into mesoscale aggregates. At room temperature, an initially amorphous phase of DNA-linked aggregates is formed upon recognition. At elevated temperatures, a melting-induced disassembly of the amorphous aggregates may be induced. Upon subsequent cooling, the system either falls back to its unordered amorphous phase or experiences spontaneous crystallization. This suggests a coexistence of reorganized densely packed particles, which may act as nuclei for crystal growth, and unassembled particles. By thermal cycling, crystalline aggregates with remarkable long-range order could be produced. This temperature-triggered reversible hybridization/dehybridization of DNA allows 3D networks of nanoparticles to be modulated and reconfigured. Reversible aggregation can provide for structural and compositional control in ordered 3D nanoassemblies [119].

In conclusion, DNA-assisted self-assembly is highly selective and offers the possibility to fabricate well-defined nanoparticle assemblies of variable interparticle distances in a variety of complex structures. Moreover, the possibility to use reconfigurable or stimuli-responsive DNA-based linkers makes this method interesting for several applications. However, this assembly method is poorly scalable and requires high cost synthesis routes, making this method less feasible in applications.

6.3.2 Protein-Assisted Assembly

In comparison to DNA and RNA, proteins are another class of biopolymers that are highly diverse and embed hierarchically many layers of molecular and structural organization. Proteins are macromolecules consisting of 21 different monomers, namely amino acids with hydrophilic and hydrophobic residues bond into a polypeptide chain. In order to minimize the conformational energy of individual amino acid residues in the chain, to maximize hydrogen bonding of polar groups, and to bury hydrophobic residues away from the aqueous environment, these chains fold and organize into highly ordered 3D architectures. The folded proteins can further self-assemble upon molecular recognition to protein superstructures and protein complexes, giving rise to unique functional entities such as transmembrane pumps and molecular motors [121].

Hence proteins and protein structures have been widely employed to assemble colloidal nanoparticles, and used either as scaffolds or recognition entities (Fischer's key–lock principle) [122] for guiding nanoparticle assemblies. In the following, we will discuss different strategies and requirements for decorating colloidal particles with proteins that can be used in their 2D and 3D assemblies. In general, there are three main principles for the protein decoration of nanoparticles that find application in protein-assisted assembly of nanoparticles: covalent binding (Figure 6.8A), peptide affinity binding (Figure 6.8B), and physisorption (Figure 6.8C).

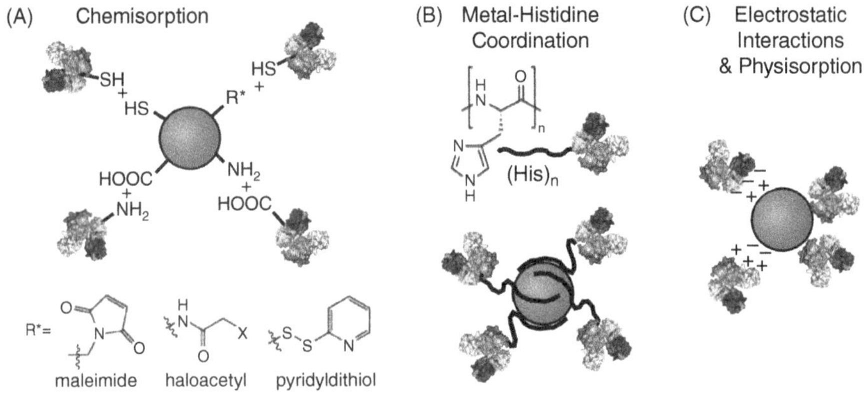

Figure 6.8 Principles of protein decoration of nanoparticles as schematic representations: (A) Chemisorption upon covalent bonding. (B) Metal-histidine coordination of proteins with polyhistidine tags. (C) Electrostatic adsorption and physisorption of proteins.

For attaching proteins covalently to nanoparticles (Figure 6.8A), the nanoparticle has to be first decorated with chemical functionalities such as carboxylates, amines, alcohols, and thiols that enable protein attachment via organic coupling reactions, yielding chemical bonds such as amides, esters, thioethers, and disulfide bonds, respectively, between the particle coating and the protein [123, 124]. However, due to the colloidal instability issues related to nanoparticles and the protein denaturation problems under reaction conditions that are beyond the ideal conditions for proteins, only few coupling reactions can be employed. The most applied reaction is the amide coupling reaction, where primary amine group (-NH_2) reacts with an activated carboxylic acid (-COOH) group in aqueous media. For the activation of the carboxylic acid group carbodiimide compounds provide the most popular and versatile method. The most readily available and commonly used carbodiimides are the water-soluble EDC (1-ethyl-3-(-3-dimethylaminopropyl) carbodiimide hydrochloride) for aqueous crosslinking and the water-insoluble DCC (*N*′,*N*′-dicyclohexylcarbodiimide) for nonaqueous organic synthesis methods. *N*-hydroxysuccinimide (NHS) or its water-soluble analogue (sulfo-NHS) is often included in carbodiimide coupling protocols to improve efficiency or create dry-stable (amine-reactive) intermediates. The amine coupling reaction is highly versatile in regard to proteins and particles types, since proteins bear both amine and carboxylic groups and nanoparticles can be easily decorated with amine or carboxylic functionalities. Besides amine-coupling reactions, sulfhydryl-reactive chemical groups such as maleimides, haloacetyls, and pyridyl disulfides (Figure 6.8A, bottom) can be used for attaching proteins covalently to nanoparticles. The proteins are then covalently attached to nanoparticles via thioether or disulfide bonds, respectively [124].

The second most applied strategy for decorating particles with proteins or peptides makes use of metal affinity binding entities such oligohistidine tags (*His-tags*), which have a high binding affinity toward metal surfaces (Figure 6.8B) [124]. Such *His-tags* are engineered biotechnologically or genetically into the protein sequence. Peptide synthesizers (solid phase) or the above-mentioned coupling reactions (liquid phase) can also be used to add a *His-tag* to peptide sequences. Finally, such *His-tag*-labeled proteins attach to nanoparticles, with the *His-tag* attaching preferentially to the particle surface. The nanoparticles that can be employed are usually metal nanoparticles, due to the high affinity of *His-tags* toward metals. However, the method is costly and yields a very low amount of labeled proteins, owed to the multiple and arduous protein engineering steps.

The third method of decorating nanoparticles with proteins comprises simply the physisorption of proteins onto nanoparticles using electrostatic and van der Waals forces (Figure 6.8C). Proteins readily adsorb onto nanoparticles forming an undefined protein layer, the so-called protein corona on any type of nanoparticle system. However, if metal nanoparticles are employed, the protein layer adsorbs strongly onto the metal surface, yielding a robust protein coating [125–128].

Once immobilized onto nanoparticle surfaces, such protein-decorated nanoparticles can be guided to self-assemble into complex structures via their inherent specific interactions. One of the prominent specific interactions of proteins is based on the bio-recognition principle, where the host protein binds a guest molecule or another protein or peptide with high specificity. This binding principle is often referred to host–guest or key–lock principle [122]. Figure 6.9 shows some of the widely applied systems, including biotin–streptavidin (Figure 6.9A) [129], enzyme-inhibitor (protein–protein) (Figure 6.9B) [130], aptamer–thrombin (oligopeptide–protein) (Figure 6.9C) [131], and carbohydrate–lectin (carbohydrate–protein) (Figure 6.9D) interactions [132]. Figure 6.9A and C show examples of gold nanorods functionalized with end-thiolated biotin or aptamer molecules at the nanorod tips, respectively. Upon addition of the multivalent host protein streptavidin or thrombin, respectively, the gold nanorods assemble tip-to-tip in longer particle chains. By attaching the guest and host molecules on different particle types, multi-component assemblies can be achieved. Figure 6.9B shows an example where a host protein (Barnase) is attached to magnetic nanoparticles and the guest protein (Barstar) to a quantum dot. Upon adding the Barstar-decorated quantum dots in excess to the Barnase-decorated nanoparticles, the quantum dots arrange on the magnetic nanoparticles in a core/satellite structure. It is worth noting that the decoration of nanoparticles with proteins and their assembly can also be achieved, via immobilizing the guest molecules (biotin, inhibitor protein, aptamer, carbohydrate; see Figure 6.9A–D) onto the nanoparticles and then locking the host protein onto the immobilized guest molecule. By adjusting the molecular ratio of the immobilized guest and free host molecules, protein decoration or particle assembly can be achieved for low or high ratios, respectively. Figure 6.9D shows an example for the latter case. Here the nanoparticles are decorated with α-D-mannosyl or α-D-glucosyl residues (guest molecule), and upon adding the host protein concanavalin A (ConA) at relative low ratios, the particles assemble into 3D nanoparticle networks.

For the colloidal assembly of nanoparticles in dispersion (3D) and on substrates (2D), it is critical that the functional coatings are robust and comply with such prerequisites as high colloidal stability, high concentration, tunable surface wettability,

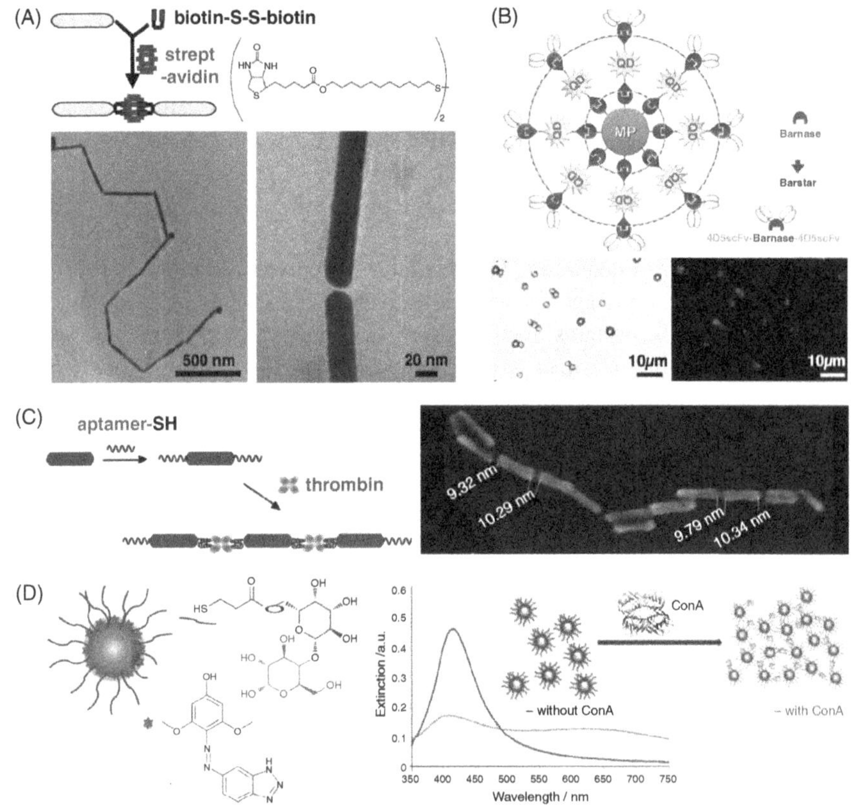

Figure 6.9 Examples of protein-mediated assembly of nanoparticles: (A) End-to-end assembly (linear aggregation) of gold nanorods tip-functionalized by biotin disulfide with streptavidin as linker. TEM images showing tip-to-tip aggregation of nanorod chains. Adapted with permission from [129]. Copyright 2003 ACS. (B) Superstructure of a magnetic core particle (MP), quantum dots (QD), and superfacial tags (4D5scFv-Bn) utilizing the key–lock type bio-recognition of two proteins (Barstar/Barnase). Bright-field (*left*) and fluorescence microscopy images (*right*) showing trifunctional nanoparticle assemblies. Adapted with permission from [130]. Copyright 2010 PNAS. (C) End-to-end assembly of gold nanorods tip-functionalized with a thiolatedaptamer with thrombin as linker. TEM image with selected tip-to-tip distances. Adapted with permission from [131]. Copyright 2009 ACS. (D) Mixed monolayer assembly by aggregation of glyco-decorated nanoparticles using a protein (ConA) as linker. UV/vis extinction before and after the addition of ConA. Adapted with permission from [132]. Copyright 2013 RSC. (*See color insert for color representation of this figure*).

and stimuli-responsiveness. Defined protein coatings on metal nanoparticles have revealed a facility to fulfill most of these requirements, thus enabling controlled modular assemblies [133] in dispersion and highly ordered, macroscopically scaled particle assemblies on substrates, for example, as shown recently by Hanske *et al.* [128] for nanospheres and Tebbe *et al.* [134, 135] for nanorods.

6.4 APPLICATION OF ASSEMBLIES

The previous sections provide scientists with a set of tools for gaining control over structural features of nanoparticle assemblies. As highlighted in Figure 6.1, optical and electronic properties correlate strongly with interparticle distances. In the following we will illustrate, with some selected examples for the case of plasmonic nanoparticles, how optical properties can be tuned in the assemblies.

In particular, plasmonic coupling between nanoparticles is highly sensitive to variations in particle spacing, orientation, and material composition. Thus, applications of macromolecular ligands as agents to direct the assembly of clusters or arrays of plasmonic nanoparticles are already found in optical sensing and nonlinear nanophotonics.

Pazos-Perez *et al.* have prepared highly symmetric plasmonic nanoparticle clusters with coordination numbers (CN) up to 7 (see Figure 6.10A), and these clusters were separated by density gradient centrifugation [136]. The authors used a macromolecular linker of poly(ethylene oxide)-*b*-poly(propylene oxide)-*b*-poly(ethylene oxide) (PEO-*b*-PPO-*b*-PEO) introduced by direct polymer ligand exchange to substitute the original surfactant shell of CTAB. The clusters were formed by gradual evaporation of toluene from emulsion droplets of aqueous nanoparticle dispersions with toluene [138]. The decrease in volume yields highly symmetric assemblies of closely assembled nanoparticles with interparticle separations of about 2 nm. Compared to DNA or other organic linkers, polymers exhibit lower Raman cross sections [139]. Therefore polymer linkers are preferable for assemblies for SERS application because analytes can be concentrated within close proximity of the particle surface while vibrational signal contamination is avoided [136]. Confocal SERS and dark-field microscopy was performed at the same positions and compared to electromagnetic modeling, revealing that the enhancement factor is proportional to the number of gaps in each cluster (see Figure 6.10A).

Another factor for functional devices is the control of orientation of anisotropic nanoparticles in their self-assemblies. Figure 6.10B shows an example of a 3D self-assembled standing array of gold nanorods on micropatterned gold/glass substrates using macromolecular ligands as directing agents [79]. The anisotropic building blocks were modified with a thiol-terminated ligands consisting of oligoethylene and oligo(ethylene glycol) blocks. Upon capillary and convective assembly, vertically aligned and hexagonally close-packed multilayer colloidal crystals are formed. The interparticle gaps of 5 ± 1 nm were determined by SAXS [79] and are consistent with the proposed thickness of the ligand shell [140]. Both the high spatial ordering and the small interparticle distances increase the hot spot density, and therefore their Raman activity. The 3D nanorod arrays exhibited a 36 times enhancement compared to a commercial SERS substrate [79].

Apart from controlling the interparticle distances also the option to perform post-modification of self-assembled particle arrays opens new pathways to controlling optical properties of nanoparticles. Müller *et al.* have shown that hexagonally close-packed monolayers of gold nanoparticles can be prepared on macroscopic substrates by spin-coating of Au/PNIPAm core/shell particles [137]. In addition,

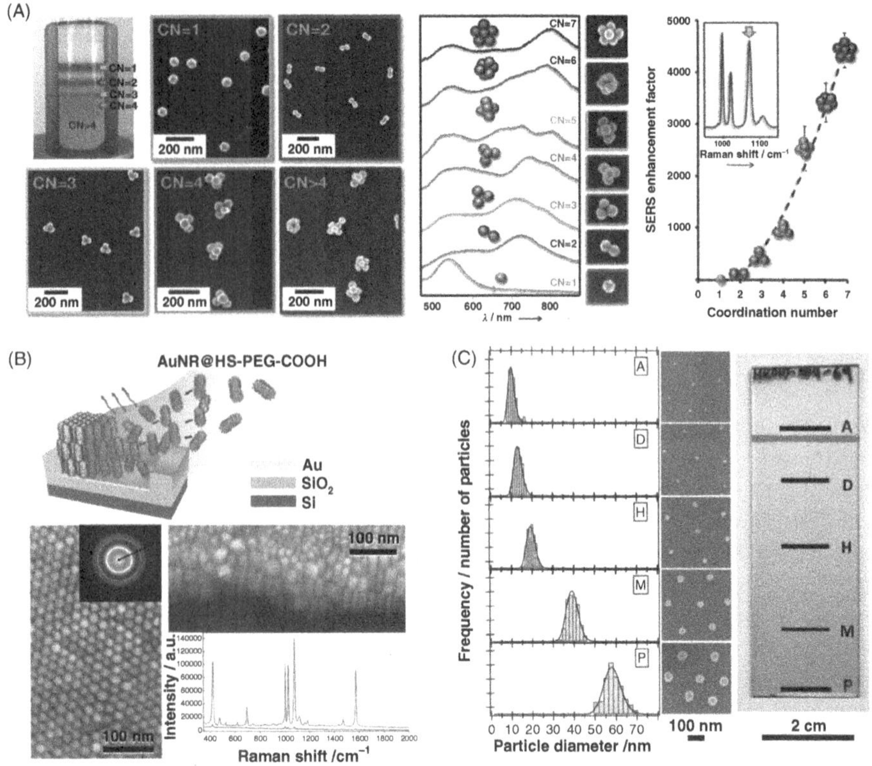

Figure 6.10 Application examples of assemblies of macromolecule-decorated nanoparticles: (A) Organized plasmonic nanoclusters assembled from gold nanoparticles with block copolymer shell. SEM images show clusters of discrete coordination numbers (CN), separated by density gradient centrifugation. Dark-field single-particle optical spectroscopy next to the corresponding clusters; SERS enhancement as a function of CNs normalize to the enhancement of a single particle. (Adapted with permission from [136]. Copyright 2012 Wiley.) (B) 3D arrays of standing gold nanorods, modified with thiol-PEO-carboxyl ligands assembled into micropatterned substrates. SEM images show top and side view of the hexagonally close packing of nanorods with interparticle gaps of 5 nm as evaluated from the transmission SAXS pattern (*see inset*). SERS enhancement was studied for benzenethiol (*see spectrum*). (Adapted with permission from [79]. Copyright 2012 Wiley.) (C) Plasmonic library based on substrate-supported gradiential plasmonic arrays with well-defined interparticle spacing: size distributions and SEM images of gold nanoparticle arrays on a macroscopic glass slide (*see photograph*). The gradual color change from nearly transparent to almost purple indicates the increasing particle sizes from top (9 nm) to bottom (57 nm) of the substrate. The monolayer assembly was achieved by spin-coating Au/PNIPAm core/shell colloids onto the substrate and subsequent overgrowth of the gold cores. (Adapted with permission from [137]. Copyright 2014 ACS.) (*See color insert for color representation of this figure*).

the authors have shown that these substrates can be used for the preparation of nanoparticle arrays with a continuous gradient in particle sizes from 9 to 57 nm (see Figure 6.10C). Consequently, this translates to a gradient in optical properties due to the size-dependent plasmonic properties of the gold particles. The gradient array was realized by immersion of a substrate-supported particle monolayer into a growing solution containing CTAB, gold ions, and a mild reducing agent (ascorbic acid). Due to the soft and porous character of the hydrogel shell of the adsorbed core/shell particles, the reactants could diffuse into the shell and allow for autocatalytic overgrowth of the gold cores. A dip-coating protocol with controlled immersion time was used to achieve a continuous gradient in particle size. Hence, the particle size became a function of the local position of the macroscopic substrate. The large separation of the gold particles, which results from the rather thick hydrogel shells, allowed avoiding aggregation and interparticle coupling of the plasmonic properties. The combination of distance control and post-modification yields a plasmonic library in form of particle arrays on macroscopic areas (see the photograph of a glass slide in Figure 6.10C) [137].

6.5 CONCLUSIONS AND OUTLOOK

This chapter gave an overview of the tools that macromolecules provide for controlling nanoparticle assemblies. At present, we can choose from a rich variety of methods for ligand exchange/coupling of both artificial and biomacromolecules. In turn, we can gain control over interparticle distances in close-packed assemblies in two and three dimensions. Linear assemblies are feasible exploiting heterogeneities in surface modification with artificial macromolecules and biomacromolecules offering access to even higher levels of complexity. Indeed structural control has already been exploited for various applications like the examples in plasmonics mentioned in the final section. Many other fields are currently emerging, such as photovoltaics [9] and biosensing [141]. Given the diversity of approaches and potential applications, it is, of course, difficult to speculate on the future of the field. Nevertheless, we want to address two remaining challenges. *Exploiting responsiveness* of macromolecular building blocks *for creating tunable nanoassemblies* is a logical next step after mastering equilibrium situations. Indeed macromolecules are or can be made sensitive to a whole range of external parameters such as solvent composition (including pH and salt concentration), temperature, and electrochemical or biochemical stimuli. Consequently the conformation/shape of nanoparticle coatings react toward those stimuli and influence the structure of nanoparticle assemblies and their structure dependent optical and electronic properties. A few works have demonstrated already the proof of concept that the responsiveness of polymer coatings can be applied to control assembly structures of nanoparticles in 2D and 3D [84, 86]. *Mechanical deformation* of the typically elastic soft polymeric coatings offers another option. While widely used for sensing and dynamic structural coloring on the scale of several hundreds of nanometers, these effects have so far been relatively little used for nanoscale distances, despite the great potential for adjusting properties. An interesting aspect is also that the electronic/optical effects that are linked to the desired structure can be directly used as a structure guiding stimulus.

Dissipative assembly offers an exciting perspective *for adaptive materials* [86]. There are, of course, classic examples for artificial systems that illustrate the potential of shear-induced structure formation [142]. But beyond that, there are many further options for influencing structure formation by dissipative processes. Especially for biological systems, structures are usually a result of equilibrium between a structure-forming and a structure-breaking process. A typical example is the constant polymerization and depolmerization of actin [143] and other fibrillar proteins as well as the constant process of bone construction/deconstruction by osteoblasts and osteoclasts [144, 145]. This dynamic equilibrium character of biological structures is crucial for their adaptivity. Indeed supramolecular chemistry provides access to similar designs, which could lead to *living materials* that are truly adaptive and not just self-healing. Of course, biomacromolecules could as well be directly used for such approaches.

These are just some examples for the broad range of options that macromolecular design in combination with colloidal building blocks offers. Indeed, the chances are very high that the macromolecular route will develop into a cornerstone of the next generation of *materials by design*.

REFERENCES

1. Fratzl, P.; Weinkamer, R. Nature's Hierarchical Materials. *Prog Mater Sci* **2007**, *52*, 1263–1334.
2. Dunlop, J. W. C.; Fratzl, P. Biological Composites. *Annu Rev Mater Res* **2010**, *40*, 1–24.
3. Dunlop, J. W. C.; Weinkamer, R.; Fratzl, P. Artful Interfaces within Biological Materials. *Mater Today* **2011**, *14*, 70–78.
4. Lu, M.-H.; Feng, L.; Chen, Y.-F. Phononic Crystals and Acoustic Metamaterials. *Mater Today* **2009**, *12*, 34–42.
5. Liu, Y.; Zhang, X. Metamaterials: a New Frontier of Science and Technology. *Chem Soc Rev* **2011**, *40*, 2494.
6. Nicolaou, Z. G.; Motter, A. E. Mechanical Metamaterials with Negative Compressibility Transitions. *Nature Nanotech* **2012**, *11*, 608–613.
7. Lee, J.-H.; Singer, J. P.; Thomas, E. L. Micro-/Nanostructured Mechanical Metamaterials. *Adv Mater* **2012**, *24*, 4782–4810.
8. Atwater, H. A.; Polman, A. Plasmonics for Improved Photovoltaic Devices. *Nat Mater* **2010**, *9*, 205–213.
9. Karg, M.; König, T. A. F.; Retsch, M.; Stelling, C.; Reichstein, P. M.; Honold, T.; Thelakkat, M.; Fery, A. Colloidal Self-Assembly Concepts for Light Management in Photovoltaics. *Mater Today* **2015**, *18*, 185–205.
10. Zheludev, N. I.; Kivshar, Y. S. From Metamaterials to Metadevices. *Nat Mater* **2012**, *11*, 917–924.
11. Pavlidou, S.; Papaspyrides, C. D. A Review on Polymer–Layered Silicate Nanocomposites. *Prog Polym Sci* **2008**, *33*, 1119–1198.

12. Paul, D. R.; Robeson, L. M. Polymer Nanotechnology: Nanocomposites. *Polymer* **2008**, *49*, 3187–3204.

13. Möller, M. W.; Handge, U. A.; Kunz, D. A.; Lunkenbein, T.; Altstädt, V.; Breu, J. Tailoring Shear-Stiff, Mica-Like Nanoplatelets. *ACS Nano* **2010**, *4*, 717–724.

14. Bieligmeyer, M.; Taheri, S. M.; German, I.; Boisson, C.; Probst, C.; Milius, W.; Altstädt, V.; Breu, J.; Schmidt, H.-W.; D'Agosto, F.; et al. Completely Miscible Polyethylene Nanocomposites. *J Am Chem Soc* **2012**, *134*, 18157–18160.

15. Lunkenbein, T.; Kamperman, M.; Li, Z.; Bojer, C.; Drechsler, M.; Förster, S.; Wiesner, U.; Müller, A. H. E.; Breu, J. Direct Synthesis of Inverse Hexagonally Ordered Diblock Copolymer/Polyoxometalate Nanocomposite Films. *J Am Chem Soc* **2012**, *134*, 12685–12692.

16. Moffitt, M. G. Self-Assembly of Polymer Brush-Functionalized Inorganic Nanoparticles: From Hairy Balls to Smart Molecular Mimics. *J Phys Chem Lett* **2013**, *4*, 3654–3666.

17. Sarikaya, M.; Tamerler, C.; Jen, A. K. Y.; Schulten, K.; Baneyx, F. Molecular Biomimetics: Nanotechnology through Biology. *Nat Mater* **2003**, *2*, 577–585.

18. Zhang, S. Fabrication of Novel Biomaterials through Molecular Self-Assembly. *Nat Biotechnol* **2003**, *21*, 1171–1178.

19. Ahmed, T. A. E.; Dare, E. V.; Hincke, M. Fibrin: a Versatile Scaffold for Tissue Engineering Applications. *Tissue Eng, Part B* **2008**, *14*, 199–215.

20. Mattoussi, H.; Mauro, J. M.; Goldman, E. R.; Anderson, G. P.; Sundar, V. C.; Mikulec, F. V.; Bawendi, M. G. Self-Assembly of CdSe–ZnS Quantum Dot Bioconjugates Using an Engineered Recombinant Protein. *J Am Chem Soc* **2000**, *122*, 12142–12150.

21. Meyers, M. A.; Chen, P.-Y.; Lin, A. Y.-M.; Seki, Y. Biological Materials: Structure and Mechanical Properties. *Prog Mater Sci* **2008**, *53*, 1–206.

22. Minko, S. Responsive Polymer Brushes. *Journal of Macromolecular Science, Part C: Polym Rev* **2006**, *46*, 397–420.

23. Stuart, M. A. C.; Huck, W. T. S.; Genzer, J.; Müller, M.; Ober, C.; Stamm, M.; Sukhorukov, G. B.; Szleifer, I.; Tsukruk, V. V.; Urban, M.; Winnik, F.; Zauscher, S.; Luzinov, I.; Minko, S. Emerging Applications of Stimuli-Responsive Polymer Materials. *Nat Mater* **2010**, *9*, 101–113.

24. Ma, X.; Tian, H. Stimuli-Responsive Supramolecular Polymers in Aqueous Solution. *Accounts Chem Res* **2014**, *47*, 1971–1981.

25. Gil, E.; Hudson, S. Stimuli-Reponsive Polymers and Their Bioconjugates. *Prog Polym Sci* **2004**, *29*, 1173–1222.

26. Ehlert, S.; Taheri, S. M.; Pirner, D.; Drechsler, M.; Schmidt, H.-W.; Förster, S. Polymer Ligand Exchange to Control Stabilization and Compatibilization of Nanocrystals. *ACS Nano* **2014**, *8*, 6114–6122.

27. Daniel, M.-C.; Astruc, D. Gold Nanoparticles: Assembly, Supramolecular Chemistry, Quantum-Size-Related Properties, and Applications Toward Biology, Catalysis, and Nanotechnology. *Chem Rev* **2004**, *104*, 293–346.

28. Lu, A.-H.; Salabas, E. L.; Schüth, F. Magnetic Nanoparticles: Synthesis, Protection, Functionalization, and Application. *Angew Chem, Int Ed* **2007**, *46*, 1222–1244.

29. Lee, H.; Dellatore, S. M.; Miller, W. M.; Messersmith, P. B. Mussel-Inspired Surface Chemistry for Multifunctional Coatings. *Science* **2007**, *318*, 426–430.

30. Chanana, M.; Jahn, S.; Georgieva, R.; Lutz, J.-F.; Bäumler, H.; Wang, D. Fabrication of Colloidal Stable, Thermosensitive, and Biocompatible Magnetite Nanoparticles and Study of Their Reversible Agglomeration in Aqueous Milieu. *Chem Mater* **2009**, *21*, 1906–1914.
31. Sodipo, B. K.; Aziz, A. A. A Sonochemical Approach to the Direct Surface Functionalization of Superparamagnetic Iron Oxide Nanoparticles with (3-Aminopropyl)Triethoxysilane. *Beilstein J Nanotechnol* **2014**, *5*, 1472–1476.
32. Francis, R.; Joy, N.; Aparna, E. P.; Vijayan, R. Polymer Grafted Inorganic Nanoparticles, Preparation, Properties, and Applications: a Review. *Polym Rev* **2014**, *54*, 268–347.
33. Kuttner, C.; Tebbe, M.; Schlaad, H.; Burgert, I.; Fery, A. Photochemical Synthesis of Polymeric Fiber Coatings and Their Embedding in Matrix Material: Morphology and Nanomechanical Properties at the Fiber-Matrix Interface. *ACS Appl Mater Interfaces* **2012**, *4*, 3485–3492.
34. Kuttner, C.; Hanisch, A.; Schmalz, H.; Eder, M.; Schlaad, H.; Burgert, I.; Fery, A. Influence of the Polymeric Interphase Design on the Interfacial Properties of (Fiber-Reinforced) Composites. *ACS Appl Mater Interfaces* **2013**, *5*, 2469–2478.
35. Hoyle, C. E.; Bowman, C. N. Thiol-Ene Click Chemistry. *Angew Chem, Int Ed* **2010**, *49*, 1540–1573.
36. Kuttner, C.; Maier, P. C.; Kunert, C.; Schlaad, H.; Fery, A. Direct Thiol-Ene Photocoating of Polyorganosiloxane Microparticles. *Langmuir* **2013**, *29*, 16119–16126.
37. Kolb, H. C.; Finn, M. G.; Sharpless, K. B. Click Chemistry: Diverse Chemical Function From a Few Good Reactions. *Angew Chem, Int Ed* **2001**, *40*, 2004–2021.
38. Hawker, C. J. The Convergence of Synthetic Organic and Polymer Chemistries. *Science* **2005**, *309*, 1200–1205.
39. Binder, W. H.; Sachsenhofer, R. "Click" Chemistry in Polymer and Materials Science. *Macromol Rapid Comm* **2007**, *28*, 15–54.
40. Moses, J. E.; Moorhouse, A. D. The Growing Applications of Click Chemistry. *Chem Soc Rev* **2007**, *36*, 1249–1262.
41. Iha, R. K.; Wooley, K. L.; Nyström, A. M.; Burke, D. J.; Kade, M. J.; Hawker, C. J. Applications of Orthogonal "Click" Chemistries in the Synthesis of Functional Soft Materials. *Chem Rev* **2009**, *109*, 5620–5686.
42. Milner, S. T. Polymer Brushes. *Science* **1991**, *251*, 905–914.
43. Zhao, B.; Brittain, W. J. Polymer Brushes: Surface-Immobilized Macromolecules. *Prog Polym Sci* **2000**, *25*, 677–710.
44. Matyjaszewski, K.; Xia, J. Atom Transfer Radical Polymerization. *Chem Rev* **2001**, *101*, 2921–2990.
45. Choi, T.-L.; Grubbs, R. H. Controlled Living Ring-Opening-Metathesis Polymerization by a Fast-Initiating Ruthenium Catalyst. *Angew Chem, Int Ed* **2003**, *42*, 1743–1746.
46. Bielawski, C. W.; Grubbs, R. H. Living Ring-Opening Metathesis Polymerization. *Prog Polym Sci* **2007**, *32*, 1–29.
47. Hawker, C. J.; Bosman, A. W.; Harth, E. New Polymer Synthesis by Nitroxide Mediated Living Radical Polymerizations. *Chem Rev* **2001**, *101*, 3661–3688.
48. Chiefari, J.; Chong, Y. K. B.; Ercole, F.; Krstina, J.; Jeffery, J.; Le, T. P. T.; Mayadunne, R. T. A.; Meijs, G. F.; Moad, C. L.; Moad, G.; et al. Living Free-Radical Polymerization by Reversible Addition-Fragmentation Chain Transfer: the RAFT Process. *Macromolecules* **1998**, *31*, 5559–5562.

49. Braunecker, W. A.; Matyjaszewski, K. Controlled/Living Radical Polymerization: Features, Developments, and Perspectives. *Prog Polym Sci* **2007**, *32*, 93–146.

50. Tebbe, M.; Kuttner, C.; Maennel, M.; Fery, A.; Chanana, M. Colloidally Stable and Surfactant-Free Protein-Coated Gold Nanorods in Biological Media. *ACS Appl Mater Interfaces* **2015**, *7*, 5984–5991.

51. Edwards, E. W.; Chanana, M.; Wang, D.; Möhwald, H. Stimuli-Responsive Reversible Transport of Nanoparticles Across Water/Oil Interfaces. *Angew Chem, Int Ed* **2008**, *47*, 320–323.

52. Stefaniu, C.; Chanana, M.; Ahrens, H.; Wang, D.; Brezesinski, G.; Möhwald, H. Conformational Induced Behaviour of Copolymer-Capped Magnetite Nanoparticles at the Air/Water Interface. *Soft Matter* **2011**, *7*, 4267.

53. Gambinossi, F.; Chanana, M.; Mylon, S. E.; Ferri, J. K. Programming Nanoparticle Aggregation Kinetics with Poly(MeO2MA-Co-OEGMA) Copolymers. *Soft Matter* **2013**, *9*, 11046–11053.

54. Hörenz, C.; Rudolph, T.; Barthel, M. J.; Günther, U.; Schacher, F. H. Amphiphilic Polyether-Based Block Copolymers as Crosslinkable Ligands for Au-Nanoparticles. *Polym Chem* **2015**, *6*, 5633–5642.

55. Edmondson, S.; Osborne, V. L.; Huck, W. T. S. Polymer Brushes via Surface-Initiated Polymerizations. *Chem Soc Rev* **2004**, *33*, 14–22.

56. Makino, K.; Yamamoto, S.; Fujimoto, K.; Kawaguchi, H.; Ohshima, H. Surface Structure of Latex Particles Covered with Temperature-Sensitive Hydrogel Layers. *J Colloid Interf Sci* **1994**, *166*, 251–258.

57. Okubo, M.; Ahmad, H. Adsorption Behaviors of Emulsifiers and Biomolecules on Temperaturesensitive Polymer Particles. *Colloid Polym Sci* **1996**, *274*, 112–116.

58. Dingenouts, N.; Norhausen, C.; Ballauff, M. Observation of the Volume Transition in Thermosensitive Core–Shell Latex Particles by Small-Angle X-Ray Scattering. *Macromolecules* **1998**, *31*, 8912–8917.

59. Zha, L. S.; Zhang, Y.; Yang, W. L.; Fu, S. K. Monodisperse Temperature-Sensitive Microcontainers. *Adv Mater* **2002**, *14*, 1090.

60. Karg, M.; Pastoriza-Santos, I.; Liz-Marzan, L. M.; Hellweg, T. A Versatile Approach for the Preparation of Thermosensitive PNIPAM Core–Shell Microgels with Nanoparticle Cores. *Chem Phys Chem* **2006**, *7*, 2298–2301.

61. Karg, M.; Wellert, S.; Pastoriza-Santos, I.; Lapp, A.; Liz-Marzan, L. M.; Hellweg, T. Thermoresponsive Core–Shell Microgels with Silica Nanoparticle Cores: Size, Structure, and Volume Phase Transition of the Polymer Shell. *Phys Chem Chem Phys* **2008**, *10*, 6708–6716.

62. Karg, M.; Wellert, S.; Prevost, S.; Schweins, R.; Dewhurst, C.; Liz-Marzan, L. M.; Hellweg, T. Well Defined Hybrid PNIPAM Core-Shell Microgels: Size Variation of the Silica Nanoparticle Core. *Colloid Polym Sci* **2010**, *289*, 699–709.

63. Kim, D. J.; Kang, S. M.; Kong, B.; Kim, W.-J.; Paik, H.-J.; Choi, H.; Choi, I. S. Formation of Thermoresponsive Gold Nanoparticle/PNIPAAm Hybrids by Surface-Initiated, Atom Transfer Radical Polymerization in Aqueous Media. *Macromol Chem Physic* **2005**, *206*, 1941–1946.

64. Contreras-Cáceres, R.; Sánchez-Iglesias, A.; Karg, M.; Pastoriza-Santos, I.; Pérez-Juste, J.; Pacifico, J.; Hellweg, T.; Fernández-Barbero, A.; Liz-Marzan, L. M. Encapsulation and Growth of Gold Nanoparticles in Thermoresponsive Microgels. *Adv Mater* **2008**, *20*, 1666–1670.

65. Contreras-Cáceres, R.; Pacifico, J.; Pastoriza-Santos, I.; Pérez-Juste, J.; Fernández-Barbero, A.; Liz-Marzan, L. M. Au@pNIPAMThermosensitive Nanostructures: Control Over Shell Cross-Linking, Overall Dimensions, and Core Growth. *Adv Funct Mater* **2009**, *19*, 3070–3076.

66. Fernández-López, C.; Pérez-Balado, C.; Pérez-Juste, J.; Pastoriza-Santos, I.; de Lera, A. R.; Liz-Marzan, L. M. A General LbL Strategy for the Growth of pNIPAMMicrogels on Au Nanoparticles with Arbitrary Shapes. *Soft Matter* **2012**, *8*, 4165.

67. Karg, M.; Jaber, S.; Hellweg, T.; Mulvaney, P. Surface Plasmon Spectroscopy of Gold–Poly- N-Isopropylacrylamide Core–Shell Particles. *Langmuir* **2011**, *27*, 820–827.

68. Dulle, M.; Jaber, S.; Rosenfeldt, S.; Radulescu, A.; Förster, S.; Mulvaney, P.; Karg, M. Plasmonic Gold–Poly(N-Isopropylacrylamide) Core–Shell Colloids with Homogeneous Density Profiles: a Small Angle Scattering Study. *Phys Chem Chem Phys* **2015**, *17*, 1354–1367.

69. Pelton, R. Temperature-Sensitive Aqueous Microgels. *Adv Colloid Interf Sci* **2000**, *85*, 1–33.

70. Nayak, S.; Lyon, L. A. Soft Nanotechnology with Soft Nanoparticles. *Angew Chem, Int Ed* **2005**, *44*, 7686–7708.

71. Das, M.; Zhang, H.; Kumacheva, E. MICROGELS: Old Materials with New Applications. *Annu Rev Mater Res* **2006**, *36*, 117–142.

72. Hoare, T. R.; Kohane, D. S. Hydrogels in Drug Delivery: Progress and Challenges. *Polymer* **2008**, *49*, 1993–2007.

73. White, E. M.; Yatvin, J.; Grubbs, J. B., III; Bilbrey, J. A.; Locklin, J. Advances in Smart Materials: Stimuli-Responsive Hydrogel Thin Films. *J Polym Sci, Polym Phys* **2013**, *51*, 1084–1099.

74. Döring, A.; Birnbaum, W.; Kuckling, D. Responsive Hydrogels – Structurally and Dimensionally Optimized Smart Frameworks for Applications in Catalysis, Micro-System Technology and Material Science. *Chem Soc Rev* **2013**, *42*, 7391.

75. Giersig, M.; Mulvaney, P. Preparation of Ordered Colloid Monolayers by Electrophoretic Deposition. *Langmuir* **1993**, *9*, 3408–3413.

76. Ebeling, B.; Vana, P. RAFT-Polymers with Single and Multiple Trithiocarbonate Groups as Uniform Gold-Nanoparticle Coatings. *Macromolecules* **2013**, *46*, 4862–4871.

77. Vogel, N.; Fernández-López, C.; Pérez-Juste, J.; Liz-Marzan, L. M.; Landfester, K.; Weiss, C. K. Ordered Arrays of Gold Nanostructures From Interfacially Assembled Au@PNIPAM Hybrid Nanoparticles. *Langmuir* **2012**, *28*, 8985–8993.

78. Martin, J. E.; Wilcoxon, J. P.; Odinek, J.; Provencio, P. Control of the Interparticle Spacing in Gold Nanoparticle Superlattices. *J Phys Chem B* **2000**, *104*, 9475–9486.

79. Thai, T.; Zheng, Y.; Ng, S. H.; Mudie, S.; Altissimo, M.; Bach, U. Self-Assembly of Vertically Aligned Gold Nanorod Arrays on Patterned Substrates. *Angew Chem, Int Ed* **2012**, *51*, 8732–8735.

80. Nishi, H.; Kobatake, S. Facile Preparation of Gold Nanoparticle with Diarylethene Polymers by Disodium Malate and Its Photoreversible Optical Properties. *Dyes Pigm* **2012**, *92*, 847–853.

81. Simancas, K. *Entropic Death of Polyelectrolyte Brushes: Chain Stretching, Bond Scission and Surfactant Complexation*; Der AndereVerlag, **2011**.

82. Pletsch, H.; Peng, L.; Mitschang, F.; Schaper, A.; Hellwig, M.; Nette, D.; Seubert, A.; Greiner, A.; Agarwal, S. Ultrasound-Mediated Synthesis of High-Molecular Weight Polystyrene-Grafted Silver Nanoparticles by Facile Ligand Exchange Reactions in Suspension. *Small* **2013**, *10*, 201–208.

83. Jaber, S.; Karg, M.; Morfa, A.; Mulvaney, P. 2D Assembly of Gold–PNIPAM Core–Shell Nanocrystals. *Phys Chem Chem Phys* **2011**, *13*, 5576–5578.

84. Clara-Rahola, J.; Contreras-Cáceres, R.; Sierra-Martin, B.; Maldonado-Valdivia, A.; Hund, M.; Fery, A.; Hellweg, T.; Fernández-Barbero, A. Structure and Plasmon Coupling of Gold-Poly(N-Isopropylacrylamide) Core–Shell Microgel Arrays with Thermally Controlled Interparticle Gap. *Colloid Surf A* **2014**, *463*, 18–27.

85. Lu, Y.; Drechsler, M. Charge-Induced Self-Assembly of 2-Dimensional Thermosensitive Microgel Particle Patterns. *Langmuir* **2009**, *25*, 13100–13105.

86. Karg, M.; Hellweg, T.; Mulvaney, P. Self-Assembly of Tunable Nanocrystal Superlattices Using Poly-(NIPAM) Spacers. *Adv Funct Mater* **2011**, *21*, 4668–4676.

87. Tao, A. R.; Ceperley, D. P.; Sinsermsuksakul, P.; Neureuther, A. R.; Yang, P. Self-Organized Silver Nanoparticles for Three-Dimensional Plasmonic Crystals. *Nano Lett* **2008**, *8*, 4033–4038.

88. Kim, S.-H.; Lee, S. Y.; Yang, S.-M.; Yi, G.-R. Self-Assembled Colloidal Structures for Photonics. *NPG Asia Mater* **2011**, *3*, 25–33.

89. Nie, Z.; Fava, D.; Kumacheva, E.; Zou, S.; Walker, G. C.; Rubinstein, M. Self-Assembly of Metal-Polymer Analogues of Amphiphilic Triblock Copolymers. *Nat Mater* **2007**, *6*, 609–614.

90. Klinkova, A.; Thérien-Aubin, H.; Ahmed, A.; Nykypanchuk, D.; Choueiri, R. M.; Gagnon, B.; Muntyanu, A.; Gang, O.; Walker, G. C.; Kumacheva, E. Structural and Optical Properties of Self-Assembled Chains of PlasmonicNanocubes. *Nano Lett* **2014**, *14*, 6314–6321.

91. Jones, M. R.; Osberg, K. D.; Macfarlane, R. J.; Langille, M. R.; Mirkin, C. A. Templated Techniques for the Synthesis and Assembly of Plasmonic Nanostructures. *Chem Rev* **2011**, *111*, 3736–3827.

92. Watson, J. D.; Crick, F. H. C. Molecular Structure of Nucleic Acids: a Structure for Deoxyribose Nucleic Acid. *Nature* **1953**, *171*, 737–738.

93. Seeman, N. C. Nucleic Acid Junctions and Lattices. *J Theor Biol* **1982**, *99*, 237–247.

94. Mirkin, C. A.; Letsinger, R. L.; Mucic, R. C.; Storhoff, J. J. A DNA-Based Method for Rationally Assembling Nanoparticles Into Macroscopic Materials. *Nature* **1996**, *382*, 607–609.

95. Alivisatos, A. P.; Johnsson, K. P.; Peng, X.; Wilson, T. E.; Loweth, C. J.; Bruchez, M. P., Jr; Schultz, P. G. Organization of "Nanocrystal Molecules" Using DNA. *Nature* **1996**, *382*, 609–611.

96. Seeman, N. C. DNA Engineering and Its Application to Nanotechnology. *Trends Biotechnol* **1999**, *17*, 437–443.

97. Li, H.; Carter, J. D.; LaBean, T. H. Nanofabrication by DNA Self-Assembly. *Mater Today* **2009**, *12*, 24–32.

98. Condon, A. Designed DNA Molecules: Principles and Applications of Molecular Nanotechnology. *Nat Rev Genet* **2006**, *7*, 565–575.

99. Willner, I.; Willner, B. Biomolecule-Based Nanomaterials and Nanostructures. *Nano Lett* **2010**, *10*, 3805–3815.

100. Aldaye, F. A.; Palmer, A. L.; Sleiman, H. F. Assembling Materials with DNA as the Guide. *Science* **2008**, *321*, 1795–1799.
101. Choi, C. L.;Alivisatos, A. P. From Artificial Atoms to Nanocrystal Molecules: Preparation and Properties of More Complex Nanostructures. *Annu Rev Phys Chem* **2010**, *61*, 369–389.
102. Lin, C.; Liu, Y.; Yan, H. Designer DNA Nanoarchitectures. *Biochemistry* **2009**, *48*, 1663–1674.
103. Sharma, J.; Chhabra, R.; Andersen, C. S.; Gothelf, K. V.; Yan, H.; Liu, Y. Toward Reliable Gold Nanoparticle Patterning on Self-Assembled DNA Nanoscaffold. *J Am Chem Soc* **2008**, *130*, 7820–7821.
104. Wen, Y.; McLaughlin, C. K.; Lo, P. K.; Yang, H.; Sleiman, H. F. Stable Gold Nanoparticle Conjugation to Internal DNA Positions: Facile Generation of Discrete Gold Nanoparticle–DNA Assemblies. *Bioconjugate Chem* **2010**, *21*, 1413–1416.
105. Claridge, S. A.; Mastroianni, A. J.; Au, Y. B.; Liang, H. W.; Micheel, C. M.; Fréchet, J. M. J.; Alivisatos, A. P. Enzymatic Ligation Creates Discrete Multinanoparticle Building Blocks for Self-Assembly. *J Am Chem Soc* **2008**, *130*, 9598–9605.
106. Aldaye, F. A.; Sleiman, H. F. Dynamic DNA Templates for Discrete Gold Nanoparticle Assemblies: Control of Geometry, Modularity, Write/Erase and Structural Switching. *J Am Chem Soc* **2007**, *129*, 4130–4131.
107. Aldaye, F. A.; Sleiman, H. F. Sequential Self-Assembly of a DNA Hexagon as a Template for the Organization of Gold Nanoparticles. *Angew Chem, Int Ed* **2006**, *45*, 2204–2209.
108. Mastroianni, A. J.; Claridge, S. A.;Alivisatos, A. P. Pyramidal and Chiral Groupings of Gold Nanocrystals Assembled Using DNA Scaffolds. *J Am Chem Soc* **2009**, *131*, 8455–8459.
109. Fu, A.; Micheel, C. M.; Cha, J.; Chang, H.; Yang, H.; Alivisatos, A. P. Discrete Nanostructures of Quantum Dots/Au with DNA. *J Am Chem Soc* **2004**, *126*, 10832–10833.
110. Sebba, D. S.; LaBean, T. H.; Lazarides, A. A. Plasmon Coupling in Binary Metal Core–Satellite Assemblies. *Appl Phys B* **2008**, *93*, 69–78.
111. Maye, M. M.; Nykypanchuk, D.; Cuisinier, M.; van der Lelie, D.; Gang, O. Stepwise Surface Encoding for High-Throughput Assembly of Nanoclusters. *Nat Mater* **2009**, *8*, 388–391.
112. Millstone, J. E.; Georganopoulou, D. G.; Xu, X.; Wei, W.; Li, S.; Mirkin, C. A. DNA-Gold Triangular Nanoprism Conjugates. *Small* **2008**, *4*, 2176–2180.
113. Xu, L.; Kuang, H.; Xu, C.; Ma, W.; Wang, L.; Kotov, N. A. RegiospecificPlasmonic Assemblies for in Situ Raman Spectroscopy in Live Cells. *J Am Chem Soc* **2012**, *134*, 1699–1709.
114. Sebba, D. S.;Lazarides, A. A. Robust Detection of Plasmon Coupling in Core-Satellite Nanoassemblies Linked by DNA. *J Phys Chem C* **2008**, *112*, 18331–18339.
115. Sebba, D. S.; Mock, J. J.; Smith, D. R.; LaBean, T. H.; Lazarides, A. A. Reconfigurable Core–Satellite Nanoassemblies as Molecularly-Driven Plasmonic Switches. *Nano Lett* **2008**, *8*, 1803–1808.
116. Pal, S.; Sharma, J.; Yan, H.; Liu, Y. Stable Silver Nanoparticle–DNA Conjugates for Directed Self-Assembly of Core-Satellite Silver–Gold Nanoclusters. *Chem Commun* **2009**, 6059–6061.
117. Cheng, W.; Campolongo, M. J.; Cha, J. J.; Tan, S. J.; Umbach, C. C.; Muller, D. A.; Luo, D. Free-Standing Nanoparticle Superlattice Sheets Controlled by DNA. *Nat Mater* **2009**, *8*, 519–525.

118. Wang, C.; Du, Y.; Wu, Q.; Xuan, S.; Zhou, J.; Song, J.; Shao, F.; Duan, H. Stimuli-Responsive Plasmonic Core–Satellite Assemblies: I-Motif DNA Linker Enabled Intracellular pH Sensing. *Chem Commun* **2013**, *49*, 5739.

119. Nykypanchuk, D.; Maye, M. M.; van der Lelie, D.; Gang, O. DNA-Guided Crystallization of Colloidal Nanoparticles. *Nature* **2008**, *451*, 549–552.

120. Zheng, Y.; Thai, T.; Reineck, P.; Qiu, L.; Guo, Y.; Bach, U. DNA-Directed Self-Assembly of Core-Satellite Plasmonic Nanostructures: a Highly Sensitive and Reproducible Near-IR SERS Sensor. *Adv Funct Mater* **2012**, *23*, 1519–1526.

121. Schliwa, M.;Woehlke, G. Molecular Motors. *Nature* **2003**, *422*, 759–765.

122. The Lock-and-Key Principle, Volume 1, the State of the Art—100 Years on; Behr, J.-P., Ed.; 1st ed.; Wiley-VCH, **1994**.

123. Sheldon, R. A. Enzyme Immobilization: the Quest for Optimum Performance. *Adv Synth Catal* **2007**, *349*, 1289–1307.

124. Liu, W.; Wang, L.; Jiang, R. Specific Enzyme Immobilization Approaches and Their Application with Nanomaterials. *Top Catal* **2012**, *55*, 1146–1156.

125. Strozyk, M. S.; Chanana, M.; Pastoriza-Santos, I.; Pérez-Juste, J.; Liz-Marzan, L. M. Protein/Polymer-Based Dual-Responsive Gold Nanoparticles with pH-Dependent Thermal Sensitivity. *Adv Funct Mater* **2012**, *22*, 1436–1444.

126. Chanana, M.; Rivera Gil, P.; Correa-Duarte, M. A.; Liz-Marzan, L. M.; Parak, W. J. Physicochemical Properties of Protein-Coated Gold Nanoparticles in Biological Fluids and Cells Before and after Proteolytic Digestion. *Angew Chem, Int Ed* **2013**, *52*, 4179–4183.

127. Chanana, M.; Correa-Duarte, M. A.; Liz-Marzan, L. M. Insulin-Coated Gold Nanoparticles: a Plasmonic Device for Studying Metal–Protein Interactions. *Small* **2011**, *7*, 2650–2660.

128. Hanske, C.; Tebbe, M.; Kuttner, C.; Bieber, V.; Tsukruk, V. V.; Chanana, M.; König, T. A. F.; Fery, A. Strongly Coupled Plasmonic Modes on Macroscopic Areas via Template-Assisted Colloidal Self-Assembly. *Nano Lett* **2014**, *14*, 6863–6871.

129. Caswell, K. K.; Wilson, J. N.; Bunz, U. H. F.; Murphy, C. J. Preferential End-to-End Assembly of Gold Nanorods by Biotin−Streptavidin Connectors. *J Am Chem Soc* **2003**, *125*, 13914–13915.

130. Nikitin, M. P.; Zdobnova, T. A.; Lukash, S. V.; Stremovskiy, O. A.; Deyev, S. M. Protein-Assisted Self-Assembly of Multifunctional Nanoparticles. *Proc Natl Acad Sci USA* **2010**, *107*, 5827–5832.

131. Zhen, S. J.; Huang, C. Z.; Wang, J.; Li, Y. F. End-to-End Assembly of Gold Nanorods on the Basis of Aptamer−Protein Recognition. *J Phys Chem C* **2009**, *113*, 21543–21547.

132. Craig, D.; Simpson, J.; Faulds, K.; Graham, D. Formation of SERS Active Nanoparticle Assemblies via Specific Carbohydrate–Protein Interactions. *Chem Commun* **2012**, *49*, 30.

133. Höller, R. P. M.; Dulle, M.; Thomä, S.; Mayer, M.; Steiner, A. M.; Förster, S.; Fery, A.; Kuttner, C.; Chanana, M. Protein-Assisted Assembly of Modular 3D Plasmonic Raspberry-like Core/Satellite Nanoclusters:Correlation of Structure and Optical Properties. *ACS Nano* **2016**, DOI: 10.1021/acsnano.5b07533.

134. Tebbe, M.; Mayer, M.; Glatz, B. A.; Hanske, C.; Probst, P. T.; Müller, M. B.; Karg, M.; Chanana, M.; König, T.; Kuttner, C.; Fery, A. Optically Anisotropic Substrates via Wrinkle-Assisted Convective Assembly of Gold Nanorods on Macroscopic Areas. *Faraday Discuss* **2015**, *181*, 243–260.

135. Mayer, M.; Tebbe, M.; Kuttner, C.; Schnepf, M. J.; König, T. A. F.; Fery, A. Template-Assisted Colloidal Self-Assembly of Macroscopic Magnetic Metasurfaces. *Faraday Discuss* **2016**, DOI: 10.1039/C6FD00013D.
136. Pazos-Perez, N.; Wagner, C. S.; Romo-Herrera, J. M.; Liz-Marzan, L. M.; Garcia de Abajo, F. J.; Wittemann, A.; Fery, A.; Alvarez-Puebla, R. A. Organized Plasmonic Clusters with High Coordination Number and Extraordinary Enhancement in Surface-Enhanced Raman Scattering (SERS). *Angew Chem, Int Ed* **2012**, *51*, 12688–12693.
137. Müller, M. B.; Kuttner, C.; König, T. A. F.; Tsukruk, V. V.; Förster, S.; Karg, M.; Fery, A. Plasmonic Library Based on Substrate-Supported Gradiential Plasmonic Arrays. *ACS Nano* **2014**, *8*, 9410–9421.
138. Plüisch, C. S.; Wittemann, A. Shape-Tailored Polymer Colloids on the Road to Become Structural Motifs for Hierarchically Organized Materials. *Macromol Rapid Comm* **2013**, *34*, 1798–1814.
139. Alvarez-Puebla, R. A.; Liz-Marzan, L. M. SERS Detection of Small Inorganic Molecules and Ions. *Angew Chem, Int Ed* **2012**, *51*, 11214–11223.
140. Jones, M. R.; Macfarlane, R. J.; Prigodich, A. E.; Patel, P. C.; Mirkin, C. A. Nanoparticle Shape Anisotropy Dictates the Collective Behavior of Surface-Bound Ligands. *J Am Chem Soc* **2011**, *133*, 18865–18869.
141. Howes, P. D.; Chandrawati, R.; Stevens, M. M. Colloidal Nanoparticles as Advanced Biological Sensors. *Science* **2014**, *346*, 1247390–1247390.
142. Schäfer, C. G.; Winter, T.; Heidt, S.; Dietz, C.; Ding, T.; Baumberg, J. J.; Gallei, M. Smart Polymer Inverse-Opal Photonic Crystal Films by Melt-Shear Organization for Hybrid Core–Shell Architectures. *J Mater Chem C* **2015**, *3*, 2204–2214.
143. Pfaendtner, J.; La Cruz, De, E. M.; Voth, G. A. Actin Filament Remodeling by Actin Depolymerization Factor/Cofilin. *Proc Natl Acad Sci USA* **2010**, *107*, 7299–7304.
144. Clarke, B. Normal Bone Anatomy and Physiology. *Clin J Am Soc Nephrol* **2008**, *3 Suppl 3*, S131–S139.
145. Kini, U.; Nandeesh, B. N. Physiology of Bone Formation, Remodeling, and Metabolism. In *Radionuclideand Hybrid Bone Imaging*; Fogelman, I.; Gnanasegaran, G.; Wall, H. V. D., Eds.; Springer Berlin Heidelberg: Berlin, Heidelberg, **2012**; pp. 29–57.

7

SELF-ASSEMBLY OF BIOHYBRID POLYMERS

Dawid Kedracki, Jancy Nixon Abraham, Enora Prado, and Corinne Nardin

University of Geneva, Sciences II, Department of Inorganic and Analytical Chemistry, Switzerland

7.1 INTRODUCTION

Self-assembly is considered to be the process through which building blocks spontaneously organize into well-defined frequently functional structures. Self-assembly is therefore thought to be attractive, cost effective, and technologically feasible. Transition from disorder to order, resemblance of the self-assembly process to the self-organization of living cells, practical ease of nanostructure production, are just a few of many factors that support this view [1, 2]. Self-assembly is omnipresent in nature from the lowest level of complexity, at the very dimerization of two components into sophisticated functional cell membranes [3].

The features of individual elements, such as their shape, charge, and mass, regulate the interactions among them and thus self-assembly. The choice of building blocks is therefore a crucial step in achieving the desired structure and in adjusting properties for potential applications. Noncovalent or weak covalent interactions [4] (i.e., van der Waals, electrostatic, hydrophobic, coordination, and hydrogen bonding) play the main role in macromolecular self-assembly [2].

From a thermodynamic point of view, two main types of self-assembly can be distinguished: static and dynamic. Static self-assembly (SSA) can achieve global or local equilibrium and dissipates energy. SSA may, however, require energy to achieve such ordered structures. SSA at the molecular level is illustrated by the formation

Macromolecular Self-Assembly, First Edition. Edited by Laurent Billon and Oleg Borisov.

of lipid bilayers, protein-folding as well as base-pairing [5]. In contrast, dynamic self-assembly (DSA) is out of equilibrium in a steady state that is preserved by a constant energy supply. Biological cell processes and structures such as protein aggregates are examples resulting from a dynamic self-assembly process. Generally speaking, DSA is mainly a feature of living organisms, although not exclusively. Most research in the self-assembly area is devoted to SSA, whereas DSA is still poorly understood [2, 6].

7.1.1 Amphiphiles

Amphiphiles contain both hydrophilic and hydrophobic parts, which are covalently linked. Amphiphiles are able to self-assemble in water to form various well-defined structures, such as spherical core-shell micelles or vesicular structures. Due to chemical incompatibility, these molecules self-organize at the air–liquid interface. The surfactant molecule, which is composed of a hydrophobic tail and a hydrophilic or polar head, is the basic example of an amphiphile. Self-assembly of amphiphilic molecules can be defined as the delicate equilibrium between the attractive (driving force) and repulsive (opposition force) intermolecular forces. It is noteworthy that there is no strong chemical bond like the covalent bond involved in self-organization. Generally, hydrophobic attraction is considered the driving force, and electrostatic repulsion and/or solvation the opposition force. The surfactant molecules are brought in close proximity to each other owing to attractive hydrophobic interaction and over the course of the process, opposition forces like hydration forces and/or electric double-layer repulsion begin to factor in [7]. Originally these opposition forces come from the charge-bearing hydrated head groups, which are comparatively short-range forces in comparison to the hydrophobic interaction [8]. The variety of structures that can be generated by self-assembly depends on a slight modification of parameters having to do with concentration, charge, solvent, composition, water content, and additives. Amphiphilic diblock copolymers, composed of at least a hydrophobic polymer segment linked to a water-soluble block, are considered promising alternatives to lipids for drug delivery [9]. This is because there are limitations arising from the use of lipids due to their low molecular weight [10], poor mechanical stability of the resulting structures, and short shelf life in comparison with more stable block-copolymer structures whose molecular weight is high. Structures that might result from the organization of an amphiphilic diblock copolymer are shown in Figure 7.1.

However, block copolymers remain relatively uncommon because they do not yet have approval from the US Food and Drug Administration (FDA) [12] unlike lipids, which are extensively used. Selecting out of many possible turns the synthetic paths to obtain self-organizing functional polymers is key to their future implementation in biomedical applications.

The tailoring of properties of such structures has drawn attention in various fields, in particular, in enzymatic cascade reactions [13], drug delivery [14], amphiphilic copolymer nanocontainers [15], and nanoreactors [16].

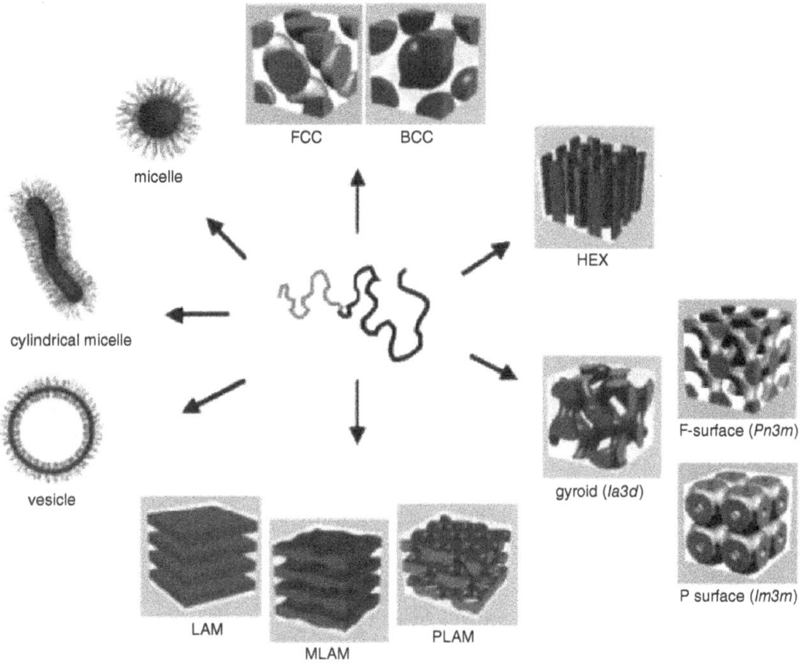

Figure 7.1 High-molecular-weight diblock-copolymer self-organized structures [11]. Adapted with permission of John Wiley and Sons.

7.1.2 Packing Parameter and Interfacial Tension

In order to predict the morphology of self-assembled structures, Israelachvili, Mitchell, and Ninham introduced a model that takes into account the geometric parameters of the surfactant molecules [17]. This model is based on a *packing parameter p*, also named the *surfactant parameter*. In this classic representation, the factor that controls the resulting self-assembled morphology is the size of the hydrophobic moiety in relation to that of the hydrophilic part. The curvature of the interface is governed by its mean curvature H and its Gaussian curvature K, which are defined by the two radii of curvature R_1 and R_2 (Figure 7.2).

The correlation between the packing parameter and the curvature is described by the following equation:

$$\frac{v}{al} = 1 + Hl + \frac{Kl^3}{3} \tag{7.1}$$

where v is the volume of the hydrophobic part of the amphiphilic molecule, l is the chain length, and a is the interfacial area. This model has been successfully applied to predict and justify the formation of structures depending on the surfactant volume fraction, such as spherical micelles ($p \approx 1/3$), vesicles ($p \approx 1$), and cylindrical micelles ($p \approx 1/2$) [19].

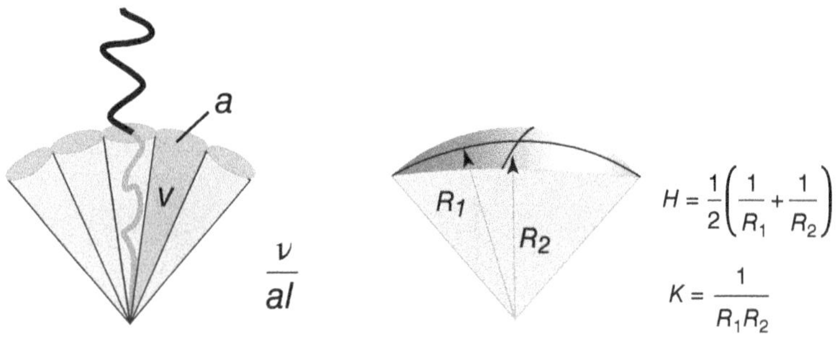

Figure 7.2 Amphiphile self-assembled morphology showing the packing parameter, interfacial curvature, and Gaussian curvature [18]. Adapted with permission from John Wiley and Sons.

The model, however, applies to molecules that can rearrange easily (e.g., surfactants). It is the case for copolymers composed of hydrophobic blocks of glass transition temperature lower than the ambient temperature of self-organization [19]. The aggregates thus remain in a dynamic equilibrium with single copolymers, so the structure rearranges while molecules attempt to reach their optimal thermodynamic equilibrium architecture [20]. Theoretical modeling of copolymer self-assembly has progressed with the self-consistent field methods and advanced with the scaling theory of polymer solution [21]. The scaling model is used to describe the configuration of the copolymer chains in the core and in the corona of diblock-copolymer micelles of different geometries (labeled by the index j): lamellar, cylindrical, and spherical micelles (Figure 7.3) [21, 22]. The morphology is governed by several interactions, as briefly described next.

7.1.3 Interaction Forces in Self-Assembly

The self-assembly of nano components into macroscopic structures depends on the balance between different interacting forces and especially on the free energy of the system to which enthalpy and entropy terms contribute. The nature of the main forces depends on many parameters such as the concentration of the components, temperature, separation distance, among many others [23]. Moreover, the nature of the interactions, the various interaction potentials, and the relative magnitudes are just few of many parameters that engage in the self-assembly process. As mentioned above, a vast majority of processes lead to nonequilibrium conditions in competition with the relaxation processes, which frequently result in the formation of amorphous phases [24, 25]. Then again, the majority of the attraction forces act exclusively at the molecular length scale [24]. It is desirable to take under consideration all these long-range interactions in comparison to the size of the building blocks in order to form well-defined structures.

The usual weak and long-range forces that play the major role in the self-assembly at all scales belong to three classes: (1) driving forces responsible for bringing the

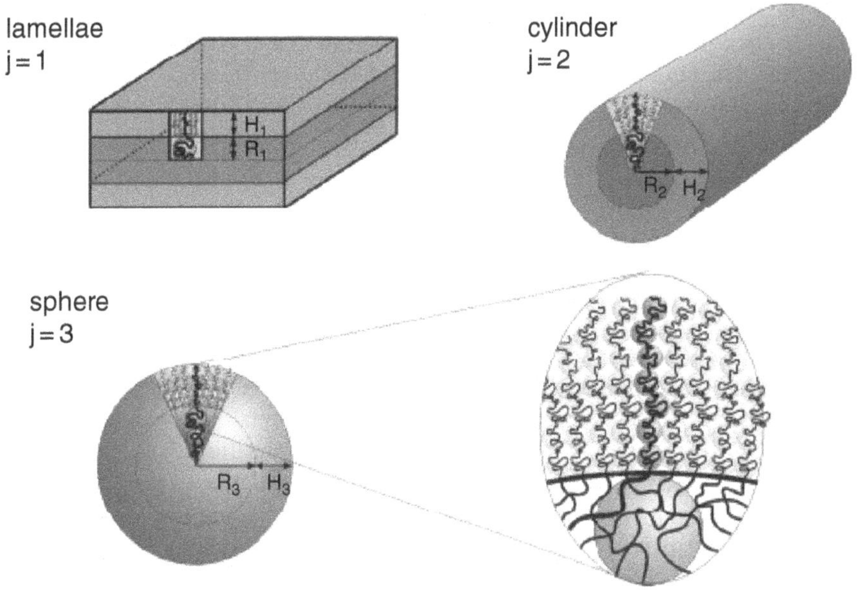

Figure 7.3 Diblock-copolymer micelles of different geometries: lamellar, cylindrical, and spherical micelles. Adapted with permission of the American Chemical Society.

building blocks in close proximity, (2) opposing forces that are in equilibrium with the driving forces, and (3) *functional forces* associated with the directionality and the functionality of self-assembled structures [8]. Depending on the self-assembly type, several kinds of interactions may be involved. However, the origins and basics of these common interactions, as well as the magnetic interactions, crosslinking interactions, and DNA base-pairing [24], are not recalled in this chapter, and the reader is referred to a publication for reference [4].

Structures resulting from the self-assembly of block copolymers composed of weak or strong polyelectrolyte blocks, as it is the case with most biohybrid polymers, can have very different features. If the corona is composed of a weak polyelectrolyte, the fraction of charged monomers is inherent to the local pH and can vary with the ionic strength and pH of the aqueous surrounding. If the copolymer is composed of a strong polyelectrolyte, the fraction of charged monomers does not depend on the environmental conditions. However, in either case, the size and shape of the self-assembly could be tuned with the ionic strength of the aqueous medium. Structures assembled from a copolymer composed of a weak polyelectrolyte are sensitive to the pH of the solution. An overview of the theories that describe the self-assembly of both nonionic and ionic micelles is nicely described in reference [21]. A noticeable outcome is that at high ionic strength, the behavior of an ionic micelle composed of a strong polyelectrolyte corona resembles that of a micelle composed of a nonionic corona. Regarding polypeptide-based biohybrid polymers, the ability of aggregates to change their size and shape due to variations of solution pH is inherent to the polypeptide secondary structure sensitivity to this stimulus. Polymer-DNA hybrid self-assemblies, in addition to responsiveness to ionic strength, are sensitive to hybridization, which renders the establishment of a theoretical framework challenging when both specific hydrogen bonding between complementary sequences and association are to be taken into consideration.

7.2 SELF-ASSEMBLY OF BIOHYBRID POLYMERS

7.2.1 Polymer-DNA Hybrids

DNA is a remarkable molecule carrying the genetic information. DNA is involved in most biological processes. However, the *in vivo* use of bare nucleotide sequences is restricted due to their low cellular penetrability and uptake ratio as well as short plasma half-life [26, 27]. Thus a DNA-polymer conjugation that enables these issues to be overcome is currently widely being investigated. Upon conjugation with a suitable polymer, the resulting biohybrids may form nanostructures that might infer cell penetrability and thus may serve as cargos for the delivery of active compounds. Moreover, the DNA higher order structures such as hybridized complementary strands or *aptamers*, which are synthetic single-stranded nucleotide sequences that undergo molecular recognition as antibodies [28–31], provide a pool for post-synthesis modifications and can ensure either specific detection or targeting. There is therefore a growing interest in designing DNA-decorated structures of potential use in drug delivery and biosensing [32, 33].

7.2.1.1 Linear Block Copolymers Coupling of a hydrophobic segment to a single-stranded nucleotide sequence results in macromolecules that are composed of the water-soluble, flexible nucleic acid strand linked to the nonsoluble segment, which could be regarded as high-molecular-weight analogues of surfactants. Along this line, Vebert *et al.* described the grafting of an aptamer, a single-stranded nucleic acid sequence that specifically recognizes the immunoglobulin E (IgE) protein, an allergy biomarker, to amino acid modified poly(2-alkyl-2-oxazoline) (POX) [27]. The resulting copolymer, of high-hydrophilic-weight fraction, exhibits surface activity that can be used in an emulsification process to stabilize submicrometer size oil-in-water emulsion droplets. Functionality of the aptameric strand engaged in the structure formation was assessed by reflectometry, confocal laser scanning microscopy, and atomic force microscopy imaging.

Depending on the DNA-copolymer composition, both chemical and physical incompatibility drives the structure formation through self-assembly. Coupling and structure formation may, however, affect the properties of the DNA strand involved in the assembly process, which stresses the need to establish a mechanism of functional structure formation through investigations of the modes of interaction of the resulting self-assemblies.

Although composed of nucleic acid strands, the examples reported reveal that linear DNA-based block copolymers do not undergo specific interactions such as intermolecular hydrogen bonding or conformal stacking of the bases or sugar rings in which the nucleic acid sequence can engage. As exemplified in the following discussion, DNA-copolymer self-assembly is primarily driven by the attraction of the hydrophobic segment coupled to the nucleic acid strand.

For instance, the conjugation of the hydrophilic *c-myc* antisense strand, which suppresses smooth muscle cell proliferation [34], to the biodegradable hydrophobic poly (D,L-lactic-*co*-glycolic acid) (PLGA) induced self-assembly into well-defined micellar structures of narrow size distribution [35]. Being micelles composed of a PLGA hydrophobic core surrounded by a hydrophilic shell of nucleic acid strands, gradual degradation of the PLGA in the intracellular medium resulted in the release of the antisense strand into the surrounding and enhanced cellular uptake subsequent to endocytosis. A general strategy to generate polystyrene-based DNA-copolymers through solid phase synthesis was reported by Mirkin *et al.* [36]. Self-assembly into spherical core-shell micelles was induced through a dialysis-mediated solvent exchange from a DMF to an aqueous solution. Functionality of the nucleic acid strands engaged in structure formation was assessed through the assembly of higher order structures by hybridization with the complementary nucleic acid strands tethered to gold nanoparticles.

However, due to the high glass transition temperature (T_g) of the hydrophobic segments (about 40°C and 100°C), PLGA- and PS-based copolymer micelles were likely kinetically "frozen" and might not have reached thermodynamic equilibrium [37]. To overcome this issue, Herrmann *et al.* conjugated polypropyleneoxide (PPO) to 22-mer long nucleic acid sequences through automated DNA synthesis [37]. Self-assembly in the aqueous solution took place. The micellar core-shell structure is composed of a PPO core surrounded by a DNA corona. Functionality of the DNA strand was assessed by hybridization with the complementary strand modified by various functional groups to carry out diverse organic reactions at the interface between the micelles and their surroundings, such as Michael addition, amide bond formation, and isoindole formation [37]. The impact of formation of specific hydrogen bonds through hybridization of the complementary nucleic acid sequences with the strand engaged in self-assembly of the DNA-*b*-PPO structures on their morphology was studied as well [38]. Scanning force microscopy (SFM) investigations revealed that hybridization with the fully complementary sequences had a very minor effect on the overall morphology of the block-copolymer micelles. However, hybridization with a much longer sequence, which encoded the complementary nucleic acid strand of the DNA-b–PPO several times, strongly affected the micellar structure. Rod-like structures were observed with a height significantly smaller than that of the initial micellar aggregates and far longer than the pristine structures, which did disintegrate. Material assembly occurred alongside the long nucleotide sequence template through hybridization, without affecting the hydrophobic core (Figure 7.4) [38].

To improve colloidal stability, inhibit nonspecific interactions and, by decreasing the electrostatic repulsion, improve the DNA micelles binding ability, Talom *et al.* reported on the synthesis and self-assembly of a DNA-triblock-copolymer hybrid obtained through the coupling of a 22-mer ssDNA with a poly(ethylene

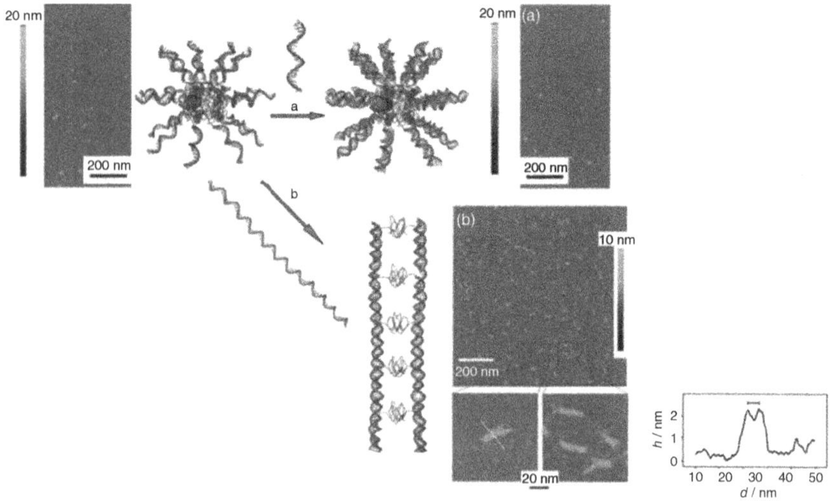

Figure 7.4 (a) Hybridization of complementary sequences does not affect the shape of the DNA-copolymer micelle. (b) Hybridization with an elongated complementary sequence results in the formation of rod-like structures [38]. Adapted with permission of Wiley-VCH. (*See color insert for color representation of this figure*)

oxide)-poly(caprolactone) (PEO-*b*-PCL) diblock copolymer [39]. Self-assembly into micelles took place in aqueous solution. Poly(caprolactone) composed the hydrophobic core, whereas the hydrophilic corona was made of PEO and DNA. Investigations using the quartz crystal microbalance (QCM) demonstrated that the micelles could be reversibly immobilized on surfaces modified with the complementary nucleotide sequence through hybridization, which revealed functional structure formation.

Synthesis of amphiphilic block copolymers to induce their self-assembly into vesicular structures was described by Vebert-Nardin *et al.* [40]. Solid phase synthesis was used to couple nucleic acid strands to highly hydrophobic moieties of low glass transition temperature, namely poly(butadiene) or poly(isobutylene). TEM (Figure 7.5), DLS, as well as encapsulation and release studies showed assembly into vesicular structures. Moreover, as assessed by circular dichroism, coupling and self-assembly did not affect the DNA chain configuration.

More recently, a very appealing *grafting-from* route of preparation of various DNA block copolymers was reported [41]. The approach involved DNA modification with an initiator and monomer addition supported by *activators generated by electron transfer* (AGET) ATRP. The authors found that this strategy allows the preparation of hydrophilic as well as amphiphilic DNA-based copolymers, which demonstrates the high versatility of this synthesis route. Due to the hydrophobicity of the polymer component, the benzyl methacrylate based DNA1-b-PBnMA-co-RMA copolymer aggregated into large DNA-latex (DTEX) particles by dialysis-mediated solvent exchange from acetone to water. Functionality of the DTEX particles was evidenced by hybridization.

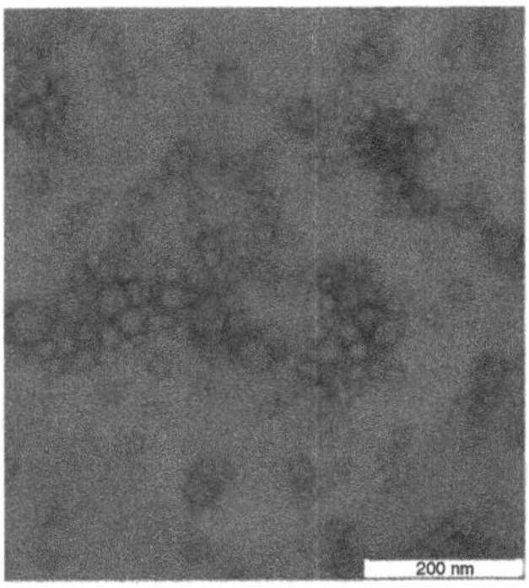

Figure 7.5 PIB_{31}-b-G_7A_5 vesicles observed by TEM. Reproduced with permission from The Royal Society of Chemistry.

7.2.1.2 ***Comb/Graft Copolymers*** Chien *et al.* reported on DNA programmable micelles that can undergo a reversible morphological transition induced enzymatically [42]. Spherical micelles of about 25 nm in diameter resulted from the self-assembly of a 19-mer long DNA brush-type amphiphilic copolymer based on (*N*-Benzyl)-5-norborene-exo-2,3-dicarboximide and (*N*-acetyloxy-2,5-pyrrolidine-dione)-5-norborene-exo-2,3-dicarboximide. Since it contained a RNA base (rA) cleavage site, the DNAzyme did cut the strand to affect the hydrophilic weight fraction. A sphere-to-cylinder shape transition could thus be induced by an enzymatic reaction (Figure 7.6). A 19-mer nucleotide sequence (In_1, Figure 7.6) added to the self-assembled DNA-copolymer formed a 9-base duplex with the DNA-copolymer to reverse the transition. To induce the sphere-to-cylinder transition back, DNA strand invasion was used. The fully complementary 19-mer sequence of In_1 (In_2, Figure 7.6) interacted competitively with the 9-mer duplex. The more thermodynamically stable $In_1 \cdot In_2$ duplex did leave the shell of the micelles, which did assemble back into cylindrical structures (Figure 7.6).

More recently, Kedracki *et al.* presented a novel synthetic approach to generate a comb/graft DNA-based copolymer of high (67%) grafting density through a one-step thiol click chemistry of high coupling efficiency (above 60%) [43]. As demonstrated by microscopy, the generated hybrids assembled into submicrometer size spherical structures in aqueous solution, seemingly of vesicular morphology. The structures were further shown to be functional through hybridization experiments, which also suggested the mechanical stability of these structures, and thus a high potential for application in diagnostic or sensing.

7.2.1.3 ***Star Copolymers*** The preparation of multifunctional star molecules was introduced by Das *et al.* [44]. The multi-arm polymer template composed of oligo(ethylene)oxide methacrylate (OEOMA) and ethylene glycol diacrylate (EGDA) on which DNA was grafted was prepared by the "arm-first" method. The presence of an azide group at each arm of the star enabled conjugation of alkyne-functionalized DNA strands through highly efficient copper-catalyzed

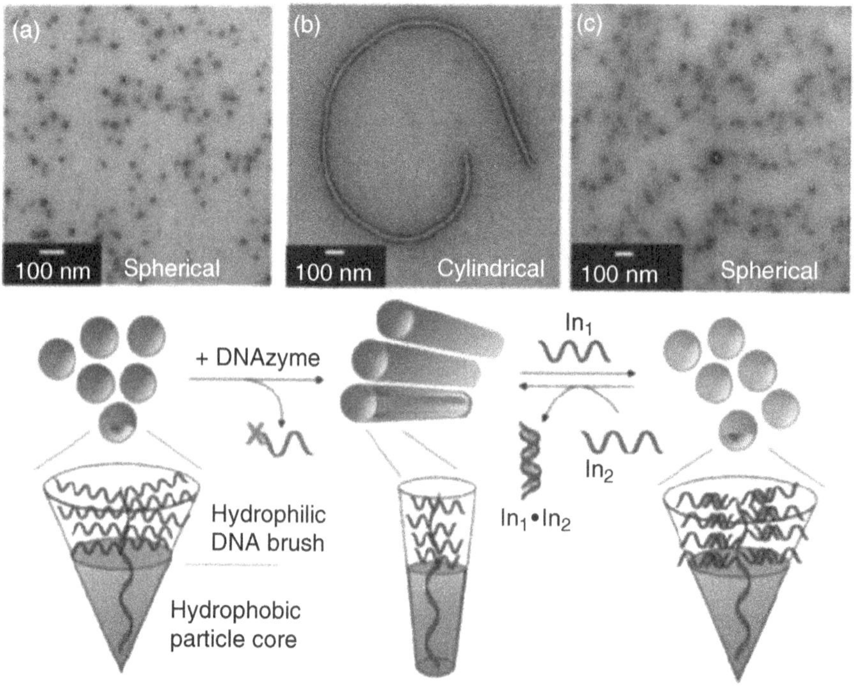

Figure 7.6 Assembly of DNA-brush copolymers: (a) Initial spherical morphology. (b) Transition to a cylindrical morphology upon the addition of a DNAzyme. (c) Hybridization with a partially matching sequence results in the micelles reassembly. Adapted with permission of Wiley-VCH.

azide-alkyne cyclo-addition (CuAAC) "click" reaction. The authors demonstrated the functionality of the obtained structures by hybridization. Two distinct populations of the polymer multi-arm structures were modified with complementary DNA sequences prior to hybridization (star-DNA1 and star-DNA 2, Figure 7.7A). As evidenced by DLS measurements after changing the molar ratio between the hybrids, it was possible to control the size of the resulting hybridized multi-arm star superstructures. The reversibility of the process was obtained by designing the duplex with an overhang ("toehold") available for an invasion of the fully complementary sequence, which caused the disassembly of the superstructure (Figure 7.7B).

In 2012 Maeda and colleagues reported on the synthesis of linear and mikto-arm star-shaped DNA diblock copolymers composed of poly(*N*-isopropylacrylamide) (PNIPAAM) and on the dependence of their assembly on various block lengths and chain architectures [45]. The PNIPAAM of different architectures were generated by ATRP and functionalized with azide groups for coupling with alkyl modified ssDNA through click-chemistry. DSC-monitored phase transition demonstrated that the PNIPAAM LCST decreased with the PNIPAAM fraction in the copolymer. DLS measurements showed lack of micellization for the copolymer with the shortest

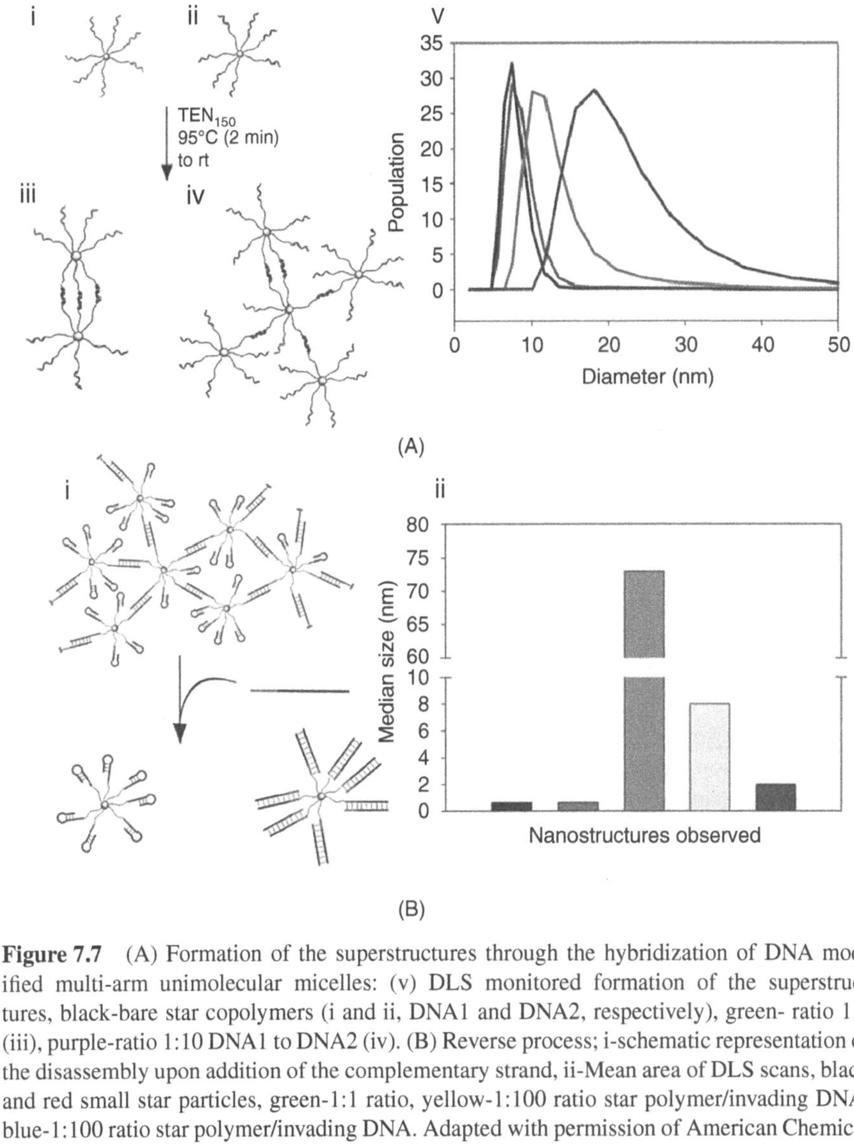

Figure 7.7 (A) Formation of the superstructures through the hybridization of DNA modified multi-arm unimolecular micelles: (v) DLS monitored formation of the superstructures, black-bare star copolymers (i and ii, DNA1 and DNA2, respectively), green- ratio 1:1 (iii), purple-ratio 1:10 DNA1 to DNA2 (iv). (B) Reverse process; i-schematic representation of the disassembly upon addition of the complementary strand, ii-Mean area of DLS scans, black and red small star particles, green-1:1 ratio, yellow-1:100 ratio star polymer/invading DNA, blue-1:100 ratio star polymer/invading DNA. Adapted with permission of American Chemical Society.

PNIPAAM chain length whereas, for the higher PNIPAAM degree of polymerization (overall target DP=50/1), the assembly of colloidal particles in the size range of few and few tens of nanometers was observed. SAXS investigations suggested the formation of hard spheres with the core composed of PNIPAAM surrounded by a DNA corona. Turbidimetric based DNA detection revealed that the increase in particle density and particle size enhanced aggregation due to van der Waals interaction between the particle cores. These studies demonstrate that DNA hybridization can thus be controlled by tuning the molecular structure.

A significant number of reports on hydrogels or networks composed of DNA-based copolymers have appeared, being structure formation induced through hybridization. A vast number of reports describe hydrogels based on polyacrylamide DNA conjugates [46–48] or PEG [49]. As this field of research is beyond the scope of this contribution, the reader is referred to the excellent review written by Xiong *et al.* [50] and an interesting article about the structure formation undergone by norbornene block copolymers composed of nucleic acid strands and ferrocenyl side chains [51].

7.2.2 Polypeptide Block Copolymers

Polypeptide-block copolymers have gained interest because the size and morphology of the self-assembled structures are responsive to solution pH, temperature, and salt concentrations [52]. This stimuli responsiveness allows for greater functions in biotechnology, as compared to the structures resulting from the self-assembly of conventional synthetic block copolymers. Moreover, the properties of polypeptide block-copolymer structures offer many advantages over other block-copolymer self-assemblies such as biocompatibility and enhanced biofunctionality. The following section focuses on how the complex three dimensional secondary structures of the polypeptide affects the self-assembly properties of peptide-based hybrid diblock copolymers.

The self-organization of poly(butadiene)-b-poly(L-glutamic acid) (PB-PGA) in aqueous solution has been studied by Lecommandoux as well as by Klok *et al.* [53]. The size and shape of the aggregates, as assessed by DLS, can be triggered by exploiting the secondary structure of poly(L-glutamic acid) with pH [54, 55]. In the PB-PGA system, the hydrophobic PB provides the driving force for self-assembly and forms the core of the micelle, while the hydrophilic polypeptide (PGA) forms the corona. The stimuli responsiveness of the structure relies on the swelling induced by changes in pH, temperature, ionic strength, and solvent type. PB_{40}-*b*-PGA_{100} is readily soluble in water and, at basic pH, well-defined vesicular structures were observed [54]. Even at high ionic strength, when the charges of the polypeptide blocks are strongly screened, the size of the aggregate could still be varied by changing the pH. CD spectroscopy studies showed that there is a transition from a compact α helical secondary structure at low pH (pH = 4.5) to a more extended random coil structure in basic media (pH = 11.5). Such pH induced size variations were completely reversible and could be regenerated numerous times [53].

Lecommandoux *et al.* further studied the impact of the polypeptide secondary structure on polypeptide-b-polymer self-assemblies by considering three different systems based on polybutadiene-*b*-poly(L-glutamic acid) (PB-*b*-PGA), polyisoprene-*b*-poly(L-lysine) (PI-*b*-PLys), and poly(L-glutamic acid)-*b*-poly(L-lysine) (PGA-*b*-PLys) [55]. Basically the self-assembly of either spherical core-shell micelles or vesicles were formed depending on the rod-(polypeptide)-to-coil block molar ratio as assessed by light scattering. For a value lower than 65%, vesicles could be obtained, whereas for greater values, stable spherical core-shell micelles were assembled.

The pH and ionic strength responsiveness of the micelles and vesicles was followed by DLS. The variation of the conformation (the transition from α helix to coil) of the polypeptide block within the self-assembled structure was studied by CD spectroscopy. This modification of the secondary structure resulted in a change of the overall size of the micelles and vesicles. The observed size variation could be attributed to the neutralization of the polypeptide block that thus did undergo a conformation transition from a charged extended coil to a neutral and compact α-helical rod in a reversible manner. This responsiveness was active and efficient even in media of high ionic strength, in which electrostatic interactions are screened [53]. PPO_{44}-$PLys_{10}$ (0.33 hydrophilic weight fraction) readily formed vesicular structures at low pH [56]. However, this species has limited solubility at pH > 5, and hence the pH responsiveness could not be determined [56]. PPO_{44}-$PLys_{62}$ and PPO_{44}-$PLys_{44}$ [57–64] are of 0.76 and 0.69 hydrophilic weight fractions, respectively. At low pH, the PLys blocks are in a fully stretched conformation. The R_G/R_H ratio estimated from light-scattering experiments was found to be 1.1 and 1.01 for PPO_{44}-$PLys_{62}$ and PPO_{44}-$PLys_{44}$, respectively. This indicates that these block copolymers did assemble into vesicular structures as evidenced further by TEM measurements. The vesicle formation at these relatively high hydrophilic weight fractions could be explained by the partial hydration of the PPO core. The PPO-PLys block-copolymer assemblies did undergo an assembly-to-unimer transition with decreasing temperature. As the temperature was below the LCST of PPO (8°C), the block copolymer became double hydrophilic and the copolymer existed as unimers. The same process did arise at low pH (~3). But, as the pH is increased above the pKa of PLys, the polypeptide became more hydrophobic. In case of PPO_{44}-$PLys_{62,}$ the PLys block became insoluble at pH >7. Contrary to the assembly occurring at lower pH, the structures formed did consist of a PLys core surrounded by a PPO corona. From DLS experiments, it was observed that the hydrodynamic radii of the vesicles increased from 10 to 75 nm, upon increasing the pH to 7. The CD data revealed an α-helical configuration at this pH. At pH 8, the solution became turbid due to the inability of the PPO blocks to stabilize the aggregates.

7.2.3 Block Copolypeptides

In the case of copolypeptide-based systems such as PGA_{15}-*b*-$PLys_{15}$, a morphological transition of the self-assembly, was observed in addition to a change in size [65–67]. This class of zwitterionic copolypeptides can reversibly self-assemble into vesicles in moderate acidic or basic aqueous solutions. They are called *schizophrenic vesicles*. Figure 7.8 schematically represents the responsiveness of the diblock copolypeptide with pH. At neutral pH, the zwitterionic diblock copolymer

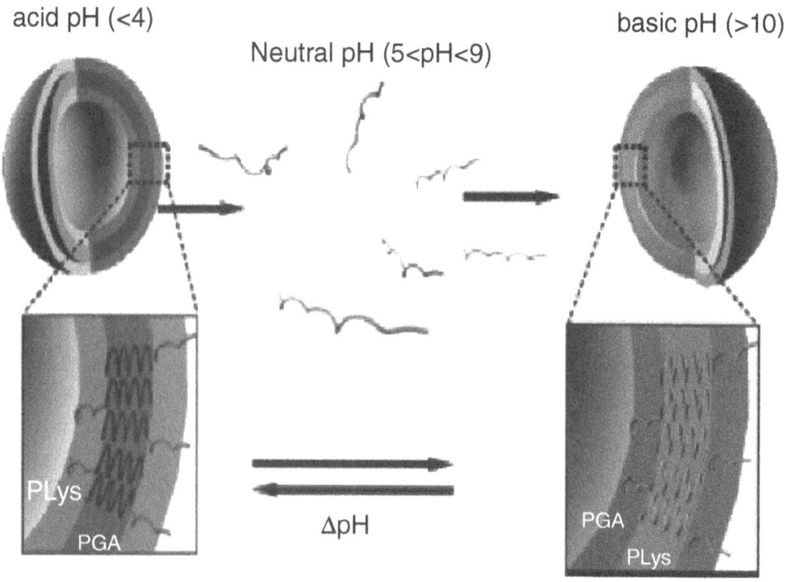

Figure 7.8 Schematic representation of the schizophrenic self-assembly into vesicular structures of PGA_{15}-b-$PLys_{15}$. Adapted with permission from American Chemical Society. (*See color insert for color representation of this figure*).

bears the positive charges of the protonated lysine and negative charges of the deprotonated glutamic acid. The dissolution of the diblock copolymer in aqueous basic and acidic solutions results in spontaneous self-assembly. At acidic pH (pH $\sim$ 3), the poly(L-glutamic acid) block is neutralized, which changes the secondary conformation from a charged coil to a more compact neutral α helical ("rod") structure. This structural variation is accompanied with a decrease in solubility, and hydrophobicity becomes the driving force for the self-assembly. The low interfacial curvature of the "rod"-like hydrophobic part induced the formation of hollow structures. The insoluble PGA did form the core of the aggregates, while the PLys block composed the shell. In basic conditions (pH $\sim$ 12), a reversible process took place. The protonated poly(L-lysine) block (NH_3^+) is transformed into a neutral (NH_2) group. These NH_2 groups are highly insoluble and participate in the formation of the hydrophobic core. The PLys block formed the shell in this case. At intermediate pH, the blocks are charged, the assembled structures disassembled, and precipitation eventually occurs. This precipitate can be re-dissolved and the process completely reversed by adding a suitable acid or base. Even at high salt concentrations, when the electrostatic interaction is expected to be screened, this reversible structural formation can be induced. This spontaneous self-assembly was proven by ^{1}H NMR and fluorescence spectroscopy. The sizes of the vesicular structures were determined by DLS. Values of 110 and 175 nm were calculated for self-assembled structures of the dipeptide in water at acidic and basic pH, respectively.

7.3 SELF-ASSEMBLY DRIVEN BY NUCLEATION POLYMERIZATION

Amyloid proteins fibril formation is suspected of playing a crucial role in several amyloid-based diseases. It is therefore of great importance to elucidate the mechanism of their aggregation [68–71]. Moreover, the formation of fibers draws attention not just to solving biological and medical issues but also important, from the perspective of potential applications in nanotechnology (scaffolding for tissue engineering), food processing and treatment of eye cataract diseases [68, 72, 73]. Amyloids are referred to proteins and peptides that exhibit a cross β-sheet structure in which the peptide backbone runs perpendicularly to the axis of the fiber, which are generally long with a diameter ranging from 40 to 120 Å [73]. Self-organization of β-sheets into proto fibers is followed by their further packing, which leads to formation of amyloid fibers (Figure 7.9). The formation of amyloid structures was identified for at least 16 various proteins and peptides [74] that share the same core structure regardless of having different composition and configuration [73]. It thus seems to be a generic feature of peptides and proteins [75].

Numerous various highly characterized self-assemblies are recognized to organize through nucleation dependent polymerization, including microtubule formation [76], sickle-cell fibril formation [77], flagellum assembly, [78], among others [79]. Frequent observation of a lag phase suggests that amyloid fiber formation also occurs according to a nucleation and growth process [70, 71, 79]. Characteristic features of these processes are depicted in Figure 7.10. Typically, a nucleation phase, which goes through series of thermodynamically unfavorable association steps to form oligomeric nuclei (with a characteristic nucleation equilibrium constant $K_n \ll 1$, and results from the fact that the entropic cost of association is not overcome by intermolecular interactions), is followed by a thermodynamically favorable growth phase (of growth equilibrium constant $K_g \gg 1$, which is associated with rapid

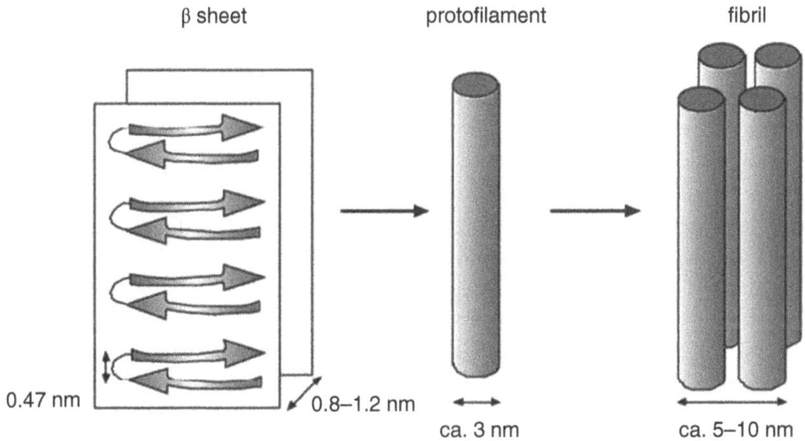

Figure 7.9 Amyloid fibers structure [68]. Adapted with permission of John Wiley and Sons.

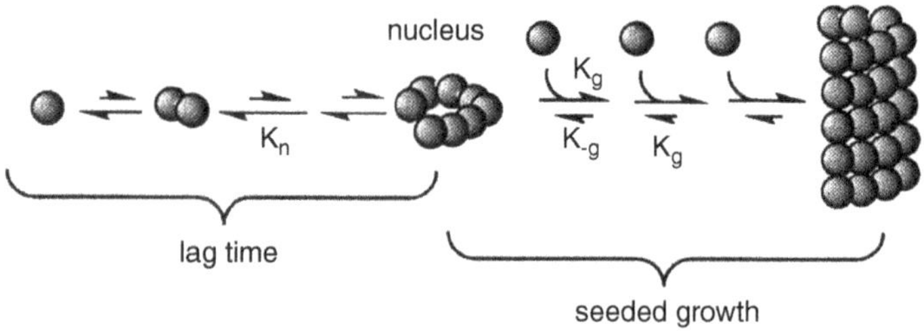

Figure 7.10 Nucleation-dependent polymerization simplified mechanism [71]. Equilibrium constants for each step are small and equal, while the growth steps' equilibrium constants K_g are large and equal [76].

polymerization occurring at multiple sites of contact of monomers with the growing polymer) and eventually a steady state phase in which the monomer and aggregate are in equilibrium [79]. The process can be considered to belong to the class of nucleation polymerization when $K_g \gg K_n$ [76].

The following properties are characteristic for nucleation polymerization: larger aggregates can be formed just above a critical concentration, which is analogous to a critical micelle concentration (CMC), lag phase after which aggregates are noticeable and the addition of seeds helps reduce the lag time [68, 71, 79].

In the common nucleation-dependent assembly, oligomers are formed when the concentration of monomer reaches a threshold value, which is defined as the critical concentration. If the concentration does not reach this level, the monomers are in the majority. The critical concentration is characterized by the growth equilibrium constant K_g, which is inherent to the protein solubility during the steady-state phase [76, 79]. Micelle generation is as well characterized by a critical concentration. However, micelles formation is much faster than amyloid fibers formation due to a larger entropic barrier to fibers formation. As a result the supersaturated protein solution is referred to as metastable or kinetically soluble [71].

The time during which the nucleus forms and solution is referred to as metastable before formation of fibers is known as lag time [71, 79]. Formation of pre-nucleus oligomers during the lag phase occurs in a quantity that might be difficult to detect (in a narrow concentration range) [79]. The slow rate of this process is a consequence of nonfavorability of the association. Protein concentration is the main factor that regulates the time of the lag phase [80].

Addition of exogenous seeds or nuclei results in intermediate polymerization, together with a reduction up to the elimination of the lag time [81]. Subsequent growth occurs according to first-order kinetics [82, 83]. Seeding depends on complementarity of the fibril seed and the substrate, which makes it highly discriminating [84]. Nevertheless, various constitution seeds can serve as heterogeneous seeds and supply complementarity between the seeds and the native substrate, exactly as in the case of the growth of inorganic crystals [71].

7.3.1 Polymer-DNA Hybrids

The synthesis of β-sheet-mimicking copolymers is of high interest in both developing new classes of materials for biomedical applications and understanding the basic rules of nanoscale assembly [85]. Reports provided up to date describe block copolymers composed of either peptide sequences able to form β-sheets or cyclic peptides forming fibrillar superstructures or nanotubes in aqueous solutions [68, 85, 86]. This self-assembly has drawn much attention because of its association with neurodegenerative diseases such as Alzheimer's, Parkinson's, Huntington's diseases, type II diabetes, and Creutzfeldt-Jacob disease [87, 88].

We reported recently on the self-assembly-driven nucleation polymerization of a DNA copolymer obtained by solid phase coupling between the amphiphilic poly[2-(3-butenyl)-2-oxazoline] (PBOX) polymer segment with *N*-protected amino acid side chains and a water-soluble nucleotide sequence [43]. To the best of our knowledge, this is the first report in which self-assembly formation of a DNA copolymer hybrid composed of a polymer segment and a DNA sequence occurs through a nucleation polymerization mechanism. Conjugation of PBOX with a short single-stranded nucleotide sequence resulted in an amphiphilic comb/graft copolymer that organizes in fibrillar structures upon direct dissolution in water. Supported by TEM, CD, and scattering data, the experimental outcomes show that the fibrils are formed through inter-and intramolecular hydrogen bonding between hydrogen accepting amide groups along the polyoxazoline backbone and the hydrogen donating nucleic acid grafts leading to the formation of hollow tubes (Figure 7.11).

7.3.2 Polymer-Peptide Hybrids

The PEGylation of peptides and proteins have various applications in biotechnology, as it often improves water solubility and stability to many hydrophobic

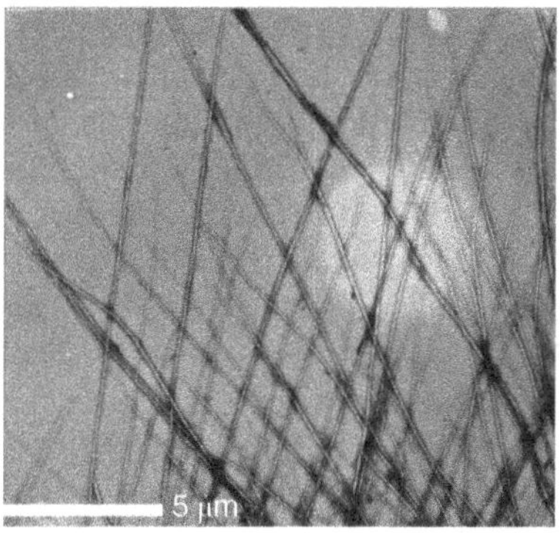

Figure 7.11 TEM imaging of PBOX-g-DNA fibers.

proteins/peptides, as well as reduces clearance through kidneys, leading to longer retention time [39, 90].

The amyloid β, a 40-43 amino acid proteolytic fragment of the β-amyloid precursor protein is the major component of plaques found in the brain of patients who suffered from Alzheimer's disease. The amyloid β (1-42) is composed of 42 amino acids, and several studies suggest that the critical sequence for fibril formation is Aβ (16-20), corresponding to KLVFF, in which LVFF is a highly hydrophobic peptide motif [68, 91–93]. The diphenylalanine dipeptide, (FF, Aβ(19, 20)) undergoes self-assembly into a range of structures from vesicles to tubes and fibers, depending on the medium and conditions. Understanding the exact mechanism of amyloid fiber formation is of importance in developing therapeutics that aim at fiber disassembly and fiber growth inhibition. The C-terminal PEG derivatization of an amyloid fragment, Aβ (10-35), to develop a PEG-peptide block copolymer was initially performed by Lynn *et al.* early in 1998 [94]. Specifically they studied the self-assembly of Aβ(10-35)-PEG3k, where PEG3k represents a poly(ethylene glycol) of 3000 Da molecular mass. Both Aβ and Aβ-PEG formed fibers composed of a similar peptide core [95]. The conjugation of Aβ(10-35) to PEG could enhance the solubility, which led to a concentration-dependent reversible fibril formation, even at high concentrations. A concentration and pH dependent study of the PEG-peptide structure formation was also performed and the following mechanism for fibril genesis was proposed [96]:

$$U_{unstructured} \rightleftharpoons \beta_{aggregate} \rightleftharpoons (\beta_{fibril})_{sol} \rightarrow (\beta_{fibril})_{precip} \tag{7.2}$$

At low concentrations, low pH and/or low ionic strength, the PEG-peptide block is unstructured ($U_{\text{unstructured}}$). As the concentration or pH was increased, the peptide self-associated to form oligomeric aggregates ($\beta_{\text{aggregates}}$), which are possibly hexamer. These $\beta_{\text{aggregates}}$ readily reversibly formed soluble fibers ($(\beta_{\text{fibril}})_{\text{sol}}$). At pH > 5.6 and long incubation times, fibril–fibril associated arrays dominate in the structure ($(\beta_{\text{fibril}})_{\text{precip}}$) [96]. The lateral aggregation of the Aβ(10-35) into linear fibrils induces the formation of insoluble β_{fibril} aggregates but could be prevented by PEGylation. PEGylation could shield the hydrophobic domain of the peptide from lateral aggregation, which inhibited irreversible fibril–fibril association. The authors did further find that the hydrophilic weight fraction of the peptide-PEG conjugate was crucial in maintaining structure formation. For example, the PEGylation at the C-terminus of the amyloid fiber forming hexapeptide (KTVIEE) could inhibit the formation of fibers [97]. Upon increasing the amphiphilic character of the peptide by introducing an additional hydrophobic group (alkyl tail) at the N-terminus, structure formation resulted in the organization of well-structured aggregates. *In situ* UV cleaving of this hydrophobic alkyl tail connected to the peptide by a UV-sensitive nitrobenzyl moiety resulted in destabilization and disassembly of the fibers.

Hamley *et al.* further studied the self-assembly of PEG conjugated to different Aβ peptide fragments [98–102]. A modified fragment of the KLVFF core peptide, FFKLVFF was used for conjugation with PEG (M_n-3300). In dilute aqueous solutions this peptide–polymer conjugate self-assembles to form core-shell fibrils being the core composed of the hydrophobic peptide in an extended β-strand configuration surrounded by a shell of the hydrophilic PEG [101]. The influence of the PEG molar mass on the self-assembly properties of FFKLVFF-PEG with PEG_{1000} and PEG_{2000} (both C-terminal) and PEG_{10000}-FFKLVFF (N-terminal) was also investigated [101]. The three hybrids formed fibrils composed of a FFKLVFF core surrounded by a PEG corona. At low concentrations, aggregates were formed through hydrophobic interactions, followed by organization into β-sheets at higher concentrations.

The effect of PEG chain length on N-terminal PEGylation of FFKLVFF, for example, $(EG)_n$-FFKLVFF-COOH (EG refers to ethylene glycol), where *n* varies from 5, 11 to 27 were also studied and the different packing motifs were examined using X-ray diffraction and *in situ* solution WAXS experiments. The $(EG)_{27}$-FFKLVFF-COOH has a different packing fraction than $(EG)_5$-FFKLVFF-COOH or $(EG)_{11}$-FFKLVFF-COOH. Fibers were observed with $(EG)_{27}$-FFKLVFF-COOH, whereas $(EG)_5$-FFKLVFF-COOH and $(EG)_{11}$-FFKLVFF-COOH formed twisted ribbons. A proposed model is shown in Figure 7.12. In the case of $(EG)_{27}$-FFKLVFF-COOH, antiparallel β sheets did offset by two residues, favored by electrostatic interaction between the positively charged lysine residue and the N-terminus.

Another example of a peptide-amphiphile is FFFF-PEG (M_n-5000), which self-assembles in aqueous solution and organizes in the solid state [103]. The peptide sequence is composed of highly hydrophobic phenylalanine residues, and there are no charges or polar residues as in the case of FFKLVFF-PEG. The end group of FFFF-PEG is charged, so the electrostatic interaction can influence the self-assembly. At low concentrations in aqueous solution, self-assembly occurred due to the hydrophobic interactions between the aromatic phenylalanine residues, while at high concentrations, β sheets developed, which resulted in the formation of "needle-shaped fibers" as evidenced by TEM (Figure 7.13). The CD spectra of this molecule in water showed a positive maximum at 220 nm assigned to n-π*

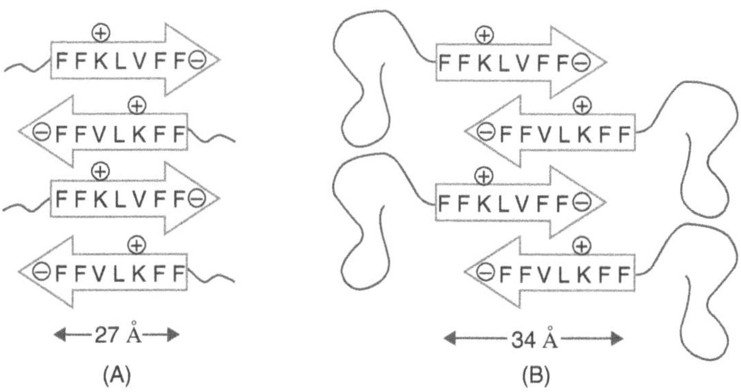

Figure 7.12 Proposed mechanism for the self-assembly of $(EG)_n$-FFKLVFF-COOH: (A) Formation of antiparallel β-sheet and (B) offset of antiparallel β-sheet due to electrostatic repulsion. Adapted with permission from American Chemical Society.

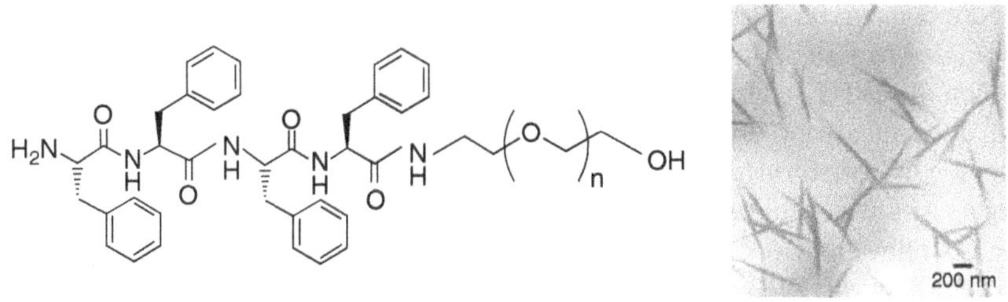

(A) (B)

Figure 7.13 (A) Structure and (B) self-assembly into fibers of FFFF-PEG. Adapted with permission from Royal Society of Chemistry.

transitions resulting from π-π staking of the phenylalanine residues. Fluorescence spectroscopy showed as well a peak at 310 nm, which was associated with the π-π stacking interactions of the aromatic residues. FTIR also showed a shoulder peak at 1670 cm^{-1} in the amide 1 region, associated with the antiparallel β-sheet structure formation.

However FF-PEG_{3000} does not undergo self-assembly, which indicates that the hydrophobic to hydrophilic balance defined by the relative number of phenylalanine repeats in respect to the PEG chain length is critical to induce structure formation.

7.3.3 DNA-Peptide Hybrids

The following discussion is based on the self-assembly of peptide-DNA hybrids, taking FF-DNA and WW-DNA as representative examples [104–107]. As discussed above, FF is a constitutive motif of the amyloid Aβ peptide. WW is a peptide characterized by intrinsic fluorescence and a stronger quadruple strength than FF [108, 109], which allows for more favorable stacking and stable structures of W containing molecules than other amino acids. Peptide-nucleic acid nanostructures are used to establish a mechanism of structure formation of DNA hybrids and for potential applications owing to biocompatibility and functionality of the amino acid or nucleic acid sequences.

The same as for other hybrids, the self-assembly of peptide-nucleic acid hybrids depends on the composition, nature of solvent, pH, temperature, and ionic strength. FF-DNA in aqueous solution was shown to assemble in vesicular structures of ~200 nm diameter in dilute aqueous solution, as assessed by an encapsulation assay [26]. The vesicular structures were shown to be composed of a spherically closed FF-DNA membrane. In contrast, WW-DNA formed spheres of ~1 μm diameter. Upon dilution, the sizes of these structures decreased. They eventually disassembled below a critical aggregation concentration, as assessed by monitoring the intrinsic fluorescence of W, which is sensitive to the polarity of the solvent. Upon increasing the concentration, fiber-like structures developed and long fibers of several micrometers length could be observed. The mode of structure formation could thus be described by nucleation polymerization, as in the case of amyloid fibril formation. At low concentrations, the formation of small nuclei takes place owing to the chemical and physical incompatibility between the hydrophobic rigid ditryptophan and the water-soluble, flexible nucleic acid sequence, prior to elongation of fibers at the suitable concentration. The hydrophobicity of the peptide moieties is the driving force for self-assembly of the nuclei, whereas β-sheets develop at high concentrations due to the stacking of the aromatic ditryptophan units in combination with intermolecular hydrogen bonding interactions enabled by the grafted nucleic acid sequences.

7.4 SELF-ASSEMBLY DRIVEN BY ELECTROSTATIC INTERACTIONS

Over the last few years, polymer scientists have focused considerable attention on the easy and straightforward design of multifunctional macromolecular architectures with a specific combination of properties required for their desired end-use applications. One of the possible and robust approaches for building up such architectures is to use co-assembly processes, which proceed through the organization of polymer-based systems built on complementary macromolecular components. In particular, one can exploit the electrostatically driven co-assembly occurring upon the simple mixing of aqueous solutions of oppositely charged polyelectrolytes [110]. Polyelectrolytes belong to a class of water-soluble macromolecules that bear high amounts of charges. Such materials are divided into either weak (annealed) or strong (quenched) polycations and polyanions. Strong polyelectrolytes are permanently charged, regardless of the pH in aqueous media. For weak polyelectrolytes, the situation is different: the charge density, and hence the chain conformation and the solubility depend on the solution pH [111]. Other than cationic or anionic polyelectrolytes, polyampholytes contain both positive and negative charges, which can be randomly distributed or arranged block-wise along the polymeric backbone [112]. In the case of polyzwitterions, each monomer unit carries at least one positive and one negative charge [113]. Two oppositely charged polyelectrolytes do co-react in solution to form interpolyelectrolyte complexes (IPECs) [114].

Interpolyelectrolyte complex is a specific term for the complex formation between two polyelectrolytes of opposite charge [115]; block ionomer complex (BIC) is another term specifically used for the complex formation between ionic/non-ionic block copolymers with one charged and one neutral segment and polyelectrolytes of opposite charge [116]. Polyion complex (PIC) [117], polyplex [18], or complex coacervate (CC), introduced for the electrostatic co-assembly of oppositely charged synthetic polyelectrolytes [119] are definitions found as well in the literature.

IPECs have attracted attention due to the large number of emerging and potential applications. Currently, various synthetic IPECs did find commercial applications [113]. Some examples include synthetic ionic polymers used in the field of water treatment [120], in the oil industry [121], and nanotechnology [122]. Other approaches focus on IPEC systems based on biomolecules (DNA, protein, polysaccharides), which are very attractive because of their numerous potential applications notably for medical purposes (targeted drug delivery, imaging) [118, 123–126]. This part of the chapter will be restricted to IPEC systems based on biomolecules, named as bio-IPECs. First, the factors influencing the complex formation will be briefly described. Later on, relevant but not exhaustive examples of bio-IPEC will be given.

As already mentioned, the mixing of oppositely charged polyelectrolytes in aqueous solution immediately induces IPECs formation. Such macromolecular co-assembly processes proceed with high rates, which are typical of diffusion-controlled reaction (millisecond range) [127]. As a result of collisions between oppositely charged macromolecules, contacts (referred to as ion-ion or salt bonds) are formed between their monomer units. Any low molecular weight counterions, which are initially located near the polymeric backbone of the involved macromolecules, are released into the solution, which causes the entropy of the system to increase. This co-assembly process is thus predominantly driven by a gain in entropy [128]. This effect was notably observed by Sato *et al.* in their investigation of interaction between DNA and a cationic comb-type copolymer (PLL-g-Dex). Double-stranded (ds) DNA has a higher density and a larger total number of anionic charges than single-strand (ss) DNA, and therefore has larger counterion condensation. These counterions are released upon IPEC formation with the cationic polymers, leading to the formation of a more stable IPEC with dsDNA than with ssDNA [129].

Hydrogen bonding might as well play an important role in complexation, notably stabilization (Figure 7.14). For example, the results obtained by Mascotti *et al.* in the study of the IPEC formation between DNA strands and oligoarginines, show that H-bonding of arginine with the phosphate backbone of the nucleic acid contributes to the increased stability of these complexes [130]. Another example is the study of IPEC based on polymethacrylate (Eudragit, E100) and sodium carboxymethylcellulose (NaCMC) achieved by Ngwuluka *et al.* In this study, computational modeling revealed that the formation of the IPEC was due to electrostatic interaction, hydrogen bonding, and van der Waals interactions [131]. Thus the hydrophilic weight fraction of the polymer is another important parameter to be taken in consideration. This parameter influences both the solubility of the polymer in aqueous media and its complex formation via hydrophobic or ionic association, hydrogen bonding, and van der Waals interaction [110, 131].

Moreover, the ratio of polyelectrolyte chain lengths and nature of the substituent (inducing screening of charges) or counterions (uni- or multivalent) can be significant in the formation of IPEC [132]. Investigations of the binding interactions between native DNA and dendrimers (Astramol) of five generations (G1, G2, G3, G4, and G5) have been carried out by Kabanov *et al.* All dendrimers interacting with DNA at an equal concentration of amine and phosphate groups did form neutral water-insoluble IPECs. However, G4 and G5 added to the DNA solution in excess formed positively charged water-soluble IPECs, contrary to G1 and G2 [133].

Environmental factors such as the medium also play a role in the complex formation process, especially pH and ionic strength (salt and polyelectrolyte concentration). The presence of salt (e.g., NaCl) drastically affects the characteristics of IPECs [135, 136]. At low salt levels, structure formation leads to the formation of rather small particles. Increased ionic strength can induce the formation of aggregated IPEC species

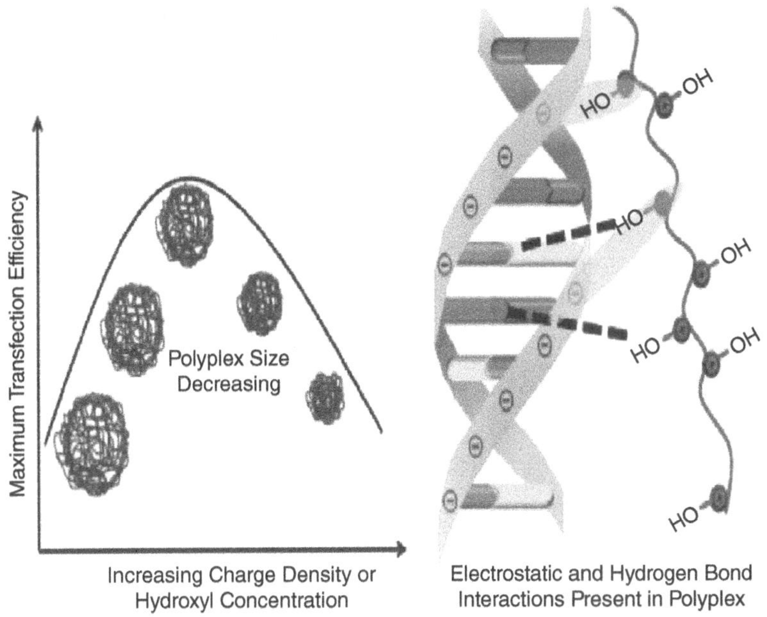

Figure 7.14 IPEC formed between a cationic polymer and condensed anionic plasmid DNA. Various factors including hydrogen bonding, impact the IPEC stability. Reprinted with the permission from [134]. (*See color insert for color representation of this figure*).

and even cause macroscopic phase separation, leading to a precipitate of an insoluble nearly stoichiometric IPEC while the prevailing supernatant solution contains either nonstoichiometric ($Z \neq 1$) IPECs or, in some cases, a pure host (excess) polyelectrolyte [111]. For example, Sung *et al.* observed that the polycation (PEG grafted PEI)/DNA complexes, in excess of the polycation, had an effective diameters of 500 nm in 10 mM NaCl, whereas, in 150 mM NaCl, complexes having size greater than 900 nm were observed [137]. At still higher salt concentrations, a threshold ionic concentration is reached, which induces dissociation of the IPEC. Therefore no complex formation was observed [138, 139]. In addition to the salt concentration, in the case of weak polyelectrolytes, the solution pH also strongly influences the amount of charges, and therefore also the extent of formation of interpolyelectrolyte complexes [131]. This factor was used by Elschner *et al.* to produce a pH-responsive IPEC based on a well soluble cellulose-based polyzwitterion with weak ionic groups. Their studies showed the formation of pH-responsive IPECs in the presence of polydiallyldimethylammonium chloride. These particles are switchable in a physiologically relevant range of pH [140].

Also of prime importance is the way that the complex formation itself is conducted, that is, in mixing parameters such as the charge-to-charge stoichiometry of the components, the addition rate, order of addition of the components (kinetic versus thermodynamic) [127], and temperature [141]. The charge-to-charge stoichiometry corresponds to the overall ratio of positive to negative charges of the oppositely charged polyelectrolytes involved. Stoichiometric IPECs are hydrophobic due to the mutual screening of electrostatic interaction and precipitate from aqueous solution, although a certain swelling due to the incorporation of water molecules cannot be excluded [135]. If nonstoichiometric mixtures are prepared, overcharging effects due to an excess of either the polycation or the polyanion occur. The solubility and stability of the IPEC then depends as well on the molecular weight of the respective polyelectrolyte added.

IPECs structure formation thus offers unique control over both properties and morphology of the resulting structures. Moreover, bio-IPECs based on biological molecules (DNA/RNA, protein, saccharide) reveal very interesting properties that include biocompatibility, swellability, stimuli responsiveness, and physicochemical stability [142]. Thus bio-IPECs belong to a very interesting and promising class of materials for several applications, notably in bio-medicine, pharmacy, and the food industry, among others. One of the most important application of bio-IPECs is as gene and/or drug delivery vectors for therapeutics [118, 125, 143, 144].

In the following discussion are given relevant examples of bio-IPECs.

7.4.1 DNA/Polymer Bio-IPECs

Gene therapy has emerged as a promising approach for the treatment or prevention of acquired and genetic diseases. The formation of bio-IPECs with DNA is encouraging for future gene therapy developments. DNA, an anionic polymer, can be complexed by cationic polymers, water-soluble or amphiphilic copolymers (depending on the hydrophilic weight fraction). The interaction between DNA and polycation can form various self-assembled structures, from spherical micelles to vesicles [145, 146]. The common polycationic polymers of natural origin are poly(L-lysine) [147, 148], and chitosan [149, 150].

Recently, bio-IPECs with potential for application have been evidenced by Nardin *et al.* [104]. In their work, they showed that the inhibition of amyloid β (1-42) fibril genesis could occur if $A\beta_{42}$ proteins interact with a polyanionic polymer, like short DNA strands. Bio-IPECs might be assembled through interaction between the protein and either synthetic polyions or nucleic acid strands through electrostatic attraction between monomers or nucleic acids and amino acids of opposite charges. Moreover, the incubation of the amyloid fibers with synthetic nucleotide sequences induced their disassembly. If the resulting complexes are less toxic than the prefibrillar oligomers or fibers, this strategy could be of particular relevance for future drug developments against amyloid-based diseases [151].

7.4.2 DNA/Copolymer Bio-IPECs

An elegant strategy, to improve the efficiency of gene delivery, is to use a diblock copolymer. Kakizawa *et al.* proposed a diblock copolymer composed of a neutral block (polyethylene glycol, PEG) and a ionic block (polyamino acid) [152]. The ionic block interacts with the phosphate groups of DNA or RNA, and thus ensures the formation of the complex. The PEG block was used to confer colloidal stability to the nanocarriers and to delay phagocytosis by prolonging the blood circulation time.

A similar strategy has been developed for protein cell targeting. However, the proteins can be neutral or charged (either positively or negatively), and several proteins are unstable in serum (parenteral route) or in the gastric and intestinal media (oral route). A solution, recently proposed by Novoa-Carballal *et al.* is the use of sulfated glycosaminoglycan (GAG) [139] (or hyaluronic acid (HA) [138])), negatively charged polysaccharides [139], as the ionic block of a PEG-based diblock copolymer (Figure 7.15) [153, 154].The main advantage of the GAGs approach is that it allows formation of complexes of neutral charge, small size, and colloidal stability at physiological (and even higher) ionic strength. All these properties contribute to prolonged blood circulation times of the formed IPECs, and thus more effective delivery to the targeted site. We can note here that PEG is the most common neutral block used for biological application of bio-IPECs due to its biocompatibility and enzymatic resistance. Furthermore, the application of various polypeptide-b-PEG copolymers as drug delivery nanocarriers (DNA, proteins, or charged anticancer drugs) has recently reached the phase III clinical trials development stage [155].

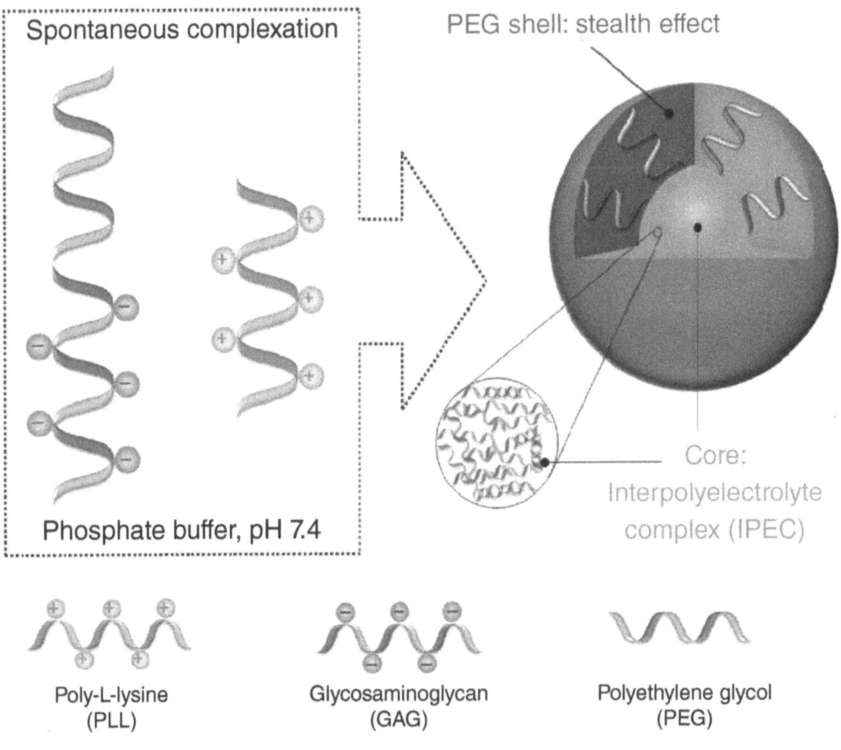

Figure 7.15 Schematic presentation of the complexation process between GAG-b-PEG and polycationic protein (poly-L-lysine (PLL)). Reprinted with the permission from [139]. (*See color insert for color representation of this figure*).

The majority of bio-IPECs strategies employed for drug delivery are very similar to those used for gene or protein delivery. For example, nanoparticles based on polysaccharide-based diblock copolymers have been developed using dextran [156] or chitosan-based copolymers [124] for the design of colon targeted ibuprofen tablets. The ternary bio-IPECS (insulin-poly(methylaminophosphazene)-dextran sulphate developed by Burova *et al.* for oral delivery of insulin is an elegant approach as well [141]. As for gene delivery, the majority of bio-IPECs structures for drug delivery assemble into a variety of structures, from spherical micelles to vesicles [114, 146].

DNA copolymers are as well investigated to induce IPECs by, for instance, the coupling of the *c-myb* antisense nucleic acid sequence with poly(ethylene glycol) (PEG) through an acid cleavable phosphoramidate linkage [157]. The resulting biocompatible diblock was complexed with a cationic peptide through electrostatic interaction in order to give rise to stable micellar aggregates. The resulting complex is composed of an inner polyelectrolyte core surrounded by a PEG chain corona. Coating of the nanocarriers with the hydrophilic, inert, and biocompatible polyethylene glycol (PEG) decreases the rate of opsonisation, leading to reduced clearance by the mononuclear phagocytic system (MPS) and prolonged circulation time but increased delivery efficiency [158, 159]. Indeed the transport of nucleic acids to cells was greatly improved in comparison to the pristine ODN sequence [157].

7.5 CONCLUSION

Our purpose, with this contribution, was to give an overview of the literature on the self-assembly of biohybrid polymers. Mainly, through this study based on the literature and our own work, we wanted to provide evidence on the mechanisms behind their structure formation.

Overall, although composed of nucleic acid sequences that might undergo several modes of interaction, namely hydrogen bonding, electrostatic interaction, and conformal stacking of sugar rings and bases, linear DNA block copolymers engage in self-assembly driven by the chemical and physical incompatibility of the covalently linked blocks, especially hydrophobic interaction. It is the same case for polypeptide block copolymers, which are in addition sensitive to the secondary structure of the constitutive polypeptide, to which the size of the self-assembly is responsive. Morphological transitions could be induced with the block copolypeptide.

Self-assembly driven nucleation polymerization was observed with a poly(oxazoline)-*graft*-DNA. Chemical and physical incompatibility between the water-soluble, flexible nucleic acid sequence and the hydrophobic polymer segment remains the driving force to the aggregation of nuclei that further assemble into fibrils according to a mechanism analogous to that of amyloid fibrils formation. Low DNA grafting density enables intra-and intermolecular hydrogen bonding to occur. At high DNA grafting density, self-assembly takes place. This mechanism is, however, different from the peptide templated assembly of comparable fibers upon conjugation of fibrils forming peptides with poly(ethelyne glycol). The polymer, which does not engage in specific interactions, does not prevent the formation of the peptide fibrils, yet hinders their aggregation into insoluble aggregates by steric hindrance. If these peptides were to be coupled with nucleic acid sequences, which might engage in intermolecular interactions, the mechanism of structure formation of the resulting hybrid would depend on the ability of the peptide motif to engage into intermolecular hydrogen bonding with the nucleotide strand.

Overall, this contribution sheds light on the mechanism of structure formation of biohybrid polymers and the crucial role of intermolecular interactions on the mechanism of structure formation. This is illustrated with nucleation polymerization of amyloid-like fibrils and the formation of interpolyelectrolyte complexes, the formation and stability of which, when based on macromolecules of biological origin, may be optimized with the several modes of interactions that these may undergo besides such electrostatic interaction as hydrogen bonding and conformal stacking.

Constant advances in polymer chemistry may enable us in the future to explore further these bio-inspired systems, in particular, the impact of composition and architecture on the modes of structure formation and interaction from the macromolecular to the ensemble scale.

REFERENCES

1. Philp, D., & Stoddart, J. F. (**1996**). Self-assembly in natural and unnatural systems. *Angewandte Chemie International Edition in English* **35**, 1154–1196.
2. Whitesides, G. M., & Grzybowski, B. (**2002**). Self-assembly at all scales. *Science* **295**, 2418–2421.
3. Wang, C., Wang, Z., & Zhang, X. (**2012**). Amphiphilic building blocks for self-assembly: From amphiphiles to supra-amphiphiles. *Accounts of Chemical Research* **45**, 608–618.
4. Israelachvili, J. N. (**2010**). *Intermolecular and Surface Forces*, Elsevier Science.
5. Isaacs, L., Chin, D.N., Bowden, N., Xia, Y., & Whitesides, G. M. (**1999**). Supramolecular Materials and Technologies. *John Wiley*, **1999**, **4**.
6. Fialkowski, M., Bishop, K. J. M., Klajn, R., Smoukov, S. K., Campbell, C. J., & Grzybowski, B. A. (**2006**). Principles and implementations of dissipative (dynamic) self-assembly. *Journal of Physical Chemistry B* **110**, 2482–2496.
7. Kralchevsky, P. A., Danov, K. D., & Basheva, E. S. (**2011**). Hydration force due to the reduced screening of the electrostatic repulsion in few-nanometer-thick films. *Current Opinion in Colloid & Interface Science* **16**, 517–524.
8. Lee, Y. S. (**2007**). *Self-Assembly and Nanotechnology: A Force Balance Approach.*
9. Dhandapani, N. V., Thapa A., Sandip G., Shrestha A., Shresta N., Bhattrai, R S. (**2013**). Liposomes as novel drug delivery system: a comprehensive review. *International Journal of Pharmaceutical Sciences* **4**, 187–193.
10. Drulis-Kawa, Z., & Dorotkiewicz-Jach, A. (**2010**). Liposomes as delivery systems for antibiotics. *International Journal of Pharmaceutics* **387**, 187–198.
11. Forster, S., & Plantenberg, T. (**2002**). From self-organizing polymers to nanohybrid and biomaterials. *Angewandte Chemie-International Edition* **41**, 689–714.
12. Adams, M. L., Lavasanifar, A., & Kwon, G. S. (**2003**). Amphiphilic block copolymers for drug delivery. *Journal of Pharmaceutical Sciences* **92**, 1343–1355.

13. Peters, R. J. R. W., Marguet, M., Marais, S., Fraaije, M. W., van Hest, J. C. M., & Lecommandoux, S. (**2014**). Cascade reactions in multicompartmentalized polymersomes. *Angewandte Chemie-International Edition* **53**, 146–150.

14. Pawar, P. V., Gohil, S. V., Jain, J. P., & Kumar, N. (**2013**). Functionalized polymersomes for biomedical applications. *Polymer Chemistry* **4**, 3160–3176.

15. Rosenkranz, T., Katranidis, A., Atta, D., Gregor, I., Enderlein, J., Grzelakowski, M., Rigler, P., Meier, W., & Fitter, J. (**2009**). Observing proteins as single molecules encapsulated in surface-tethered polymeric nanocontainers. *Chembiochem* **10**, 702–709.

16. Spulber, M., Baumann, P., Saxer, S. S., Pieles, U., Meier, W., & Bruns, N. (**2014**). Poly(N-vinylpyrrolidone)-poly(dimethylsiloxane)-based polymersome nanoreactors for laccase-catalyzed biotransformations. *Biomacromolecules* **15**, 1469–1475.

17. Israelachvili, J. N., Mitchell, D. J., & Ninham, B. W. (**1976**). Theory of self-assembly of hydrocarbon amphiphiles into micelles and bilayers. *Journal of the Chemical Society, Faraday Transactions 2: Molecular and Chemical Physics* **72**, 1525–1568.

18. Antonietti, M., & Forster, S. (**2003**). Vesicles and liposomes: a self-assembly principle beyond lipids. *Advanced Materials* **15**, 1323–1333.

19. Holder, S. J., & Sommerdijk, N. A. J. M. (**2011**). New micellar morphologies from amphiphilic block copolymers: disks, toroids and bicontinuous micelles. *Polymer Chemistry* **2**, 1018–1028.

20. Nicolai, T., Colombani, O., & Chassenieux, C. (**2010**). Dynamic polymeric micelles versus frozen nanoparticles formed by block copolymers. *Soft Matter* **6**, 3111–3118.

21. Zhulina, E. B., & Borisov, O. V. (**2012**). Theory of Block Polymer Micelles: Recent Advances and Current Challenges. *Macromolecules* **45**, 4429–4440.

22. van Hest, J. C., Delnoye, D. A., Baars, M. W., van Genderen, M. H., & Meijer, E. W. (**1995**). Polystyrene-dendrimer amphiphilic block copolymers with a generation-dependent aggregation. *Science* **268**, 1592–1595.

23. Thiruvengadathan, R., Korampally, V., Ghosh, A., Chanda, N., Gangopadhyay, K., & Gangopadhyay, S. (**2013**). Nanomaterial processing using self-assembly-bottom-up chemical and biological approaches. *Reports on Progress in Physics* **76**, 066501 066554.

24. Bishop, K. J. M., Wilmer, C. E., Soh, S., & Grzybowski, B. A. (**2009**). Nanoscale forces and their uses in self-assembly. *Small* **5**, 1600–1630.

25. Foffi, G., McCullagh, G. D., Lawlor, A., Zaccarelli, E., Dawson, K. A., Sciortino, F., Tartaglia, P., Pini, D., & Stell, G. (**2002**). Phase equilibria and glass transition in colloidal systems with short-ranged attractive interactions: Application to protein crystallization. *Physical Review E* **65**, 031407 1–17.

26. Gour, N., Kedracki, D., Safir, I., Ngo, K. X., & Vebert-Nardin, C. (**2012**). Self-assembling DNA-peptide hybrids: morphological consequences of oligonucleotide grafting to a pathogenic amyloid fibrils forming dipeptide. *Chemical Communications* **48**, 5440–5442.

27. Kedracki, D., Safir, I., Gour, N., Kien Xuan, N., & Vebert-Nardin, C. (**2013**). DNA-Polymer Conjugates: From Synthesis, Through Complex Formation and Self-assembly to Applications. *Bio-Synthetic Polymer Conjugates* **253**, 115–149.

28. Peyrin, E. (**2009**). Nucleic acid aptamer molecular recognition principles and application in liquid chromatography and capillary electrophoresis. *Journal of Separation Science* **32**, 1531–1536.

29. Osborne, S. E., Matsumura, I., & Ellington, A. D. (**1997**). Aptamers as therapeutic and diagnostic reagents: problems and prospects. *Current Opinion in Chemical Biology* **1**, 5–9.

30. Proske, D., Blank, M., Buhmann, R., & Resch, A. (**2005**). Aptamers—basic research, drug development, and clinical applications. *Applied Microbiology and Biotechnology* **69**, 367–374.

31. Nimjee, S. M., Rusconi, C. P., & Sullenger, B. A. (**2005**). Aptamers: an emerging class of therapeutics. In *Annual Review of Medicine*, Vol. **56**, pp. 555-+. Annual Reviews, Palo Alto.

32. Schnitzler, T., & Herrmann, A. (**2012**). DNA block copolymers: functional materials for nanoscience and biomedicine. *Accounts of Chemical Research* **45**, 1419–1430.

33. Kwak, M., & Herrmann, A. (**2006**). Nucleic acid polymers and DNA synthetic polymer hybrid materials generated by molecular biology Techniques. In *Materials Science and Technology*. Wiley-VCH Verlag GmbH & Co. KGaA.

34. Biro, S., Fu, Y. M., Yu, Z. X., & Epstein, S. E. (**1993**). Inhibitory effects of antisense oligodeoxynucleotides targeting c-myc mRNA on smooth muscle cell proliferation and migration. *Proceedings of the National Academy of Sciences of the United States of America* **90**, 654–658.

35. Jeong, J. H., & Park, T. G. (**2001**). Novel polymer-DNA hybrid polymeric micelles composed of hydrophobic poly(D,L-lactic-co-glycolic acid) and hydrophilic oligonucleotides. *Bioconjugate Chemistry* **12**, 917–923.

36. Li, Z., Zhang, Y., Fullhart, P., & Mirkin, C. A. (**2004**). Reversible and chemically programmable micelle assembly with DNA block-copolymer amphiphiles. *Nano Letters* **4**, 1055–1058.

37. Alemdaroglu, F. E., Ding, K., Berger, R., & Herrmann, A. (**2006**). DNA-templated synthesis in three dimensions: introducing a micellar scaffold for organic reactions. *Angewandte Chemie International Edition* **45**, 4206–4210.

38. Ding, K., Alemdaroglu, F. E., Börsch, M., Berger, R., & Herrmann, A. (**2007**). Engineering the structural properties of DNA block copolymer micelles by molecular recognition. *Angewandte Chemie International Edition* **46**, 1172–1175.

39. Talom, R. M., Fuks, G., Kaps, L., Oberdisse, J., Cerclier, C., Gaillard, C., Mingotaud, C., & Gauffre, F. (**2011**). DNA–polymer micelles as nanoparticles with recognition ability. *Chemistry – A European Journal* **17**, 13495–13501.

40. Cottenye, N., Anselme, K., Ploux, L., & Vebert-Nardin, C. (**2012**). Vesicular structures self-assembled from oligonucleotide-polymer hybrids: mechanical prevention of bacterial colonization upon their surface tethering through hybridization. *Advanced Functional Materials* **22**, 4891–4898.

41. Averick, S. E., Dey, S. K., Grahacharya, D., Matyjaszewski, K., & Das, S. R. (**2014**). Solid-phase incorporation of an ATRP initiator for polymer–DNA biohybrids. *Angewandte Chemie* **126**, 2777–2782.

42. Chien, M.-P., Rush, A. M., Thompson, M. P., & Gianneschi, N. C. (**2010**). Programmable shape-shifting micelles. *Angewandte Chemie International Edition* **49**, 5076–5080.

43. Kedracki, D., Filippov, S. K., Gour, N., Schlaad, H., & Nardin, C. (**2015**). Formation of DNA-copolymer fibrils through an amyloid-like nucleation polymerization mechanism. *Macromolecular Rapid Communications* **36**, 768–773.

44. Averick, S., Paredes, E., Li, W., Matyjaszewski, K., & Das, S. R. (**2011**). Direct DNA conjugation to star polymers for controlled reversible assemblies. *Bioconjugate Chemistry* **22**, 2030–2037.
45. Pan, P., Fujita, M., Ooi, W.-Y., Sudesh, K., Takarada, T., Goto, A., & Maeda, M. (**2012**). Thermoresponsive micellization and micellar stability of poly(*N*-isopropylacrylamide)-*b*-DNA diblock and miktoarm star polymers. *Langmuir* **28**, 14347–14356.
46. Kang, H., Trondoli, A. C., Zhu, G., Chen, Y., Chang, Y.-J., Liu, H., Huang, Y.-F., Zhang, X., & Tan, W. (**2011**). Near-infrared light-responsive core–shell nanogels for targeted drug delivery. *Acs Nano* **5**, 5094–5099.
47. Lin, H. X., Zou, Y., Huang, Y. S., Chen, J., Zhang, W. Y., Zhuang, Z. X., Jenkins, G. & Yang, C. J. (**2011**). DNAzyme crosslinked hydrogel: a new platform for visual detection of metal ions. *Chemical Communications* **47**, 9312–9314.
48. Yang, H., Liu, H., Kang, H., & Tan, W. (**2008**). Engineering target-responsive hydrogels based on aptamer–target interactions. *Journal of the American Chemical Society* **130**, 6320–6321.
49. Lewis, C. L., Choi, C. H., Lin, Y., Lee, C. S., & Yi, H. (**2010**). Fabrication of uniform DNA-conjugated hydrogel microparticles via replica molding for facile nucleic acid hybridization assays. *Analytical Chemistry* **82**, 5851–5858.
50. Xiong, X. L., Wu, C. C., Zhou, C. S., Zhu, G. Z., Chen, Z., & Tan, W. H. (**2013**). Responsive DNA-based hydrogels and their applications. *Macromolecular Rapid Communications* **34**, 1271–1283.
51. Watson, K. J., Park, S. J., Im, J. H., Nguyen, S. T., & Mirkin, C. A. (**2001**). DNA-block copolymer conjugates. *Journal of the American Chemical Society* **123**, 5592–5593.
52. Ray, J. G., Johnson, A. J., & Savin, D. A. (**2013**). Self-assembly and responsiveness of polypeptide-based block copolymers: how "smart" behavior and topological complexity yield unique assembly in aqueous media. *Journal of Polymer Science Part B: Polymer Physics* **51**, 508–523.
53. Checot, F., Lecommandoux, S., Gnanou, Y., & Klok, H. A. (**2002**). Water-soluble stimuli-responsive vesicles from peptide-based diblock copolymers. *Angewandte Chemie-International Edition* **41**, 1339–1343.
54. Checot, F., Brulet, A., Oberdisse, J., Gnanou, Y., Mondain-Monval, O., & Lecommandoux, S. (**2005**). Structure of polypeptide-based diblock copolymers in solution: stimuli-responsive vesicles and micelles. *Langmuir* **21**, 4308–4315.
55. Checot, F., Rodriguez-Hernandez, J., Gnanou, Y., & Lecommandoux, S. (**2007**). pH-responsive micelles and vesicles nanocapsules based on polypeptide diblock copolymers. *Biomolecular Engineering* **24**, 81–85.
56. Naik, S. S., Ray, J. G., & Savin, D. A. (**2011**). Temperature- and pH-responsive self-assembly of poly(propylene oxide)-*b*-poly(lysine) block copolymers in aqueous solution. *Langmuir* **27**, 7231–7240.
57. Agut, W., Taton, D., & Lecommandoux, S. (**2007**). A versatile synthetic approach to polypeptide based rod-coil block copolymers by click chemistry. *Macromolecules* **40**, 5653–5661.
58. Altintas, O., Hizal, G. & Tunca, U. (**2006**). ABC-type hetero-arm star terpolymers through "click" chemistry. *Journal of Polymer Science Part a-Polymer Chemistry* **44**, 5699–5707.

59. Gao, H., & Matyjaszewski, K. (**2007**). Arm-first method as a simple and general method for synthesis of miktoarm star copolymers. *Journal of the American Chemical Society* **129**, 11828–11834.

60. Sanda, F., Gao, G. Z., & Masuda, T. (**2004**). Helical polymer carrying helical grafts from peptide-based acetylene macromonomers: synthesis. *Macromolecular Bioscience* **4**, 570–574.

61. Naik, S. S., & Savin, D. A. (**2009**). Poly(Z-lysine)-based organogels: effect of interfacial frustration on gel strength. *Macromolecules* **42**, 7114–7121.

62. Vogt, A. P., & Sumerlin, B. S. (**2006**). An efficient route to macromonomers via ATRP and click chemistry. *Macromolecules* **39**, 5286–5292.

63. Eckhardt, D., Groenewolt, M., Krause, E., & Börner, H. G. (**2005**). Rational design of oligopeptide organizers for the formation of poly(ethylene oxide) nanofibers. *Chemical Communications*, 2814–2816.

64. Tzokova, N., Fernyhough, C. M., Butler, M. F., Armes, S. P., Ryan, A. J., Topham, P. D., & Adams, D. J. (**2009**). The effect of PEO length on the self-assembly of poly(ethylene oxide)-tetrapeptide conjugates prepared by "click" chemistry. *Langmuir* **25**, 11082–11089.

65. Rodriguez-Hernandez, J., Babin, J., Zappone, B., & Lecommandoux, S. (**2005**). Preparation of shell cross-linked nano-objects from hybrid-peptide block copolymers. *Biomacromolecules* **6**, 2213–2220.

66. Rodriguez-Hernandez, J., Checot, F., Gnanou, Y., & Lecommandoux, S. (**2005**). Toward "smart" nano-objects by self-assembly of block copolymers in solution. *Progress in Polymer Science* **30**, 691–724.

67. Rodriguez-Hernandez, J., & Lecommandoux, S. (**2005**). Reversible inside-out micellization of pH-responsive and water-soluble vesicles based on polypeptide diblock copolymers. *Journal of the American Chemical Society* **127**, 2026–2027.

68. Hamley, I. W. (**2007**). Peptide fibrillization. *Angewandte Chemie-International Edition* **46**, 8128–8147.

69. Jeong, J. S., Ansaloni, A., Mezzenga, R., Lashuel, H. A., & Dietler, G. (**2013**). Novel mechanistic insight into the molecular basis of amyloid polymorphism and secondary nucleation during amyloid formation. *Journal of Molecular Biology* **425**, 1765–1781.

70. Kad, N. M., Thomson, N. H., Smith, D. P., Smith, D. A., & Radford, S. E. (**2001**). β2-microglobulin and its deamidated variant, N17D form amyloid fibrils with a range of morphologies in vitro. *Journal of Molecular Biology* **313**, 559–571.

71. Harper, J. D., & Lansbury, P. T. (**1997**). Models of amyloid seeding in Alzheimier's disease and scrapie: Mechanistic truths and physiological consequences of the time-dependent solubility of amyloid proteins. *Annual Review of Biochemistry* **66**, 385–407.

72. Adamcik, J., Jung, J.-M., Flakowski, J., De Los Rios, P., Dietler, G., & Mezzenga, R. (**2010**). Understanding amyloid aggregation by statistical analysis of atomic force microscopy images. *Nat Nano* **5**, 423–428.

73. Koga, T., Taguchi, K., Kobuke, Y., Kinoshita, T., & Higuchi, M. (**2003**). Structural regulation of a peptide-conjugated graft copolymer: A simple model for amyloid formation. *Chemistry-a European Journal* **9**, 1146–1156.

74. Kelly, J. W. (**1998**). The alternative conformations of amyloidogenic proteins and their multi-step assembly pathways. *Current Opinion in Structural Biology* **8**, 101–106.

75. Dobson, C. M. (**2003**). Protein folding and misfolding. *Nature* **426**, 884–890.
76. Andreu, J. M., & Timasheff, S. N. (**1986**). The measurement of cooperative protein self-assembly by turbidity and other techniques. *Methods in Enzymology* **130**, 47–59.
77. Eaton, W. A., & Hofrichter, J. (**1995**). The biophysics of sickle cell hydroxyurea therapy. *Science* **268**, 1142–1143.
78. Asakura, S. (**1970**). Polymerization of flagellin and polymorphism of flagella. *Advances in Biophysics* **1**, 99–155.
79. Jarrett, J. T., & Lansbury Jr, P. T. (**1993**). Seeding "one-dimensional crystallization" of amyloid: a pathogenic mechanism in Alzheimer's disease and scrapie? *Cell* **73**, 1055–1058.
80. Hofrichter, J., Ross, P. D., & Eaton, W. A. (**1974**). Kinetics and mechanism of deoxyhemoglobin-S gelatin—new approach to understanding sickle-cell disease. *Proceedings of the National Academy of Sciences of the United States of America* **71**, 4864–4868.
81. Jarrett, J. T., Berger, E. P., & Lansbury, P. T. (**1993**). The carboxy terminus of the Beta-Amyloid protein is critical for the seeding of amyloid formation- Implications for the Pathogenesis of Alzhemier's disease. *Biochemistry* **32**, 4693–4697.
82. Naiki, H., & Nakakuki, K. (**1996**). First-order kinetic model of Alzheimer's beta-amyloid fibril extension in vitro. *Laboratory Investigation* **74**, 374–383.
83. Esler, W. P., Stimson, E. R., Ghilardi, J. R., Lu, Y. A., Felix, A. M., Vinters, H. V., Mantyh, P. W., Lee, J. P.– & Maggio, J. E. (**1996**). Point substitution in the central hydrophobic cluster of a human beta-amyloid congener disrupts peptide folding and abolishes plaque competence. *Biochemistry* **35**, 13914–13921.
84. Jarrett, J. T., & Lansbury, P. T. (**1992**). Amyloid fibril formation requires a chemically discriminating nuleation event—studies of an amyloidogenic sequence from bacterial protein OSMB *Biochemistry* **31**, 12345–12352.
85. Koenig, H. M., & Kilbinger, A. F. M. (**2007**). Learning from nature: beta-sheet-mimicking copolymers get organized. *Angewandte Chemie-International Edition* **46**, 8334-8340.
86. Borner, H. G., & Schlaad, H. (**2007**). Bioinspired functional block copolymers. *Soft Matter* **3**, 394–408.
87. Glabe, C. G. (**2006**). Common mechanisms of amyloid oligomer pathogenesis in degenerative disease. *Neurobiology of Aging* **27**, 570–575.
88. Lashuel, H. A., Hartley, D., Petre, B. M., Walz, T., & Lansbury, P. T. (**2002**). Neurodegenerative disease—amyloid pores from pathogenic mutations. *Nature* **418**, 291–291.
89. Healy, J. M., Lewis, S. D., Kurz, M., Boomer, R. M., Thompson, K. M., Wilson, C., & McCauley, T. G. (**2004**). Pharmacokinetics and biodistribution of novel aptamer compositions. *Pharm Res* **21**, 2234–2246.
90. Collier, J. H., & Messersmith, P. B. (**2004**). Self-assembling polymer-peptide conjugates: Nanostructural tailoring. *Advanced Materials* **16**, 907–910.
91. Xie, L. G., Luo, Y., & Wei, G. H. (**2013**). A beta(16-22) peptides can assemble into ordered beta-barrels and bilayer beta-sheets, while substitution of phenylalanine 19 by tryptophan increases the population of disordered aggregates. *Journal of Physical Chemistry B* **117**, 10149–10160.
92. Tjernberg, L. O., Lilliehook, C., Callaway, D. J. E., Naslund, J., Hahne, S., Thyberg, J., Terenius, L., & Nordstedt, C. (**1997**). Controlling amyloid beta-peptide fibril formation with protease-stable ligands. *Journal of Biological Chemistry* **272**, 17894–17894.

93. Tjernberg, L. O., Naslund, J., Thyberg, J., Gandy, S. E., Terenius, L., & Nordstedt, C. (**1997**). Generation of Alzheimer amyloid beta peptide through nonspecific proteolysis. *Journal of Biological Chemistry* **272**, 1870–1875.

94. Burkoth, T. S., Benzinger, T. L. S., Jones, D. N. M., Hallenga, K., Meredith, S. C., & Lynn, D. G. (**1998**). C-terminal PEG blocks the irreversible step in beta-amyloid(10-35) fibrillogenesis. *Journal of the American Chemical Society* **120**, 7655–7656.

95. Burkoth, T. S., Benzinger, T. L. S., Urban, V., Lynn, D. G., Meredith, S. C., & Thiyagarajan, P. (**1999**). Self-assembly of A beta((10-35))-PEG block copolymer fibrils. *Journal of the American Chemical Society* **121**, 7429–7430.

96. Thiyagarajan, P., Burkoth, T. S., Urban, V., Seifert, S., Benzinger, T. L. S., Morgan, D. M., Gordon, D., Meredith, S. C., & Lynn, D. G. (**2000**). pH dependent self assembly of beta-amyloid(10-35) and beta-amyloid(10-35)PEG3000. *Journal of Applied Crystallography* **33**, 535–539.

97. Meijer, J. T., Henckens, M. J. A. G., Minten, I. J., Lowik, D. W. P. M., & van Hest, J. C. M. (**2007**). Disassembling peptide-based fibres by switching the hydrophobic-hydrophilic balance. *Soft Matter* **3**, 1135–1137.

98. Hamley, I. & Hamley, I. (**2010**). Liquid crystal phase formation by biopolymers. *Soft Matter* **6**, 1863.

99. Hamley, I. W. (**2010**). Liquid crystal phase formation by biopolymers. *Soft Matter* **6**, 1863–1871.

100. Hamley, I. W., Krysmann, M. J., Castelletto, V., & Noirez, L. (**2008**). Multiple lyotropic polymorphism of a poly(ethylene glycol)-peptide conjugate in aqueous solution. *Advanced Materials* **20**, 4394–4397.

101. Hamley, I. W., Krysmann, M. J., Kelarakis, A., Castelletto, V., Noirez, L., Hule, R. A., & Pochan, D. J. (**2008**). Nematic and columnar ordering of a PEG-peptide conjugate in aqueous solution. *Chemistry—A European Journal* **14**, 11268–11268.

102. Hamley, I. W., Krysmann, M. J., Newby, G. E., Castelletto, V., & Noirez, L. (**2008**). Orientational ordering in the nematic phase of a polyethylene glycol-peptide conjugate in aqueous solution. *Physical Review E* **77**, 062901.

103. Castelletto, V., & Hamley, I. W. (**2009**). Self assembly of a model amphiphilic phenylalanine peptide/polyethylene glycol block copolymer in aqueous solution. *Biophysical Chemistry* **141**, 169–174.

104. Guo, C., Luo, Y., Zhou, R., & Wei, G. (**2012**). Probing the self-assembly mechanism of diphenylalanine-based peptide nanovesicles and nanotubes. *ACS Nano* **6**, 3907–3918.

105. Na, N., Mu, X., Liu, Q., Wen, J., Wang, F., & Ouyang, J. (**2013**). Self-assembly of diphenylalanine peptides into microtubes with "turn on" fluorescence using an aggregation-induced emission molecule. *Chemical Communications* **49**, 10076–10078.

106. Reches, M., & Gazit, E. (**2005**). Self-assembly of peptide nanotubes and amyloid-like structures by charged-termini capped diphenylalanine peptide analogues. *Febs Journal* **272**, 388–388.

107. Reches, M., & Gazit, E. (**2005**). Self-assembly of peptide nanotubes and amyloid-like structures by charged-termini-capped diphenylalanine peptide analogues. *Israel Journal of Chemistry* **45**, 363–371.

108. Joshi, K. B., & Verma, S. (**2008**). Ditryptophan conjugation triggers conversion of biotin fibers into soft spherical structures. *Angewandte Chemie-International Edition* **47**, 2860–2863.

109. Schweizer, S., & Reed, J. (**2008**). Effect of variation of the strength of the aromatic interactions of tryptophan on the cooperative structural refolding behavior of a peptide from HIV 1. *Biophysical Journal* **95**, 3381–3390.
110. Pergushov, D., Borisov, O., Zezin, A.– & Müller, A. E. (**2011**). Interpolyelectrolyte complexes based on polyionic species of branched topology. In *Self Organized Nanostructures of Amphiphilic Block Copolymers I* (Müller, A. H. E. & Borisov, O., eds.), Vol. **241**, pp. 131–161. Springer Berlin Heidelberg.
111. Pergushov, D. V., Müller, A. H. E., & Schacher, F. H. (**2012**). Micellar interpolyelectrolyte complexes. *Chemical Society Reviews* **41**, 6888–6901.
112. Ciferri, A., & Kudaibergenov, S. (**2007**). Natural and Synthetic Polyampholytes, 1. *Macromolecular Rapid Communications* **28**, 1953–1968.
113. Lowe, A. B., & McCormick, C. L. (**2002**). Synthesis and solution properties of zwitterionic polymers†. *Chemical Reviews* **102**, 4177–4190.
114. Thünemann, A., Müller, M., Dautzenberg, H., Joanny, J.-F., & Löwen, H. (**2004**). Polyelectrolyte complexes. In *Polyelectrolytes with Defined Molecular Architecture II* (Schmidt, M., ed.), Vol. **166**, pp. 113–171. Springer Berlin Heidelberg.
115. Zezin, A. B., Izumrudov, V. A., & Kabanov, V. A. (**1989**). Interpolyelectrolyte complexes as a new family of enzyme carriers. Makromolekulare Chemie. *Macromolecular Symposia* **26**, 249–264.
116. Kabanov, A. V., Bronich, T. K., Kabanov, V. A., Yu, K., & Eisenberg, A. (**1996**). Soluble stoichiometric complexes from poly(N-ethyl-4-vinylpyridinium) Cations and Poly(ethylene oxide)-block-polymethacrylate Anions. *Macromolecules* **29**, 6797–6802.
117. Harada, A., & Kataoka, K. (**1995**). Formation of polyion complex micelles in an aqueous milieu from a pair of oppositely-charged block copolymers with poly(ethylene glycol) segments. *Macromolecules* **28**, 5294–5299.
118. Tros de Ilarduya, C., Sun, Y., & Düzgüneş, N. (**2010**). Gene delivery by lipoplexes and polyplexes. *European Journal of Pharmaceutical Sciences* **40**, 159–170.
119. Voets, I. K., de Keizer, A., & Cohen Stuart, M. A. (**2009**). Complex coacervate core micelles. *Advances in Colloid and Interface Science* **147-148**, 300-318.
120. Bolto, B., & Gregory, J. (**2007**). *Water Res.* **41**, 2301–2324.
121. Kudaibergenov, S. E., & Ciferri, A. (**2007**). *Macromol. Rapid Commun.* **28**, 1969–1986.
122. Ballauff, M., & Borisov, O. (**2006**). *Curr. Opin. Colloid Interface Sci.* **11**, 316–323.
123. Hoffman, A. S. (**2002**). Hydrogels for biomedical applications. *Advanced Drug Delivery Reviews* **54**, 3–12.
124. Ofokansi, K. C., & Kenechukwu, F. C. (**2013**). Formulation development and evaluation of drug release kinetics from colon-targeted ibuprofen tablets based on Eudragit RL 100-chitosan interpolyelectrolyte complexes. *ISRN Pharmaceutics* **2013**, 838403, 1–8.
125. Lee, Y., & Kataoka, K. (**2009**). Biosignal-sensitive polyion complex micelles for the delivery of biopharmaceuticals. *Soft Matter* **5**, 3810–3817.
126. Samoilova, N., Tikhonov, V., Krayukhina, M., & Yamskov, I. (**2014**). Interpolyelectrolyte complexes of maleic acid copolymers and chitosan for stabilization and functionalization of magnetite nano- and microparticles. *Journal of Applied Polymer Science* **131**, 39663–39675.
127. Bakeev, K. N., Izumrudov, V. A., Kuchanov, S. I., Zezin, A. B., & Kabanov, V. A. (**1992**). Kinetics and mechanism of interpolyelectrolyte exchange and addition reactions. *Macromolecules* **25**, 4249–4254.

128. Pergushov, D., Zezin, A., Zezin, A., & Müller, A. E. (**2014**). Advanced functional structures based on interpolyelectrolyte complexes. In *Polyelectrolyte Complexes in the Dispersed and Solid State I* (Müller, M., ed.), Vol. **255**, pp. 173–225. Springer Berlin Heidelberg.

129. Sato, Y., Moriyama, R., Choi, S. W., Kano, A., & Maruyama, A. (**2006**). Spectroscopic investigation of cationic comb-type copolymers/DNA interaction: interpolyelectrolyte complex enhancement synchronized with DNA hybridization†. *Langmuir* **23**, 65–69.

130. Mascotti, D. P., & Lohman, T. M. (**1997**). Thermodynamics of oligoarginines binding to RNA and DNA. *Biochemistry* **36**, 7272–7279.

131. Ngwuluka, N. C., Choonara, Y. E., Kumar, P., du Toit, L. C., Khan, R. A., & Pillay, V. (**2014**). A novel pH-responsive interpolyelectrolyte hydrogel complex for the oral delivery of levodopa. I. IPEC modeling and synthesis. *Journal of Biomedical Materials Research Part A*, 1077–1084.

132. Sukhishvili, S. A., Kharlampieva, E., & Izumrudov, V. (**2006**). Where polyelectrolyte multilayers and polyelectrolyte complexes meet. *Macromolecules* **39**, 8873–8881.

133. Kabanov, V. A., Sergeyev, V. G., Pyshkina, O. A., Zinchenko, A. A., Zezin, A. B., Joosten, J. G. H., Brackman, J., & Yoshikawa, K. (**2000**). Interpolyelectrolyte complexes formed by DNA and astramol poly(propylene imine) dendrimers. *Macromolecules* **33**, 9587–9593.

134. Allen, M. H., Green, M. D., Getaneh, H. K., Miller, K. M., & Long, T. E. (**2011**). Tailoring charge density and hydrogen bonding of imidazolium copolymers for efficient gene delivery. *Biomacromolecules* **12**, 2243–2250.

135. Kabanov, V. A. (**2005**). Polyelectrolyte complexes in solution and in bulk. *Russian Chemical Reviews* **74**, 3–20.

136. Dautzenberg, H., & Kriz, J. (**2003**). Response of polyelectrolyte complexes to subsequent addition of salts with different cations. *Langmuir* **19**, 5204–5211.

137. Sung, S. J., Min, S. H., Cho, K. Y., Lee, S., Min, Y. J., Yeom, Y. I., & Park, J. K. (**2003**). Effect of polyethylene glycol on gene delivery of polyethylenimine. *Biol Pharm Bull* **26**, 492–500.

138. Novoa-Carballal, R., Pergushov, D. V., & Müller, A. H. E. (**2013**). Interpolyelectrolyte complexes based on hyaluronic acid-block-poly(ethylene glycol) and poly-l-lysine. *Soft Matter* **9**, 4297–4303.

139. Novoa-Carballal, R., Silva, C., Moller, S., Schnabelrauch, M., Reis, R. L. & Pashkuleva, I. (**2014**). Tunable nano-carriers from clicked glycosaminoglycan block copolymers. *Journal of Materials Chemistry B* **2**, 4177–4184.

140. Elschner, T. & Heinze, T. (**2014**). A promising cellulose-based polyzwitterion with pH-sensitive charges. *Beilstein Journal of Organic Chemistry* **10**, 1549–1556.

141. Burova, T. V., Grinberg, N. V., Tur, D. R., Papkov, V. S., Dubovik, A. S., Shibanova, E. D., Bairamashvili, D. I., Grinberg, V. Y., & Khokhlov, A. R. (**2013**). Ternary interpolyelectrolyte complexes insulin-poly(methylaminophosphazene)-dextran sulfate for oral delivery of insulin. *Langmuir* **29**, 2273–2281.

142. Moustafine, R. I., Margulis, E. B., Sibgatullina, L. F., Kemenova, V. A., & Mooter, G. V. d. (**2008**). Comparative evaluation of interpolyelectrolyte complexes of chitosan with Eudragit® L100 and Eudragit® L100-55 as potential carriers for oral controlled drug delivery. *European Journal of Pharmaceutics and Biopharmaceutics* **70**, 215–225.

143. Bertin, A. (**2014**). Polyelectrolyte complexes of DNA and polycations as gene delivery vectors. In *Polyelectrolyte Complexes in the Dispersed and Solid State II* (Müller, M., ed.), Vol. **256**, pp. 103–195. Springer Berlin Heidelberg.

144. Krayukhina, M. A., Samoilova, N. A., & Yamskov, I. A. (**2008**). Polyelectrolyte complexes of chitosan: formation, properties and applications. *Russian Chemical Reviews* **77**, 799.

145. Iatrou, H., Frielinghaus, H., Hanski, S., Ferderigos, N., Ruokolainen, J., Ikkala, O., Richter, D., Mays, J., & Hadjichristidis, N. (**2007**). Architecturally induced multiresponsive vesicles from well-defined polypeptides: formation of gene vehicles. *Biomacromolecules* **8**, 2173–2181.

146. Müller, M. (**2014**). Sizing, shaping and pharmaceutical applications of polyelectrolyte complex nanoparticles. In *Polyelectrolyte Complexes in the Dispersed and Solid State II* (Müller, M., ed.), Vol. **256**, pp. 197–260. Springer Berlin Heidelberg.

147. Ramsay, E., Hadgraft, J., Birchall, J., & Gumbleton, M. (**2000**). Examination of the biophysical interaction between plasmid DNA and the polycations, polylysine and polyornithine, as a basis for their differential gene transfection in-vitro. *International Journal of Pharmaceutics* **210**, 97–107.

148. Miyoshi, D., Ueda, Y. M., Shimada, N., Nakano, S., Sugimoto, N., & Maruyama, A. (**2014**). Drastic stabilization of parallel DNA hybridizations by a polylysine comb-type copolymer with hydrophilic graft chain. *ChemMedChem* **9**, 2156–2163.

149. Vimal, S., Abdul Majeed, S., Taju, G., Nambi, K. S. N., Sundar Raj, N., Madan, N., Farook, M. A., Rajkumar, T., Gopinath, D., & Sahul Hameed, A. S. (**2013**). Chitosan tripolyphosphate (CS/TPP) nanoparticles: preparation, characterization and application for gene delivery in shrimp. *Acta Tropica* **128**, 486–493.

150. Dang, J. M., & Leong, K. W. (**2006**). Natural polymers for gene delivery and tissue engineering. *Advanced Drug Delivery Reviews* **58**, 487–499.

151. Abraham, J. N., Kedracki, D., Prado, E., Gourmel, C., Maroni, P., & Nardin, C. (**2014**). Effect of the interaction of the amyloid β (1–42) peptide with short single-stranded synthetic nucleotide sequences: morphological characterization of the inhibition of fibrils formation and fibrils disassembly. *Biomacromolecules* **15**, 3253–3258.

152. Kakizawa, Y., & Kataoka, K. (**2002**). Block copolymer micelles for delivery of gene and related compounds. *Advanced Drug Delivery Reviews* **54**, 203–222.

153. Amorim, S., Pires, R. A., Costa, D. S. d., Reis, R. L., & Pashkuleva, I. (**2013**). Interactions between exogenous FGF-2 and sulfonic groups: in situ characterization and impact on the morphology of human adipose-derived stem cells. *Langmuir* **29**, 7983–7992.

154. Pashkuleva, I., & Reis, R. L. (**2010**). Sugars: burden or biomaterials of the future? *Journal of Materials Chemistry* **20**, 8803–8818.

155. Miyata, K., Christie, R. J. & Kataoka, K. (**2011**). Polymeric micelles for nano-scale drug delivery. *Reactive and Functional Polymers* **71**, 227–234.

156. Novoa-Carballal, R., Pfaff, A., & Müller, A. H. E. (**2013**). Interpolyelectrolyte complexes with a polysaccharide corona from dextran-block-PDMAEMA diblock copolymers. *Polymer Chemistry* **4**, 2278–2285.

157. Jeong, J. H., Kim, S. W., & Park, T. G. (**2003**). Novel intracellular delivery system of antisense oligonucleotide by self-assembled hybrid micelles composed of DNA/PEG conjugate and cationic fusogenic peptide. *Bioconjugate Chemistry* **14**, 473–479.

158. Aravind, A., Jeyamohan, P., Nair, R., Veeranarayanan, S., Nagaoka, Y., Yoshida, Y., Maekawa, T., & Kumar, D. S. (**2012**). AS1411 aptamer tagged PLGA-lecithin-PEG nanoparticles for tumor cell targeting and drug delivery. *Biotechnology and Bioengineering* **109**, 2920–2931.

159. Bensaid, F., Thillaye du Boullay, O., Amgoune, A., Pradel, C., Harivardhan Reddy, L., Didier, E., Sablé, S., Louit, G., Bazile, D., & Bourissou, D. (**2013**). Y-shaped mPEG-PLA cabazitaxel conjugates: well-controlled synthesis by organocatalytic approach and self-assembly into interface drug-loaded core–corona nanoparticles. *Biomacromolecules* **14**, 1189–1198.

8

BIOMEDICAL APPLICATION OF BLOCK COPOLYMERS

MARTIN HRUBÝ, SERGEY K. FILIPPOV, AND PETR ŠTĚPÁNEK

Institute of Macromolecular Chemistry AS CR, Prague, Czech Republic

8.1 INTRODUCTION

Nanotechnology is the driving force of current research on new functional materials both in materials science and in biological, bioanalogous, and biomimetic applications. Nanomaterials can be created either by a top-down approach when the desired structure is produced in an already existing material or by a bottom-up approach when the desired structure is induced by self-assembly of selected building blocks using various physical and physicochemical interactions.

Self-assembly of block copolymers into higher organized structures is a result of the amphiphilic character of the blocks composing the polymer. Block copolymer self-assembly in bulk or in solution has been the subject of numerous reviews [1, 2]. In biomedical applications, it is implied that inherently in order to generate (or preserve) a self-assembled structure, the copolymers in contact with aqueous environment will have at least one block that is hydrophilic and one that is hydrophobic. Self-assembly of block copolymers in water has been recently reviewed by Aseyev *et al.* [3], where also other more specialized reviews of polymer self-association in aqueous environment are reviewed.

The commonly used structure-forming polymeric materials are block copolymers, and in particular, diblock copolymers where the long-range repulsion interaction is given by the incompatibility of the blocks and the short-range attraction is provided by the covalent bond between the blocks. In the presence of a solvent in the system,

Macromolecular Self-Assembly, First Edition. Edited by Laurent Billon and Oleg Borisov.

several types of self-assembled structures arise whose structures are governed by the thermodynamic interactions between the solvent and the blocks of the copolymer. A classic example is themicellar structure of a diblock copolymer solution occurring when the solvent is selective (thermodynamically good for one block and bad for the second block); other types of possible common arrangements are liposomes, vesicles, and layered structures. The most common application of these structured materials is to encapsulate a bioactive substance in the hydrophobic parts of the object and to release it at a later time in a different environment.

The various reported morphologies are primarily a result of the variable intra-molecular solvophilic/solvophobic balance in terms of polymerization degrees and volumes of soluble and insoluble blocks (Figure 8.1). Spherical micelles are favored when soluble block has comparable or larger length than the insoluble one. Other morphologies are thermodynamically stable in the case of strongly asymmetric copolymers with long insoluble blocks. Such copolymers form so-called crew cut micelles or vesicles where the size of the dense core is larger than the span of the solvated coronal domain formed by extended soluble blocks.

A block copolymer chain may have various sequences of monomers, depending on how many different monomers are used for its construction and how many blocks are created. The simplest case is a diblock copolymer A-B, a more complex is a triblock copolymer A-B-A or A-B-C, and the increasing complexity of the block copolymer architecture can be imagined in an infinite number of ways. In this chapter we include examples of statistical copolymers of A and B that can be considered block copolymers with irregular sizes of blocks, and graft copolymers with longer side chains.

When a block copolymer self-assembles in an organized structure in solution, such as a micelle, it means that one of the constituent monomers is put into conditions where it is not soluble in the surrounding solvent (water). In order to create well-defined structures (micelles, nanoparticles), polymeric chains have to be first individually dissolved and subsequently the thermodynamic conditions have to be changed to make one component of the copolymer insoluble. Dispersing an

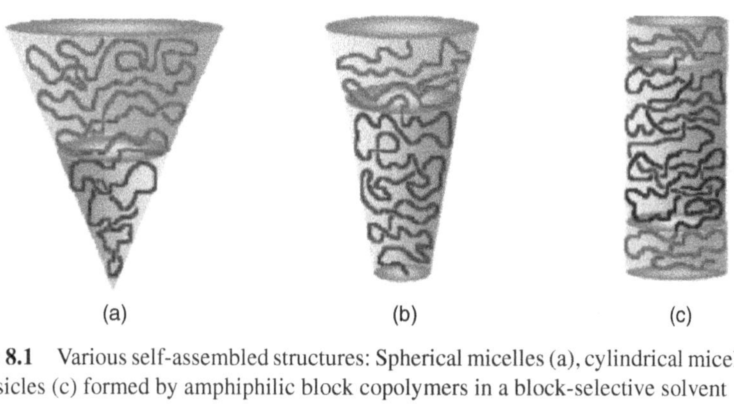

Figure 8.1 Various self-assembled structures: Spherical micelles (a), cylindrical micelles (b), and vesicles (c) formed by amphiphilic block copolymers in a block-selective solvent [4].

amphiphilic copolymer directly in an aqueous solvent may produce small particles, but their size, shape, and stability will depend on details of the dispersion procedure and may not be reproducible. In other cases direct dispersion of a bioapplicable polymer in water may not be possible in the same way, for example, as it is not possible to prepare micelles of a polystyrene-polyacrylic acid diblock copolymer in water by direct dissolution. This is because the crystalline polystyrene core is in glassy state (glass transition temperature of polystyrene is 100°C), so the micelles are "frozen" with severely restricted unimer–micelle exchange equilibration kinetics. In this case a dialysis procedure has to be used [4].

In all cases, however, the hydrophobic block must not be miscible with fatty acids. The opposite case would lead to membrane toxicity [5]. An example of a convenient popular polymer that satisfies this condition is the FDA-approved Pluronic F127. It is a triblock copolymer consisting of a central hydrophobic block of poly(propylene oxide) with two hydrophilic blocks of polyethylene glycol at each end. The approximate lengths of the two PEG blocks is 101 repeating units while the approximate length of the poly(propyleneoxide) block is 56 repeating units [6]. From a general point of view, self-assembly in solution involves phase separation of (at least) one component of the block copolymer. Controlling this phase separation is an essential step in preparing well-defined self-assembled systems, as needed for biomedical applications, for which a precisely defined size and low size dispersity are essential. Phase separation in a system occurs as a result of a change in an external variable. This parameter can be temperature, pH, or an added co-solvent. In principle, it could also be a change in pressure, but the latter is not very practical because usually large pressure changes are needed to achieve relatively small changes in phase diagrams.

A phase diagram for a polymer A/polymer B or polymer/solvent system is schematically represented in Figure 8.2. In the classic case [7] the energy of the system is given by enthalpic and entropic contributions, and the interaction parameter

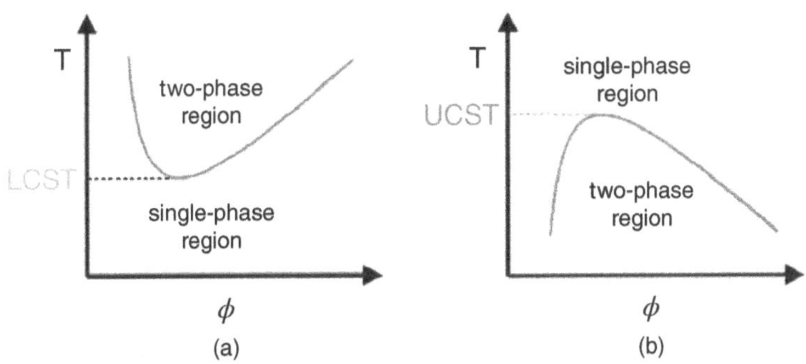

Figure 8.2 Schematic temperature–composition phase diagram for a polymer A/polymer B or polymer/solvent. ϕ is the volume fraction of the first component of the system. LCST (UCST) stands for lower (upper) critical solution temperature of the mixture. At concentrations other than that corresponding to LCST (UCST), the transition from the single-phase to the two-phase region is characterized by a cloud point temperature, CPT.

χ is given by the Flory–Huggins relation $\chi = a/T + b$, where T is absolute temperature and a and b are specific for the polymer/polymer or polymer/solvent pair. For more complex systems like several of those discussed here, a third term χ_s has to be included:

$$\chi = a/T + b + \chi_s$$

In this simple representation the third χ_s term includes all additional interactions in the system, and in particular, the effect of different temperature expansions of the system components and that of various specific interactions in the system (hydrogen bonds, ionic interactions, etc.) that may be dominant compared to the enthalpic/entropic terms a, b.

Depending on the various parameters in equation (8.1), which is to say, mostly on the types of monomers and the solvent, the system has a lower or upper critical solution temperature, LCST or UCST. For aqueous environments related to biomedical applications, almost exclusively LCST-type polymer systems are used. The self-assembled system may be either permanent or "smart"—meaning responsive to external stimuli, most often to changes in temperature and/or pH, in some cases also to changes in illumination, redox potential, or magnetic field.

8.2 DIBLOCK AND TRIBLOCK COPOLYMERS

Interfacial activity accounts for the high utility of amphiphilic block copolymers [8] in the biomedical field as colloidal dispersants, micellar drug carriers [9], and surface modifiers, influencing block copolymer adsorption [10, 11] on solid surfaces.

A vast majority of diblock and triblock copolymers are used for the creation of micellar systems that are loaded with a biologically active substance and transport it, in a protected way, to the application site where the substance is then released at a controlled rate. Polymer micelles are prospective anticancer drug and radionuclide delivery systems, enabling efficient tumor accumulation due to their high apparent molecular weight. The passive tumor accumulation mechanism is called the enhanced permeability and retention (EPR) effect [12–14]; see Figure 8.3. The high-molecular-weight species are accumulated in the solid tumor lesions as a consequence of the low-quality fenestrated neovasculature accompanied by poor or missing lymphatic drainage in such solid tumors.

Micelles exist in dynamic equilibrium with individual surfactant molecules (unimers) that are constantly being exchanged between the solvent and the micelles [5, 16]. The micelles themselves are also continuously disintegrating and reassembling. These two processes are characterized by relaxation times of the order of microseconds and milliseconds, respectively [17]. Unimers may have molecular weight below renal filtration threshold [18–20] (ca. 40–50 kDa depending on type of polymer) and thus may be eliminated rapidly from the organism by filtration in the kidneys after disassembly of the micellar system that fulfilled its task as a drug carrier.

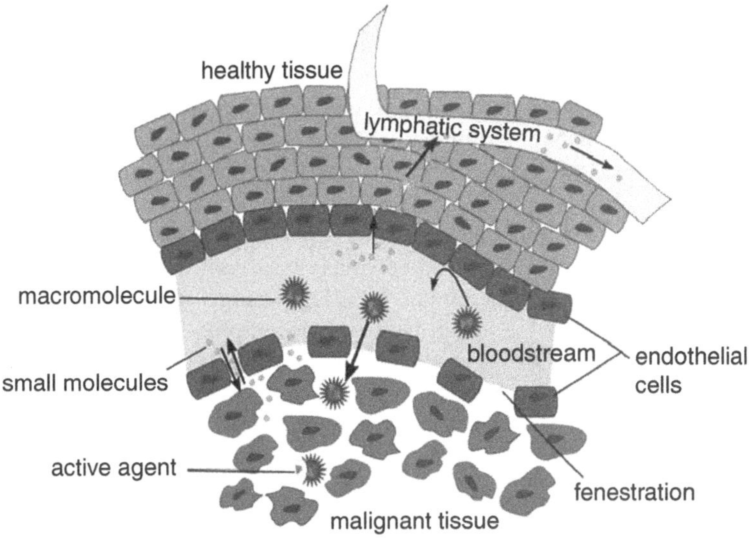

Figure 8.3 EPR effect [13]. Macromolecules and nanoparticles do not penetrate the compact endothelium but can enter the defective malignant tissue. (*See color insert for color representation of this figure*).

A very important parameter in the block copolymer self-association process is the critical micelle concentration, *cmc*. It is the concentration that indicates the usually narrow range of concentrations separating the limits below which most of the copolymer is in the unimolecular state and above which virtually all additional copolymers enter the micellar state [21]. For block copolymers, the *cmc* is very low compared to low-molecular-weight surfactants, which is advantageous for biomedical applications where the presence of a free nonassociated polymer must be avoided as much as possible [22]. For example, the *cmc* of a low-molecular-weight surfactant sodium dodecyl sulphate (SDS) is $cmc = 8 \times 10^{-3}$mol/L [23], for the F127 copolymer it is $cmc = 1 \times 10^{-4}$mol/L [24], and for block copolymers with longer hydrophobic blocks, *cmc* decreases exponentially with chain length [25], and is, for example, 10^{-7} to 10^{-8}mol/L for diblock copolymers with ~500–600 monomeric units [26]. The *cmc* is also influenced by the ratio of block lengths, and the polydispersity and heterogeneity of the block copolymer. A low *cmc* favors therapeutical applications of polymeric micellar systems, since the system would remain in a micellar state even after the critical dilution results from the injections into the bloodstream. Last, but not least, it is important whether or not the hydrophobic block(s) of the copolymer is/are compatible/miscible with phospholipids forming cell membranes. On the one hand, if there is miscibility, one can expect cytotoxicity of such a macromolecular detergent due to membrane damage. On the other hand, block copolymers with a hydrophobic block that is immiscible with the aliphatic part of the phospholipids—which is the case, for example, of Pluronic F127 with a poly(propylene oxide) hydrophobic central block—have only very little if any cytotoxicity.

The size of block copolymer micelles is dependent on the polymer characteristics and ranges usually from 10 to 100 nm [27]. Therefore lipophilic agents, such as hydrophobic drugs, can by solubilized in the micellar core, significantly affecting their concentration in aqueous media. The encapsulation efficiency is dependent on several parameters, including drug and polymer solubility parameters, size, shape, and physical state of the hydrophobic guest molecule, block copolymer length and volume ratio.

Many micelles with a thermoresponsive core and hydrophilic corona have the additional advantage in that they can be simply prepared by heating the thermoresponsive polymer in aqueous solution to a temperature above their cloud point temperature, CPT [28–30]. Among the polymeric micellar drug delivery systems, poly(ethylene oxide), PEO, is most often employed because of its high hydrophilicity and biocompatibility, long experimental record, and FDA approval [31]. However, careful studies of PEO-derived systems over recent decades have revealed some drawbacks, namely interactions with various immunological entities and superfluous accumulation in the body [32]. Another important reason for the search for alternatives is the wide range of patent coverage for nearly all PEO drug carriers and therapeutics. The PEO alternatives poly[*N*-(2-hydroxypropyl)methacrylamide], HPMA, polyvinylpyrrolidone, PVP, and poly(2-alkyl-2-oxazoline)s, POX, have been examined [33].

Because of their unique self-assembly properties, such micellar systems have been found suitable for a variety of applications such as in gene and drug delivery, chemosensors/biosensors, cell adhesion, protein-ligand recognition, and molecular separation, and their functions have been evaluated *in vitro* and *in vivo* [34–36]. Special attention has been devoted to micelles formed from hydrophilic/hydrophilic copolymers (also called double-hydrophilic copolymers, or DHCs) consisting of two or more water-soluble monomers with different chemical properties [37, 38]. The DHCs typically exhibit properties of reversible micellization in water upon temperature variation when one of the blocks presents a lower critical solution temperature (LCST) above room temperature. In addition, there are cases where one of the blocks promotes their dissolution but the other block interacts with a substrate or is sensitive to the influence of an external stimulus such as changes in temperature, pH, or ionic strength [36, 39].

Because the micelle is supramolecular aggregate of tens to hundreds of macromolecules, one can easily construct hybrid micelles containing several types of moieties and polymer blocks in the desired ratio by simply mixing several appropriate block copolymers and forming the mixed micelle afterward from the mixture. The only prerequisite for the formation of such macromolecular modules is compatibility of the copolymers with each other, which may be ensured, for example, by using an identical hydrophobic core-forming block.

A novel polymeric micellar pH-sensitive system for delivery of doxorubicin (Dox) was prepared by the self-assembly of amphiphilic diblock copolymers in aqueous solutions [40, 41]. The copolymers consist of a biocompatible hydrophilic poly(ethylene oxide) (PEO) block and a hydrophobic block containing covalently bound anthracycline antibiotic Dox. The starting block copolymers poly(ethylene oxide)-*block*-poly(allylglycidyl ether) (PEO-PAGE) with a very narrow molecular-weight distribution (M_w/M_n ca. 1.05) were prepared by anionic ring-opening polymerization using a sodium salt of poly(ethylene oxide) monomethyl ether as the macro initiator and allyl glycidyl ether as the functional monomer; the double bond of the allyl glycidyl ether moiety was further functionalized to form the polymer hydrazide. The hydrazide was coupled with Dox, yielding pH-sensitive hydrazone bonds between the drug and carrier. The resulting conjugate containing ca. 3 wt% Dox formed micelles with R_h ~100 nm in phosphate-buffered saline.

Thermoresponsive polymer micelles for applications as drug and radionuclide carriers [41] were prepared [1] from ABA triblock copolymers poly[2-methyl-2-oxazoline-*block*-(2-isopropyl-2-oxazoline-*co*-2-butyl-2-oxazoline)-*block*-2-methyl-2-oxazoline]. These polymers are molecularly dissolved in an aqueous milieu below the cloud point temperature (CPT) of the thermoresponsive central (2-isopropyl-2-oxazoline-*co*-2-butyl-2-oxazoline) block, and above the CPT they form polymeric micelles with a diameter of ~200 nm. The phenolic moiety introduced into the copolymer allowed radionuclide labeling with iodine-125 with sufficient *in vitro* stability under model conditions. Various iodine radioisotopes are suitable for both the diagnosis and the radionuclide therapy of the solid tumors. The composition of these triblock copolymers is more complex than that of the vast majority of poly(2-alkyl-2-oxazoline)s: a statistical thermoresponsive (iPrOx) and hydrophobic (BuOx) central block with terminal hydrophilic blocks (MeOx). As the temperature increases, nanoparticles form in a process starting with single molecules that become loose aggregates and end with the formation of compact nanoparticles. It has been proved [42] that the iPrOx/MeOx ratio determines the value of the cloud point temperature, whereas the different BuOx–iPrOx blocks determine the character of the process.

Polymers that are pH-sensitive have attracted considerable research interest because the differences in pH between normal tissue and cancer tissues create an opportunity to design pH-sensitive drug delivery systems that can target tumors and release loaded drugs at the tumor site [43, 44]. Much effort is being devoted to biodegradable polymeric micelles because their degradability prevents micelle accumulation that can result in toxicity in the long term. The employment of pH-responsive micelles is another approach because the polymers can be triggered and destabilized in response to the slightly acidic tumor microenvironment, enabling rapid release of active agents upon arrival at the desired site and simplifying renal clearance. In such way, we recently devised block copolymers to produce nanoparticles comprising a PDPA (poly[2-diisopropylaminoethyl methacrylate]) core (Figure 8.4). PDPA is a promising smart material for the construction of tumor-targeting drug delivery polymeric nanocarriers because it is able to encapsulate hydrophobic anticancer drugs and it undergoes a sharp hydrophobic-hydrophilic pH-induced transition within a pH range that is desirable for tumor-targeting drug delivery [45]. PDPA is one of the few polymers that can quickly release hydrophobic guest molecules at specific tumor sites via a pH-triggered pathway at slightly acidic conditions [46, 47].

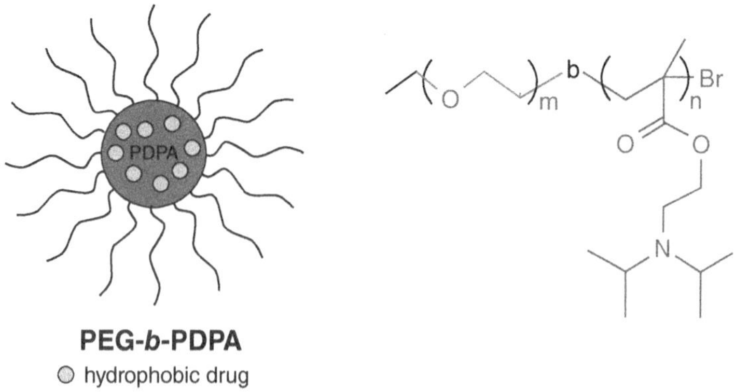

Figure 8.4 Schematic representation of a pH-responsive PEG-b-PDPA drug-loaded block copolymer micelle and the molecular structure of the block copolymer PEG-*b*-PDPA.

The block copolymer poly(ethylene oxide)-*b*-poly(glycerol monomethacrylate)-*b*-poly[2-diisopropylaminoethyl methacrylate] self-assembles in a phosphate buffered saline (PBS) into highly regular spherical micelles. The micellar size (diameter ~42 nm) and micellar molecular weight (M_w > 100 kDa) are in the range intended to avoid renal clearance in order to provide for a long blood circulation time. Their size is below the cutoff size of the leaky pathological vasculature (d < 200 nm), making them candidates for the use in cancer therapy based on the EPR effect. The pH-responsive PDPA core could be loaded with the poorly water-soluble anticancer drug paclitaxel (PTX) whose encapsulation efficiency is ~70% and drug loading content ~7%. Release experiments have evidenced that approximately 90% of the encapsulated PTX is sustained at the PDPA micellar core within the first 9 hours at pH 7.4, while only 18 hours were required for complete drug release at pH 5.0. Thus the micellar dissociation could be triggered at the slightly acidic tumoral extracellular environments. In contact with human plasma or human serum albumin (HSA) diluted in PBS, the micelles are stable for up to 48 hours; long blood circulation time of the nanoparticles at serum environments is a prerequisite for drug delivery applications. Cell viability experiments also demonstrated that such (drug-free) block copolymer micelles are nontoxic.

The stability and interaction of block copolymer micelles with model proteins (e.g., lysozyme or bovine serum albumine, BSA) was studied [48] for micellar systems featuring a hydrophobic (poly[2-diisopropylaminoethyl methacrylate]) (PDPA) core and hydrophilic coronas comprising poly(ethylene oxide)/poly(glycerol monomethacrylate) (PEO-b-PG2MA) or poly[2-(methacryloyloxy)ethyl phosphoryl choline] (PMPC). The results revealed that protein size and hydrophilic chain density play important roles in the observed interactions. The nanoparticles proved to be stable, and protein adsorption was prevented at all the investigated protein environments. The successful protein-repellent characteristic of these nanoparticles is attributed to a highly hydrophilic surface chain density (>0.1 chains per nm^2) and to the length of the hydrophilic chains.

The fouling properties of nanoparticles based on block copolymer of poly(lactic acid), PLA and poly(ε-caprolactone), PCL, have been evaluated [49] by placing them in contact with the model proteins. Block copolymer NPs were produced through the self-assembly of PEO-*b*-PLA and PEO-*b*-PCL. This procedure yielded nanosized objects with distinct structural features, depending on the length of the hydrophobic and hydrophilic blocks and the volume ratio. The protein adsorption events were examined in relation to size, chain length, surface curvature, and hydrophilic chain density. Fouling by BSA and lysozyme was considerably reduced as the length of the hydrophilic PEO-stabilizing shell increased. In contrast to the case of hydrophilic polymer-grafted planar surfaces, the hydrophilic chain density did not seem to markedly influence protein fouling. The protein adsorption took place at the outer surface of the NPs, since neither BSA nor lysozyme was able to diffuse within the hydrophilic layer due to geometric restrictions. Protein binding is an exothermic process, and it is modulated mainly by polymer features.

A new biocompatible and biodegradable diblock copolymer [50] that contains a specific acid-labile degradable linkage (acyclic ketal group) between the hydrophobic poly(ε-caprolactone) (PCL) and the hydrophilic poly(ethylene oxide monomethyl ether) (MPEO) block was prepared. Upon dissolution in a mild organic solvent, the MPEO-*b*-PCL block copolymer self-assembled in water–PBS into regular, spherical, stable nanoparticles. Furthermore, the presence of the acid-labile ketal linker enabled the disassembly of these nanoparticles in a buffer that simulated acidic cytosolic or endosomal conditions in tumor cells. This disassembly led to hydrolysis profiles that resulted in neutral degradation products.

The diblock copolymer monomethoxy-PEG-*b*-poly(D,L-lactic acid) (MPEG-PDLLA) is used in the approved micellar formulation of the anticancer drug paclitaxel for the treatment of cancer in clinical studies [1]. In a similar way, paclitaxel-loaded micellar systems prepared from PEG-poly(aspartic acid) (PEG-*b*-Pas) are used in clinical trial of gastric cancer [1]. Diblock copolymers made of 6-*O*-methacryloyl-1,2:3,4-di-*O*-isopropylidene-D-galactopyranose (MAIGal) and 2-dimethylaminoethyl methacrylate (DMAEMA) were used to construct spherical brushes with a crosslinked polystyrene core [53]. The variation of pH makes it possible to reversibly expose to the external medium the inner block attached to the nanoparticle surface. This property makes these nanoparticles a very useful model of a delivery system specifically targeted to the liver for treatment of hepatocellular carcinoma. In the bloodstream at pH $\sim$ 7.4, the PDMAEMA block is collapsed; the glycopolymer block is then exposed to the outside and is selectively accumulated in liver. After contact with the tumor at pH $\sim$ 5, the PDMAEMA block will expand (Figure 8.5), and its newly exposed cationic surface will damage the anionic surface of the cancer cells.

Melatonin-loaded vesicular nanocarriers were prepared [54] by interfacial deposition using a blend of an amphiphilic diblock copolymer, poly(methyl methacrylate)-block-poly(2-dimethylaminoethyl methacrylate), PMMA-*b*-PDMAEMA, with poly(ε-caprolactone), PCL. Dynamic light scattering showed the nanocarriers to have hydrodynamic radii between 100 and 180 nm, with unimodal particle size distribution for each formulation. The standard TEM for nanocapsules

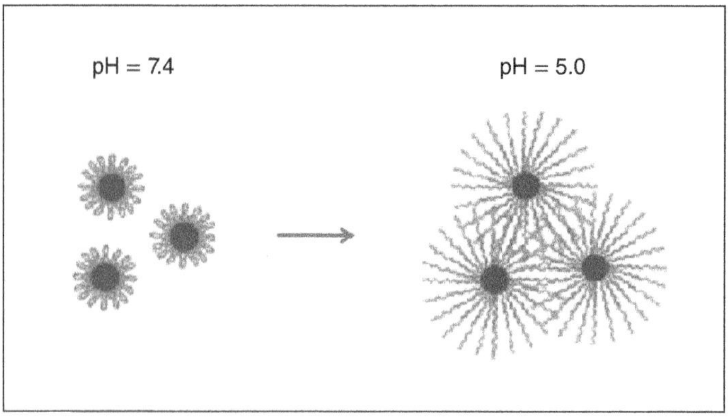

Figure 8.5 Response of the MAIGal-*b*-DMAEMA diblock copolymer attached to the surface of a polystyrene nanoparticle to change of pH.

showed an oily core surrounded by a thin layer composed by PCL/PMMA-*b*-PDMAEMA. Encapsulation efficiencies of ca. 25% were determined by assaying the nanoparticles using HPLC. Platinum nanoparticles were incorporated into the nanocarrier as evidenced by the TEM, which opens up the possibility for promising applications like monitoring the encapsulated drug in the body.

Polymer vesicles called also polymersomes [55–58] are an important type of self-assembled block copolymer system and hold promise for the development of robust nanocontainers for both hydrophobic and hydrophilic bioactive molecules. Figure 8.6 schematically shows a polymersome structure where the hydrophilic drug is loaded into the aqueous core of the polymersome and the hydrophobic drug is inserted in the lipophilic interlayer [62].

Block copolymer polymersomes can be also prepared with compartimentalized inner structure that can be considered as a model of more complex biological environment [60–62].

8.3 GRAFT AND STATISTICAL COPOLYMERS

Graft and statistical copolymers are considered special cases of block copolymers. In the first case, B blocks are attached to the A backbone either regularly or randomly. In the second case, the polymer is a linear chain whose alternating A and B blocks are short and have random lengths. These polymers often do not from micelles or other self-assembled systems spontaneously but can be used to create nanoparticles for biomedical purposes by different physicochemical procedures such as emulsification [63], complexation [64], and nanoprecipitation [65]. The latter technique has become quite popular in the past few years, since it allows a controlled preparation of well-defined nanoparticles.

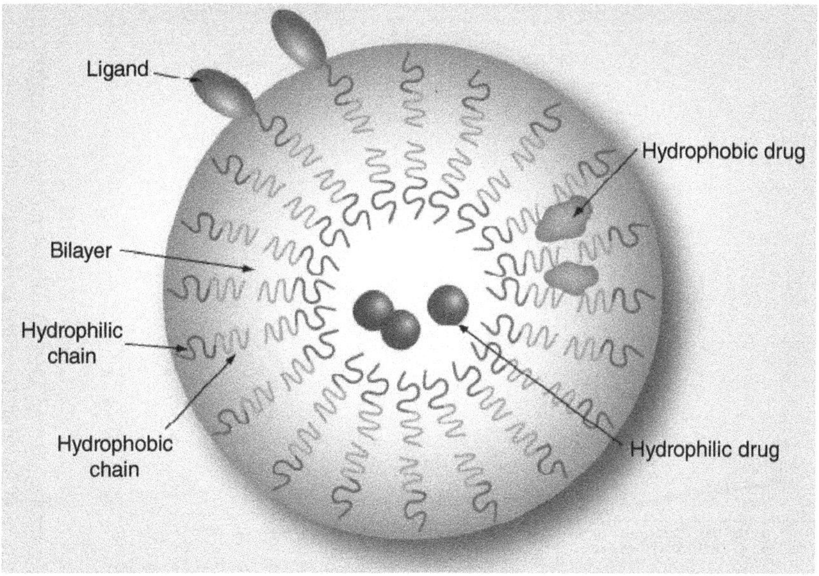

Figure 8.6 Structure of a polymersome with both a hydrophobic and hydrophilic drug incorporated [59]. *(See color insert for color representation of this figure).*

Polymer nanoparticles prepared by self-assembling thermoresponsive poly(*N*-isopropylacrylamide)- *graft*-poly[*N*-(2-hydroxypropyl)methacrylamide] copolymers with hydrolytically degradable *N*-glycosylamine groups between the polymer blocks were prepared [66] for delivery of diagnostic and therapeutic radionuclides into solid tumors. The nanoparticles are formed by fast heating an aqueous solution of the copolymer to 37.8°C. They have a hydrodynamic diameter of 128 nm measured by dynamic light scattering, and they slowly degrade during incubation in aqueous buffer at pH = 7.4. Labeling with both ^{131}I and ^{90}Y proceeds with high yields (>85%). The unlabeled polymers are not cytotoxic for any of the tested murine and human cell lines.

Statistical graft copolymers of poly[*N*-(2-hydroxypropyl)methacrylamide and a different amount of cholesterol moieties attached to the backbone by a noncleavable spacer have been studied previously [67, 68]. Additionally, some of the copolymers were modified with the anticancer drug doxorubicin (Dox), which was bound via the pH-sensitive hydrazone bond. Dilute solutions of conjugates in phosphate buffers at pH = 5.0 and 7.2 were examined by dynamic and static light scattering (DLS/SLS), small-angle X-ray, and neutron scattering—SAXS (Figure 8.7a) and SANS. The presence of any amount of cholesterol proved to result in the formation of anisotropic nanoparticles. The obtained results show that the size and anisotropy of the nanoparticles grow with increasing amounts of cholesterol moieties (Figure 8.7b).

The SAXS and SANS experiments led to the determination of the 3D structure of the nanoparticles as composed of poly[N-(2-hydroxypropyl)methacrylamide-*graft*-

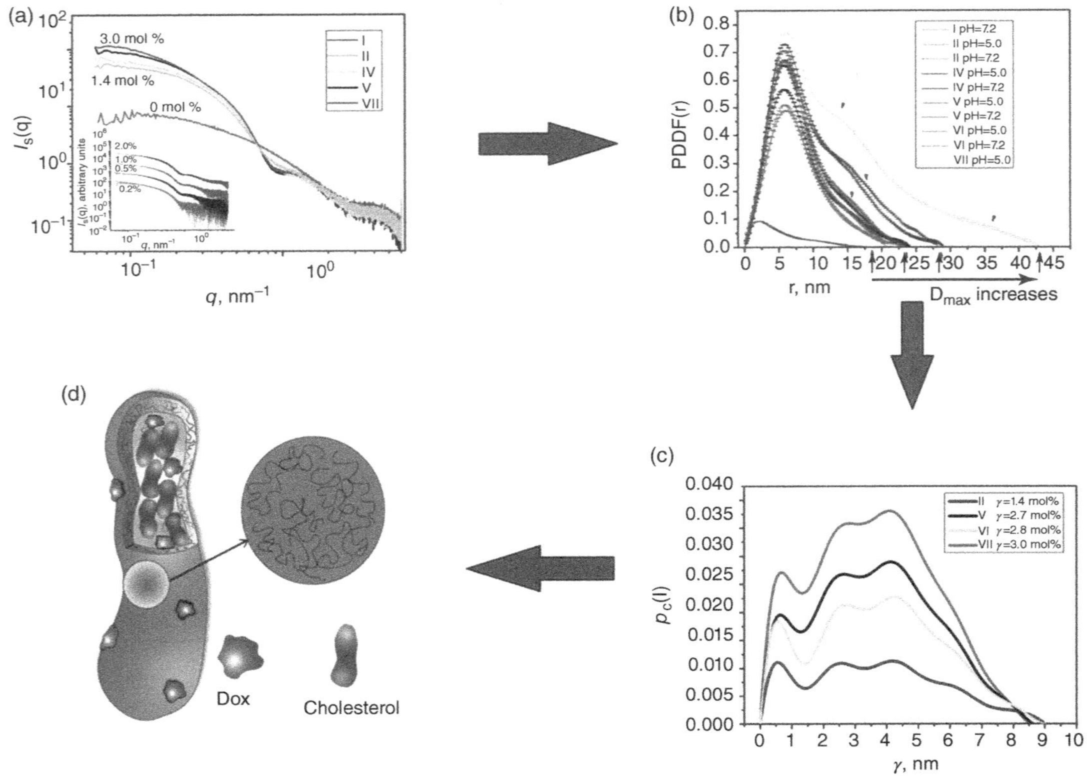

Figure 8.7 (a) SAXS intensity profiles for conjugates with different amounts of cholesterol: I (0 mol%)—red, II (1.4 mol%)—cyan; IV (2.3 mol%)—green, V (2.7 mol%)—blue, and VII (3.0 mol%)—black. (b) Pair-distance-distribution function for some conjugates. (c) Cross-sectional pair-distance distribution functions, $p_c(r)$, for some conjugates. (d) 3D structure of HPMA-Dox-cholesterol NP with bound Dox [69]. (*See color insert for color representation of this figure*).

cholesterol conjugates. From *ab initio* calculations and the shape of pair-distance distribution function (Figure 8.7b), together with the shape of cross-sectional pair-distance distribution function (Figure 8.7c), it was concluded that the most probable structure of nanoparticles is the pearl-necklace type whereby the ellipsoidal pearls composed of cholesterol are mainly connected by bridges composed of hydrophilic HPMA copolymer chains (Figure 8.7d).

Self-assembled polymeric chelate nanoparticles were prepared [69, 70] as potential theranostic agents for improving cancer diagnostics and therapy. The graft copolymers consist of a backbone containing 8-hydroxyquinoline-5-sulfonic acid-chelating groups and poly(ethylene oxide) hydrophilic grafts. The polymers assemble and form nanoparticles after the addition of a multivalent metal cation, such as Fe^{3+} or Cu^{2+} (Figure 8.8).

The obtained nanoparticles exhibit a hydrodynamic diameter of around 25 nm and a stability of at least several hours, which are counted as essential parameters for biomedical purposes. Their biodegradability was confirmed by degradation with deferoxamine. Convenient usage of these nanoparticles for nuclear medicine purposes was demonstrated by their high radiolabeling efficiency with ^{64}Cu.

Thermoresponsive statistical polyoxazolines, poly[(2-butyl-2-oxazoline)-*stat*-(2-isopropyl-2-oxazoline)] [pBuOx-*co*-piPrOx] with different hydrophobic moieties were combined with the FDA-approved triblock copolymer Pluronic F127 (PEO-PPO-PEO) as a template system for the creation of thermosensitive nanoparticles [71]. It was shown that the presence of the thermosensitive F127 triblock copolymer in solution reduces the nanoparticle size and polydispersity. When the temperature is raised above the CPT, nanoparticles composed of polyoxazolines and F127 are dominant in the solution. The molecular weight and hydrophobicity of the polymer do not influence the outer radius and only slightly change the inner radius of the nanoparticles. Poly(2-oxazoline) molecules were fully incorporated inside of F127 micelles, and this result is very promising for the successful application of such systems in the delivery of radionuclides and bioactive compounds. Pluronic F127 and some similar polyamphiphiles are also known to be inhibitors of ABC cassette

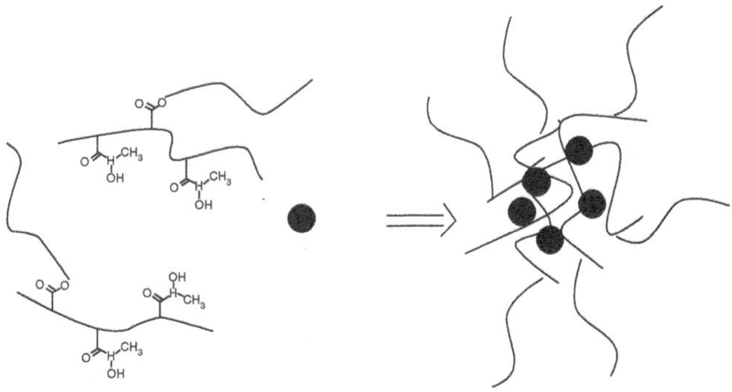

Figure 8.8 Scheme of reaction of the graft-polymer (PMMHA–PEO) with metal cation (Fe^{3+} or Cu^{2+}) [71].

transporters such as P-glycoprotein, whose overexpression causes multidrug resistance of tumor cells. Therefore such copolymers are able to restore the sensitivity of the cancer cells toward chemotherapy.

An interesting group of thermoresponsive graft copolymers was prepared by Hruby *et al.* [72] and is based on glycogen-*graft*-poly(2-alkyl-2-oxazolines). These nanostructured hybrid dendrimeric stimuli-responsive polymers combine the body's own biodegradable polysaccharidic dendrimer glycogen with the widely tunable thermoresponsive poly(2-alkyl-2-oxazoline)s, the latter being polypeptide analogues and are known to be biocompatible. Glycogen-*graft*-poly(2-alkyl-2-oxazoline)s were prepared by a simple one-pot, two-step procedure involving cationic ring-opening polymerization of 2-alkyl-2-oxazolines followed by termination of the living cationic ends with sodium glycogenate. As confirmed by light and X-ray scattering, as well as by cryotransmission electron microscopy, the grafted dendrimer structure allows easy adjustment of the cloud point temperature and the nanostructure of the self-assembled phase separated polymer by the graft composition, length, and the grafting density.

Aliphatic biodegradable copolyesters are a very promising group of copolymers for biologically applicable nanoparticles. Surfactant-free, narrowly distributed, nano-sized spherical particles (R_h < 60 nm) have been produced from the poly(butylene succinate-*co*-butylene dilinoleate) (PBS/PBDL) by a single-step nanoprecipitation method [73]. Combined SLS and DLS measurements suggested that the nanoparticles comprise a porous core (Figure 8.9). The nanoparticles loaded with 6–7% paclitaxel had a pronounced stability and relatively rapid degradation. Cell viability experiments

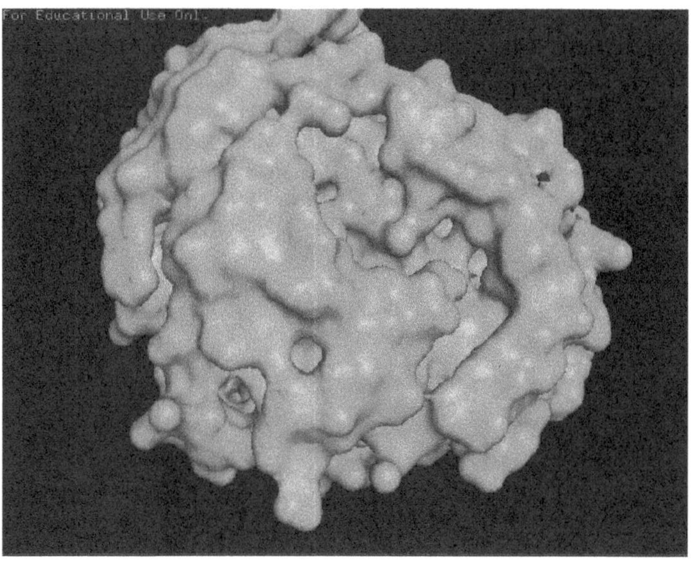

Figure 8.9 Porosity of PBS/PBDL nanoparticles. Schematic structure reconstructed from SAXS by DAMMIF procedure [74].

have demonstrated that the nanoparticles are fully biocompatible and nontoxic, making them useful for biomedical applications.

Their porosity enables water to be entrapped, which is responsible for their pronounced stability and relatively fast degradation as followed by size exclusion chromatography (SEC). The polymeric nanoparticles could be loaded with the hydrophobic model drug paclitaxel (PTX) with an encapsulation efficiency of ~95% and drug loading content of 6–7%. The drug encapsulation and release modifies the inner structure of the nanoparticles, which holds a large amount of entrapped water in the drug-free condition. PTX encapsulation leads to replacement of the entrapped water by the hydrophobic model drug and to shrinking of the nanoparticles, due to favorable drug–polymer hydrophobic interactions. Cell viability experiments demonstrated that the nanoparticles are biocompatible and nontoxic, making them potentially useful for applications in nanomedicine.

8.4 CONCLUDING REMARKS

Block copolymers forming supramolecular assemblies of the micellar or polymerosomal type in aqueous media represent materials that are extremely useful for the construction of drug delivery systems especially for cancer applications. The micellar and polymerosomal formulations suppress unwanted physicochemical properties of the encapsulated drugs, modify biodistribution of the drugs toward targeted delivery into tissues of interest, and allow triggered release of the active cargo.

ACKNOWLEDGMENT

The work was supported by the Ministry of Education, Youth and Sports of CR within the National Sustainability Program I (NPU I), Project POLYMAT LO1507.

REFERENCES

1. Lazzari, M.; Lin, G.; Lecommandoux, S. *Block Copolymers in Nanoscience.* **2006**, Wiley-VCH, Weinheim.
2. Hamley, I. W. *Block Copolymers in Solution: Fundamentals and Applications.* **2005**, Wiley, Chichester.
3. Aseyev, V.; Tenhu, H.; Winnik, F. M. Non-ionic Thermoresponsive Polymers in Water. *Adv Polym Sci* **2011**, *242*, 29–89.
4. Li, G. H.; Yang, P. P.; Gao, Z. S.; Zku, Y. Q. Synthesis and Micellar Behavior of Poly(acrylic acid-b-styrene) Block Copolymers. *Colloid Polym Sci* **2012**, *290*, 1825–1831.
5. Bhattacharjee, S.; Ershov, D.; Fytianos, K.; van der Gucht, J.;Alink, G. M.; Rietjens, I. M. C. M.; Marcelis, A. T. M.; Zuilhof, H. Cytotoxicity and Cellular Uptake of Tri-block Copolymer Nanoparticles with Different Size and Surface Characteristics. *Part Fibre Toxicol* **2012**, *9*, Article Number 11.

6. http://en.wikipedia.org/wiki/Poloxamer_407
7. Flory, P. J. *Principles of Polymer Chemistry*, Cornell University Press,**1953**
8. Tuzar, Z.; Kratochvil, P. Micelles of Block and Graft Copolymers in Solutions. *Surf Colloid Sci* **1993**, *15*, 1–83.
9. Cammas-Marion, S.; Okano, T.; Kataoka, K. Functional and Site-Specific Macromolecular Micelles as High Potential Drug Carriers. *Colloids Surf B-Biointerfaces* **1999**, *16*, 207–215.
10. Vladkova, T.; Krasteva, N.; Kostadinova, A.; Altankov, G. Preparation of PEG-coated Surfaces and a Study for Their Interaction with Living Cells. *J Biomater Sci-Polym Ed* **1999**, *10*, 609–620.
11. Kramarenko, E.; Yu.; Potemkin, I. I.; Khokhlov, A. R.; Winkler, R. G.; Reineker, P. Surface Micellar Nanopattern Formation of Adsorbed Diblock Copolymer System. *Macromolecules* **1999**, *32*, 3495–3501.
12. Maeda, H.; Wu, J.;Sawa, T.; Matsumura, Y.; Hori, K. Tumor Vascular Permeability and the EPR Effect in Macromolecular Therapeutics: A Review. *J Control Release* **2000**, *1–2*, 271–284.
13. Stockhofe, K.; Postema, J. M.; Schieferstein, H.; Ross, T. L. Radiolabeling of Nanoparticles and Polymers for PET Imaging. *Pharmaceuticals* **2014**, *7*, 392-418.
14. Greish, K. Enhanced Permeability and Retention (EPR) Effect for Anticancer Nanomedicine Drug Targeting. *Methods Mol Biol* **2010**, *624*, 25–37. doi: 10.1007/978-1-60761-609-2_3.
15. Sushant, S. Kulthe; Yogesh M. Choudhari; Nazma N. Inamdar; Vishnukant Mourya. Polymeric Micelles: Authoritative Aspects for Drug Delivery. *Des Monomers Polym* **2012**, *15*, 465–521.
16. Lund, R.; Willner, L.; Richter, D. Kinetics of Block Copolymer Micelles Studied by Small-Angle Scattering Methods. *Adv Polym Sci* **2013**,*259*, 51–158.
17. Patist, A.; Kanicky, J. R.; Shukla, P. K.; Shah, D. O. Importance of Micellar Kinetics in Relation to Technological Processes. *J Colloid Interface Sci* **2002**, *245*, 1–15.
18. Chen, B.; Jerger, K.; Frechet, J. M. J.; Szoka, F. C. The Influence of Polymer Topology on Pharmacokinetics: Differences between Cyclic and Linear PEGylatedPoly(acrylic acid) Comb Polymers. *J Control Release* **2009**,*140*, 203–209.
19. Wang, Y. L.; Ye, F. R.; Jeong, E. K.; Sun, Y.; Parker, D. L.; Lu, Z. R. Noninvasive Visualization of Pharmacokinetics, Biodistribution and Tumor Targeting of Poly N-(2-hydroxypropyl)methacrylamide in Mice Using Contrast Enhanced MRI. *Pharm Res* **2007**, *24*, 1208–1216.
20. Yang, Y.; Pan, D. Y.; Luo, K.; Li, L.; Gu, Z. W. Biodegradable and Amphiphilic Block Copolymer–Doxorubicin Conjugate as Polymeric Nanoscale Drug Delivery Vehicle for Breast Cancer Therapy. *Biomaterials* **2013**, *34*, 8430–8443.
21. Desando, M. A.; Reeves, L. W. The Demicellization Temperature of Potassium Normal-Octanoate in Deuterium-Oxide as Estimated from H-1 and C-13 Nuclear-Magnetic-Resonance Spectra. *Can J Chem* **1986**, *64*, 1817–1822.
22. Chernitsky, E. A.; Rozin, V. V.; Senkovich, O. A. Influence of pH of the Medium on Parameters of Detergent-Induced Hemolysis and Vesiculation of Erythrocytes. *Biol Membr* **2000**, *17*, 420–426.
23. Cifuentes, A.; Bernal, J. L.; Diez-Masa, J. C. Determination of Critical Micelle Concentration Values Using Capillary Electrophoresis Instrumentation. *Anal Chem* **1997**, *69*, 4271–4274.

24. Singh, V.; Khullar, P.; Dave, P. N.; Kaur, N. Micelles, Mixed Micelles, and Applications of Polyoxypropylene (Ppo)-Polyoxyethylene (Peo)-Polyoxypropylene (Ppo) Triblock Polymers.*International J Indust Chem* **2013**, *4*, art. 12.

25. Nikoubashman, A.; Panagiotopoulos, A. Z. Effect of Solvophobic Block Length on Critical Micelle Concentration in Model Surfactant Systems. *J Chem Phys* **2014**, *141*, 041101.

26. Khougaz, K.; Gao, Z.; Eisenberg, A. Determination of the Critical Micelle Concentration of Block Copolymer Micelles by Static Light Scattering. *Macromolecules* **1994**, *27*, 6341–6346.

27. Kataoka, K.; Harada, A.; Nagasaki, Y. Block Copolymer Micelles for Drug Delivery: Design, Characterization and Biological Significance. *Adv Drug Del Rev* **2001**, *47*, 113–131.

28. Hruby, M.; Konak, C.; Kucka, J.; Vetrik, M.; Filippov, S. K.; Vetvicka, D. *et al.* Thermoresponsive, Hydrolytically Degradable Polymer Micelles Intended for Radionuclide Delivery. *Macromol Biosci* **2009**, *9*, 1016–1027.

29. Hruby, M.; Filippov, S. K.; Panek, J.; Novakova, M.; Mackova, H.; Kucka, J., *et al.* Thermoresponsive Micelles for Radionuclide Delivery. *J Control Release* **2010**, *148*, E60–E62.

30. Hruby, M.; Filippov, S. K.; Panek, J.; Novakova, M.; Mackova, H.; Kucka, J., *et al.* Polyoxazoline Thermoresponsive Micelles as Radionuclide Delivery Systems. *Macromol Biosci* **2010**, *10*, 916–924.

31. Torchilin, V. P. PEG-based Micelles as Carriers of Contrast Agents for Different Imaging Modalities. *Adv Drug Deliv Rev* **2002**, *54*, 235–252.

32. Knop, K.; Hoogenboom, R.; Fischer, D.; Schubert, U. S. Poly(ethylene glycol) in Drug Delivery: Pros and Cons as Well as Potential Alternatives. *Angew Chem Int Edit* **2010**, *49*, 6288–6308/

33. Sedlacek, O.; Monnery, B. D.; Filippov, S. K.; Hoogenboom, R.; Hruby, M. Poly(2-Oxazoline)s—Are They More Advantageous for Biomedical Applications than Other Polymers? *Macromol Rapid Commun* **2012**, *33*, 1648–1662.

34. Huh, K. M.; Min, H. S.; Lee, S. C.; Lee, H. J.; Kim, S.; Park, K. A New Hydrotropic Block Copolymer Micelle System for Aqueous Solubilization of Paclitaxel. *J Control Release* **2008**, *126*, 122–129.

35. Ruan, G.; Feng, S.-S.Preparation and Characterization of Poly(lactic acid)-poly(ethylene glycol)-poly(lactic acid) (PLA-PEG-PLA) Microspheres for Controlled Release of Paclitaxel. *Biomaterials* **2003**, *24*, 5037–5044.

36. Nakayama, M.; Kawahara, Y.; Akimoto, J.; Kanazawa, H.; Okano, T. pH-induced Phase Transition Control of Thermoresponsive Nano-Micelles Possessing Outermost Surface Sulfonamide Moieties. *Colloids Surf B-Biointerfaces* **2012**, *99*, 12–19.

37. Tang, X.; Pan, C.-Y. Double Hydrophilic Block Copolymers PEO-*b*-PGA: Synthesis, Application as Potential Drug Carrier and Drug Release via pH-sensitive Linkage. *J Biomed Mater Res A* **2008**, *86A*, 428–438.

38. Lou, S.-F.; Zhang, H.; Williams, R. G.; Branford-White, C.; Nie, H.-L.; Quana, J., *et al.* Fabrication and Aggregation of Thermoresponsive Glucose-Functionalized Double Hydrophilic Copolymers. *Colloids Surf B-Biointerfaces* **2013**, *105*, 180–186.

39. Colfen, H. Double-Hydrophilic Block Copolymers: Synthesis and Application as Novel Surfactants and Crystal Growth Modifiers. *Macromol Rapid Commun* **2001**, *22*, 219–252.

40. Hrubý, M., Koňák, Č.; Ulbrich, K. Polymeric Micellar pH-sensitive Drug Delivery System for Doxorubicin. *J Control Release* **2005**,*103*, 137–148.

41. Hrubý, M.; Koňák, Č.; Ulbrich, K. Poly(allylglycidyl ether)-block-poly(ethylene oxide): A Novel Promising Polymeric Intermediate for the Preparation of Micellar Drug Delivery Systems. *J Appl Polym Sci* **2005**,*95*, 201–211.

42. Bogomolova, A.; Filippov, S. K.; Starovoytova, L.; Angelov, B.; Konarev, P.; Sedlacek, O.; Hruby, M.; Stepanek, P. Study of Complex Thermosensitive Amphiphilic Polyoxazolines and Their Interaction with Ionic Surfactants: Are Hydrophobic, Thermosensitive, and Hydrophilic Moieties Equally Important? *J Phys Chem B*, **2014**, *118*, 4940–4950.

43. Liua, Y.; Wanga, W.; Yanga, J.;Zhoua, Ch.; Sun, J. pH-sensitive Polymeric Micelles Triggered Drug Release for Extracellular and Intracellular Drug Targeting Delivery. *Asian J Pharm Sci* **2013**, *8*, 159–167.

44. Zhang, Z.; Sun, Q.; Zhong, J.; Yang, Q.; Li, H.; Du, C.; Liang, B.; Shuai, X. Magnetic Resonance Imaging-Visible and pH-Sensitive Polymeric Micelles for Tumor Targeted Drug Delivery. *J Biomed Nanotechnol* **2014**, *10*, 216–226.

45. Giacomelli, F. C.; Štpánek, P.; Giacomelli, C.; Schmidt, V.; Jäger, E.; Jäger, A.; Ulbrich, K. pH-Triggered Block Copolymer Micelles Based on a pH-responsive PDPA (poly[2-diisopropylaminoethyl methacrylate]) Inner Core and a PEO (poly(ethylene oxide)) Outer Shell as a Potential Tool for the Cancer Therapy. *Soft Matter* **2011**, *7*, 9316–9325.

46. Pegoraro, C.; Cecchin, D.; Gracia, L. S.; Warren, N.; Madsen, J.; Armes, S. P.; Lewis, A.; MacNeil, S.; Battaglia, G. Enhanced Drug Delivery to Melanoma Cells Using PMPC-PDPA Polymersomes. *Cancer Lett* **2013**, *334*, 328–337.

47. Wang, Y.; Zhou, K.; Huang, G.; Hensley, C.; Huang, X.; Ma, X.; Zhao, T.; Sumer, B. D.; DeBerardinis, R. J.; Gao, J. A Nanoparticle Based Strategy for the Imaging of a Broad Range of Tumours by Nonlinear Amplification of Microenvironment Signals. *Nat Mater* **2014**, *13*, 204–212.

48. Giacomelli, F. C.; Stepánek, P.; Schmidt, V.; Jäger, E.; Jäger, A.; Giacomelli, C. Light Scattering Evidence of Selective Protein Fouling on Biocompatible Block Copolymer Micelles. *Nanoscale* **2012**, *4*, 4504–4514.

49. de Castro, C. E.; Mattei, B.; Riske, K. A.; Jäger, E.; Jäger, A.; Stepánek, P.; Giacomelli, F. Understanding the Structural Parameters of Biocompatible Nanoparticles Dictating Protein Fouling. *Langmuir* **2014**, *30*, 9770–9779.

50. Petrova, S.; Jager, E.; Konefal, R.; Jager, A.; De Garcia Venturini, C.; Spěváček, J.; Štěpánek, P. Novel Poly(ethylene oxide monomethyl ether)-b-poly (.epsilon.-caprolactone) Diblock Copolymers Containing a pH-Acid Labile Ketal Group as a Block Linkage. *Polym Chem* **2014**, *5*, 3884–3893.

51. Cabral, H.; Kataoka, K. Progress of Drug-loaded Polymeric Micelles into Clinical Studies. *J Control Release* **2014**, *190*, 465–476.

52. Kato, K.; Chin, K.; Yoshikawa, T.; Yamaguchi, K.; Tsuji, Y.; Esaki, T.; Sakai, K.; Kimura, M.; Hamaguchi, T.; Shimada, Y.; Matsumara, Y.; Ikeda, R. Phase II Study of NK105, a Paclitaxel Incorporating Micellar Nanoparticle, for Previously Treated Advanced or Recurrent Gastric Cancer. *Invest New Drugs* **2012**, 30, 1621–1627.

53. Penott-Chang, E.; Walther, A.; Millard, P.; Jäger, A.; Jäger, E.; Müller, A. H.; Guterres, S. S.; Pohlmann, A. R. Amphiphilic Diblock Copolymer and Polycaprolactone Blends to Produce New Vesicular Nanocarriers. *J Biomed Nanotechnol* **2012**, *8*, 272–279.

54. Garanger, E.; Lecommandoux, S. Towards Bioactive NanovehiclesBased on ProteinPolymers. *Angew Chem Int Edit* **2012**, *51*, 3060–3062.

55. Sanson, C.; Schatz, C.; Le Meins, J. F.; Soum, A.; Thevenot, J.; Garanger, E.; Lecommandoux, S. A Simple Method to Achieve High Doxorubicin Loading in Biodegradable Polymersomes. *J Control Release* **2010**, *147*, 428–435.

56. Sanson, C.; Schatz, C.; Le Meins, J. F.; Brulet, A.; Soum, A.; Lecommandoux, S. Biocompatible and Biodegradable Poly(trimethylene carbonate)-b-Poly (L-glutamicacid) Polymersomes: Size Control and Stability. *Langmuir* **2010**, *26*, 2751–2760.

57. Sanson, C.; Le Meins, J. F.; Schatz, C.; Soum, A.; Lecommandoux, S. Temperature Responsive Poly(trimethylene carbonate)-block-poly(L-glutamicacid) Copolymer: Polymersomes Fusion and Fission. *Soft Matter* **2010**, *6*, 1722–1730.

58. Bei, D.; Meng, J.; Youan, B. C. Engineering Nanomedicines for Improved Melanoma Therapy: Progress and Promises. *Nanomedicine* **2010**, *5*, 1385–1399.

59. Marguet, M.; Edembe, L.; Lecommandoux, S. Polymersomes in Polymersomes: Multiple Loading and Permeability Control. *Angew Chem Int Edit* **2012**, *51*, 1173–1176.

60. Marguet, M.; Sandre, O.; Lecommandoux, S. Polymersomes in "Gelly" Polymersomes: Toward Structural Cell Mimicry. *Langmuir* **2012**, *28*, 2035–2043.

61. Marguet, M.; Bonduelle, C.; Lecommandoux, S. Multicompartmentalized Polymeric Systems: Towards Biomimetic Cellular Structure and Function. *Chem Soc Rev* **2013**, *42*, 512–529.

62. Weikun, L.; Shanqin, L.; Renhua, D.; Jintao, Z. Encapsulation of Nanoparticles in Block Copolymer Micellar Aggregates by Directed Supramolecular Assembly. *Angew Chem Int Edit* **2011**, *50*, 5865–5868.

63. *Polyelectrolyte Complexes in the Dispersed and Solid State*. Mueller, M.; Ed. Springer Verlag, Heidelberg **2014**.

64. Schubert, S.; Delaney, J. T. Jr.; Schubert, U. S. Nanoprecipitation and Nanoformulation of Polymers: From History to Powerful Possibilities beyond Poly(lactic acid). *Soft Matter* **2011**, *7*, 1581–1588.

65. Hruby, M.; Konak, C.; Kucka, J.; Vetrik, M.; Filippov, S. K.; Vetvicka, D.;Mackova, H.; Karlsson, G.; Edwards, K.; Rihova, B.; Ulbrich, K. Thermoresponsive, Hydrolytically Degradable Polymer Micelles Intended for Radionuclide Delivery. *Macromol Biosci* **2009**, *9*, 1016–1027.

66. Filippov, S. K.; Chytil, P.; Dyakonova, M.; Papadakis, C. M.; Jigounov, A.; Plestil, J.; Stepanek, P.; Etrych, T.; Ulbrich, K.; Svergun, D. I. Macromolecular HPMA-Based Nanoparticles with Cholesterol for Solid-Tumor Targeting: Detailed Study of the Inner Structure of a Highly Efficient Drug Delivery System. *Biomacromolecules* **2012**, *13*, 2594–2604.

67. Filippov, S. K.; Franklin, J. M.; Konarev, P. V.; Chytil, P.; Etrych, T.; Bogomolova, A.; Dyakonova, M.; Papadakis, C. M.; Radulescu, A.; Ulbrich, K.; Stepanek, P.; Svergun, D. I. Hydrolytically Degradable Polymer Micelles for Drug Delivery: A SAXS/SANS Kinetic Study. *Biomacromolecules* **2013**, *14*, 4061–4070.

68. Škodová, M.; Černoch, P.; Štěpánek, P.; Chánová, E.; Kučka, J.; Kálalová, Z.; Kaňková, D.; Hrubý, M. Self-Assembled Polymeric Chelate Nanoparticles as Potential Theranostic Agents. *Chem Phys Chem* **2012**, *13*, 4244–4250.

69. Skodova, M.; Hruby, M.; Filippov, S. K.; Karlsson, G.; Mackova, H.; Spirkova, M.; Kankova, D.; Steinhart, M.; Stepanek, P.; Ulbrich K. Novel Polymeric Nanoparticles Assembled by Metal Ion Addition. *Mac Chem Phys* **2011**, *212*, 2339–2348.

70. Bogomolova, A.; Hruby, M.; Panek, J.; Rabyk, M.; Turner, S.; Bals, S.; Steinhart, M.; Zhigunov, A.; Sedlacek, O.; Stepanek, P.; Filippov, S. K. Small-Angle X-ray Scattering and Light Scattering Study of Hybrid Nanoparticles Composed of Thermoresponsive Triblock Copolymer F127 and Thermoresponsive Statistical Polyoxazolines with Hydrophobic Moieties. *J Appl Crystallogr* **2013**, *46*, 1690–1698.

71. Pospisilova, A.; Filippov, S. K.; Bogomolova, A.; Turner, S.; Sedlacek, O.; Matushkin, N.; Cernochova, Z.; Stepanek, P.; Hruby, M. Glycogen-graft-poly(2-alkyl-2-oxazolines)—The New Versatile Biopolymer-Based Thermoresponsive Macromolecular Toolbox. *RSC Adv* **2014**, *4*, 61580–61588.

72. Jäger, A.; Gromadzki, D.; Jäger, E.; Giacomelli, F. C.; Kozlowska, A.; Kobera, L.; Brus, J.; Říhová, B.; El Fray, M.; Ulbrich, K.; Štpánek, P. Novel "Soft" Biodegradable Nanoparticles Prepared from Aliphatic Based Monomers as a Potential Drug Delivery System. *Soft Matter* **2012**, *8*, 4343–4354.

73. Franke, D.; Svergun, D. I. DAMMIF, A Program for Rapid *ab-initio* Shape Determination in Small-Angle Scattering. *J Appl Crystallogr* **2009**, *42*, 342–346.

INDEX

Macromolecular Self-Assembly, First Edition. Edited by Laurent Billon and Oleg Borisov.